CASCADE SEPARATION OF POWDERS

CASCADE SEPARATION OF POWDERS

E. Barsky and M. Barsky

Viva Books

New Delhi | Mumbai | Chennai | Kolkata | Bangalore | Hyderabad | Kochi

First Indian Edition 2008

VIVA BOOKS PRIVATE LIMITED

- 4737/23 Ansari Road, Daryaganj, New Delhi 110 002
  E-mail: vivadelhi@vivagroupindia.net, Tel. 42242200
- Plot No. 76, Service Industries, Shirvane, Sector 1, Nerul, Navi Mumbai 400 706
  E-mail: vivamumbai@vivagroupindia.net, Tel. 27721273, 27721274
- Jamals Fazal Chambers, 26 Greams Road, Chennai 600 006
  E-mail: vivachennai@vivagroupindia.net, Tel. 28290304, 28294241
- B-103, Jindal Towers, 21/1A/3 Darga Road, Kolkata 700 017
  E-mail: vivakolkata@vivagroupindia.net, Tel. 22836381, 22816713
- 7, Sovereign Park Apartments, 56-58, K. R. Road, Basavanagudi, Bangalore 560 004
  E-mail: vivabangalore@vivagroupindia.net, Tel. 26607409, 26607410
- 101-102 Mughal Marc Apartments, 3-4-637 to 641, Narayanguda, Hyderabad 500 029
  E-mail: vivahyderabad@vivagroupindia.net, Tel. 27564481, 27564482
- First Floor, Beevi Towers, SRM Road, Kaloor, Kochi 682 018, Kerala
  E-mail: vivakochi@vivagroupindia.net, Tel: 0484-2403055, 2403056

www.vivagroupindia.com

Published by arrangement with

Cambridge International Science Publishing
7, Meadow Walk, Great Abington
Cambridge CB1 6AZ, UK

ISBN-10: 81-309-0945-6
ISBN-13: 978-81-309-0945-5

Published by Vinod Vasishtha for Viva Books Private Limited, 4737/23, Ansari Road, Daryaganj, New Delhi-110 002. Printed and bound by Anand Sons, Delhi - 110 092.

PREFACE

Industry imposes stringent requirements on the quality of powder materials used in many areas of technology. To satisfy these requirements, it is necessary to overcome technical problems and find solutions of the problems, in most cases by the application of highly efficient separation processes.

The following aims are followed in the fractionation of powder materials:

- the production of dust-free products in relation to the given boundary grain size. Small amounts of the fine classes are permitted in these products.
- the production of pure fine products as a result of the removal of coarse particles. This task is inverse to the first task. The apparatus facilities in the process greatly differ from the devices used in the first task. Usually, the first task is realised with a loss of part of the course product together with dust, and the second task with the loss of the final product with the coarse particles. In most cases, these losses are large. The situation has been aggravatedby the absence of highly efficient separation systems:
- separation of bulk (loose) material on the basis of the density of particles irrespective of particle size;
- the separation of the polydispersed material with a wide range of the grain size composition into fractions with a narrower range of the grain size;
- the production of powders whose grain size characteristic is specified in advance for the entire grain size range of the initial material.

At present, the fractionation of materials in technology is carried out using different systems and different classification methods. These methods may be divided into the following groups:

- screening and vibroseparation on flat surfaces (perforated and smooth)
- Hydraulic classification in a moving or stationary liquid medium;
- dry classification in gas flows.

In most cases, fractionation carried out by screening methods. In this case, classification sections are represented by different technological lines with a large number of hoppers, feeders and conveyeres. In most cases, they do not ensure that the required efficiency of the process. The main shortcoming of screening is that its separating capacity greatly decreases when the boundary grain size of separation approaches 1 mm and practically approaches zero in the regions of separation in respect of classes finer than 0.5 mm which are mostly characteristic of modern industrial technology. The powders of this grain size can be separated most efficiently in moving flows.

At present, the methods of hydraulic classification are used most widely. These methods have been developed and studied for a long time and the available experience creates favourable conditions for extensive application of the methods. However, the application of hydraulic classification results in difficult-to-solve technological problems. The main problems are associated with the disruption of the principles of environmental protection and also with a high consumption of water which causes considerable difficulties. In addition to this, a relatively large number of materials can be separated by the wet method owing to the fact that in wetting they changed their physical properties or bonder together.

It should also be mentioned that the technology using hydraulic method of fractionation is characterised by high energy consumption because after the separation operations the powders must be often dried because further processing (dosing, mixing, shaping, etc.) is possible only in the dehydrated condition.

Dry separation methods are more efficient. These methods are realised in most cases in equipment with air flows or, if necessary, flows of inert, flue and other gases. Their efficiency is indicated by the current tendency of transition to the dry methods of production.

Therefore, without reducing the significance of the hydraulic classification methods and discussing methods of improving them, in the book special attention is paid to the analysis of the results of investigations and main relationships of the dry fractionation methods.

It should be mentioned that there is no principal difference in the physics of the process of hydraulic and pneumatic classification. Naturally, the difference in the density and viscosity of water and air assumes different orders of the rates of the process and this is reflected in the design features of separation equipment. However, in both cases, the process is based on the ratio of the forces of natural or artificial, for example centrifugal, gravity of particles to the value of their hydrodynamic resistance in a moving medium. Regardless of the design of equipment and the separation medium, only this factor predetermines the nature of phenomena taking place during fractionation.

The presence in the Arsenal of modern technology of sufficiently reliable dust cleaning systems and also the possibility of carrying out separation in a closed cycle create favourable conditions for the extensive application of air classification methods.

Until recently, it was generally recognised that pneumatic classification does not result in acceptable efficiency of the process, and equipment used for this method should be very large. These assumptions are basically confirmed only for the conventional methods of organisation of the process based on the principle of equalisation by the gas flows of the particles of the boundary separation size. At the same time, the possibilities of pneumatic classification are not exhausted only by this principle. In the examination of the mechanism of separation of bulk materials in the flows, it has become possible to use this method with high efficiency.

The role of the processes of separation of bulk materials increases at the present time owing to the fact that, firstly, the requirements on the quality of powders and intermediate products continuously increase and, secondly, because of the increase of the volume of production larger and larger quantities of low-quality starting material are used in processing.

It should be mentioned that, regardless of the extensive application of classification systems used for the separation of bulk materials, no significant advances have been made in the design of these systems with the exception of, possibly, the construction of cascade separation systems in the last couple of decades.

The main but not only reason explaining the given situation is that no accurate methods of comparison of the separation capacity of the classification systems and qualitative parameters of the process have been developed. This prevents the effective definition of advanced design of separation systems and suppresses the tendency in the development of this group of systems.

Work on the development of the criteria of quality for the evaluation of the separation processes started at the beginning of the 20th century. However, in addition to correct concepts, these developments have been based erroneous concents which prevented a solution of the given problem. More than 100 years have passed since the publication of Hancock's studies in which the generalised quality criteron was formulated for the first time. In this period, approximately 100 different dependences for expressing the efficiency of classification have been proposed. The large number of the criterial methods, the absence of unity in the problem of the method of evaluation and optimisation of separation have resulted in an uncertain situation in the selection of classification systems and evaluation of the quality of their operation. Therefore, in the majority of cases the design of new production processes and optimisation of existing equipment have not been carried out on a strictly scientific basis but on the basis of experience obtained in the service of related equipment, intuition and the 'courage' of designers.

Attempts have been made to systematise the entire range of the methods of optimisation of separation. However, investigators could not link clearly the criteria parameters of the process with its physical nature and, in the majority of cases, they did not even formulate this task.

In the middle of the 30s of the previous century, methods of optimisation of separation, based on the Tromp curve, were introduced in enrichment practice. The curve was used for formulating a group of parameters which, however, also have significant shortcomings.

The investigations carried out in recent years have shown that the methods of objective and unambiguous evaluation of the classification processes must be based on the relationship between the separating capacity of the system and the physical fundamentals of the investigated processes.

Recently, it has become necessary to develop new methods of organisation of separation of powders. They include the separation of powders in a single system into more than two products, and the production of bulk media with the defined grain size characteristic. Multiproduct separation differs principally from two-product separation in both the methods of physical organisation and the methods of evaluating the quality of realisation.

New concepts are also being proposed in the solution of the problem of the special purity two-product separation of powders. The main concept is based on the application of combined classification schemes.

All these problems are reflected in the book in which special attention is given to the examination of cascade principles of organisation of separation.

It should be mentioned in particular that the main relationships obtained for the cascade processes are of the phenomenological nature. They may be used for various cascade processes, such as rectification, extraction, isotope separation, cascade drying of bulk materials, etc.

In the book, the authors generalise the results of experimental and theoretical investigations carried out by them in Russia (Ural Polytechnical Institute, Department of Silicate Technology) and in Israel (Institute for Applied Research and Department of Mathematics of the Ben Gurion University of the Negev and Department of Industrial engineering of the Sami Shamoon College of Engineering).

CONTENTS

Chapter I

GRAIN SIZE COMPOSITION OF BULK MATERIALS

1. METHODS OF DETERMINATION OF THE PARTICLE SIZE

The refining of materials in modern technology is carried out by different methods. In a large majority of cases, the products of refining consist of particles of irregular geometrical shapes and different sizes. In most cases, the dimensions of some grains in the products of refining are hundreds and thousand times larger than the dimensions of other grains. The difference in the size of the grains of bulk materials of natural origin (for example, river or sea and) is slightly smaller than that of the products of refining but their composition is polydisperse. Strictly speaking, in nature, there are no monodisperse materials.

In most cases, the experimental determination of the grain size composition of the products of refining and classification is carried using the methods of sieve, microscopic and sedimentation analysis. Sieve analysis gives a satisfactory result only for fractions larger than 0.04 mm. For particles smaller than 0.04 mm, the grain size composition is determined by the sedimentation or centrifuging methods. These methods are based on different rates of settling of particles of different sizes. The size of the smallest grains (smaller than 5 μm) is determined by laser scanning.

In classification in moving media, the separation of the material depends on the hydrodynamic properties of particles such as size, shape, surface condition, density, elasticity. The generally accepted analysis of the results of separation processes using sieves provides only a characteristic of the size of separation products and this is clearly insufficient for processes. However, the determination of the sizes of particles of refined materials, regardless of the apparently simpler procedure, is a relatively difficult task.

The particle size is evaluated using different characteristics of the particles. The concept of the radius or diameter of the particle is used in theoretical calculations. For particles of irregular shapes, the diameter greatly differs from the characteristic size. The concept of the Feret diameter or the Martin diameter is used in certain cases. This concept is introduced in laser or photo projection of particles on a plane. The Feret diameter is the maximum distance between the edges of a single particle in projection, and the Martin diameter is the length of a straight line which divides a particle into two equal parts on the projection area. These determinations are relatively complicated and cumbersome and, evidently, competent, if they are averaged out for a large number of particles and carried out using the same procedure. In this case, it is possible to calculate the mean size of the particles only by assuming that their orientation is random. Other characteristics are also used, for example, the largest or smallest particle size, the difference between the largest and smallest sizes, the mean size, specific surface, *etc.*

The combined characteristics of the size and shape of the particles include the concept of the 'equivalent' d_e, and sedimentometric d_s diameters.

The mean equivalent size of the particles is determined as the mean arithmetic value of three mutually orthogonal measurements of the particles:

$$d_e = \frac{a+b+c}{3}$$

or as the mean geometrical value

$$d_e = \sqrt[3]{abc}$$

where a, b, c are the sizes of the particles of irregular shape in three mutually perpendicular directions.

The mean equivalent diameter of a specific particle of irregular shape, determined by this procedure, may differ depending on the selected direction of the measurement axes.

The mean sedimentometric diameter, calculated from the hovering velocity, is also not unambiguous because for the laminar region of settling

$$d_s = \sqrt{\frac{18\mu v}{g(\rho - \rho_0)}}$$

and for the developed self-modelling region

$$d_s = \frac{0.33v^2\rho_0}{g(\rho_t - \rho_0)}$$

where μ is the kinematic coefficient of the viscosity of the medium, kg/m s; v is the velocity of settling of the particle, m/s; ρ is the density of the material, kg/m^3; ρ_0 is the density of the medium, kg/m^3; v is dynamic viscosity, m^2/s^2.

The equivalent diameter may be calculated if it is assumed that an equivalent sphere and a particle have the same volume:

$$d_e = \sqrt[3]{\frac{6V}{\pi}}$$

Here V is the volume of the particle. However, it is assumed that the specific surfaces of the particles and the equivalent sphere are equal:

$$d_e = \sqrt{\frac{S}{\pi}}$$

where S is the surface area of the particle.

It should be mentioned that the degree of deviation of the shape of real particles from the equivalent sphere is characterised by the shape factor which is the ratio of the surface of the sphere, equivalent in relation to the particle as regards volume, to the surface of the particle:

$$\psi = \frac{S_v}{S}$$

Another parameter, characterising the degree of deviation of the particle from the spherical shape, is the isometric coefficient, which is the ratio of three dimensions of the particle (the largest, medium and smallest), taken in three mutually perpendicular axes: a:b:c. In addition to the geometrical coefficient of the shape, there is also the dynamic coefficient, which takes into account the differences in the resistance of the particle and the equivalent sphere:

$$\psi_d = \frac{\lambda_p}{\lambda_s}$$

Here λ_p, λ_s are the coefficients of the resistance of the particle and the equivalent sphere, respectively.

If it is required to determine the mean size for a narrow class range of the particles, the equivalent diameter may be determined from the mean geometrical diameter from the size of the cells of adjacent sieves

$$d_e = \sqrt{X_i X_{i+1}}$$

For the material of narrow classes, for which the X_i / X_{i+1} ratio is small, it may be assumed with a sufficient degree of accuracy that

$$d_e = \frac{X_i + X_{i+1}}{2}$$

There also other equations which can be used, depending on which feature of dispersion is regarded as controlling:

The mean arithmetic diameter (mean-weighted)

$$d_e = \frac{\sum \gamma_i d_i}{\sum \gamma_i}$$

where γ_i is the fraction of the particle of the *i*-th class.

The harmonic diameter (based on the number of particles)

$$d_e = \frac{\sum \frac{\gamma_i}{d_i^3}}{\sum \frac{\gamma_i}{d_i^4}}$$

The mean logarithmic diameter

$$\lg d_e = \frac{\sum \gamma_i \lg d_i}{\sum \gamma_i}$$

The mean diameter with respect to the volume

$$d_e = \sqrt[3]{\frac{\sum \gamma_i d_i^3}{\sum \gamma_i}}$$

The mean diameter calculated as the ratio of the volume of the particles to the surface is

$$d_e = \frac{\sum \gamma_i d_i^{\ 3}}{\sum \gamma_i d_i^2}, \text{ and so on.}$$

All this indicates that the determination of the mean diameter of a single particle and, even more so, of a narrow range of the particles, is far from unambiguous. In practice, it is not possible to obtain the unambiguous and accurate numerical value of the size of one or a group of particles, forming a narrow fraction. Therefore, using the terminology of mathematical statistics, it should be accepted that the diameter of the particles of the products of refining and classification should be treated as an unidimensional random quantity.

In fact, if it is assumed that we have been successful in counting all the particles of the material and measuring the size of each particle in three mutually perpendicular directions, it is impossible to think that such detailed information on the product would be useless.

Therefore, it is usually necessary to accept the main idea of statistics which is averaging. The practical result of averaging is reduced to the fact that in this case it is necessary to use probabilities instead of reliabilities. Within the framework of this approach, it is not possible to talk about the specific shape and size of the particles and we should consider only the probability realisation of these parameters.

2. THE SIZE DISTRIBUTION OF PARTICLES

In some cases, a spot indicator is used to evaluate the size of particles. It is either the mean-weighted or median diameter. The median diameter of particles can be determined by recording the diameters of all particles in the order of their increase, with the determination of the diameter which divides these series into halves. Although these estimates are very simple, they are very inaccurate, because they completely ignore the size distribution of the particles.

The dispersion of the comminuted particles is characterised most accurately by the grain size composition. In this characterisation, it is necessary to determine not only the previously mentioned parameters but also the percent content of particles of each size.

The curves graphically predict the grain size composition of the material and the grain size characteristics. In order to understand the main concepts of this approach, further discussion will be carried out using a specific example associated with the determination of the dispersion of casting sands (Table 1).

Table 1. Dispersion characteristics of casting sands

Characteristic	Notation	Mesh size of the sieve, mm											
		2.5	1.6	1.0	0.63	0.40	0.315	0.20	0.16	0.1	0.063	0.05	Bottom
Partial residues	r,%	2.35	37.58	25.95	12.38	10.64	3.2	4.35	1.33	1.24	0.71	0.18	0
Total residues	R,%	2.35	39.93	65.88	78.26	88.9	92.1	96.45	97.78	99.02	99.73	99.91	100
Total passes	D,%	97.65	60.07	34.12	21.74	11.1	7.9	3.55	2.22	0.98	0.27	0.09	0

The size characteristics can be efficiently described by the distribution functions $D(x)$ of the mass of the material or by the associated function $R(x)$.

Function $D(x)$ is a total (cumulative) characteristic of dispersion, expressing in percent the ratio of the mass of all particles with a diameter smaller than x to the total mass of the refined material.

Function $R(x)$ is determined as the cumulative characteristic expressing the ratio (expressed in percent) of the mass of all particles with a diameter is larger than x, to the total mass of the material. Figure I-1 shows the curves plotted on the basis of Table 1.

Since at any point $D + R = 100\%$, the curves $D(x)$ and $R(x)$ intersect at a point where $D = R = 50\%$. This value ($D = 50\%$) is the parameter of distribution of the grains of the material or the distribution mode.

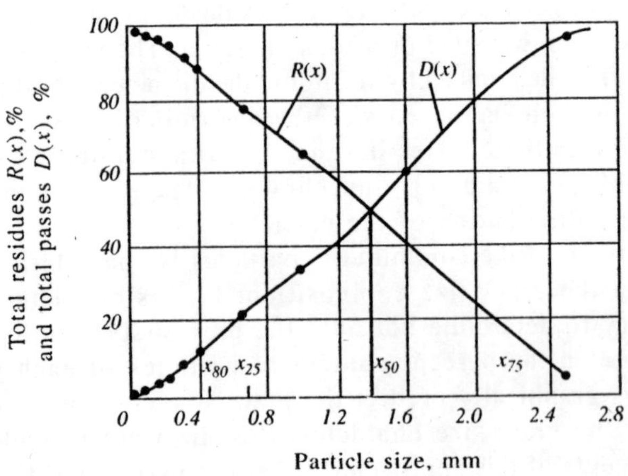

Fig. I-1. Grain size characteristics of casting sands in full residues $R(x)$ and full passes $D(x)$.

Another characteristic of the distribution is the value $D = 80\%$. This parameter represents the grain size with 80% (by mass) of all particles of the material smaller than this size. In some cases, the characteristic of the grain size composition is determined by dividing it into equal parts, *i.e.* determining, in addition to $D = 50\%$, also $D = 25\%$ and $D = 75\%$. These parameters should not be confused with the mean probability deviation, characteristic of the integral Gauss curve. These parameters provide general information on the nature of the grain size distribution of the disperse material.

The values of the distribution functions for all particle sizes could not be determined by experiments, and $D(x)$ and $R(x)$ are determined only for a limited number of points on the size axis $x_1 < x_2 < x_3 ... < x_n$, in which the function $D(x)$ has positive 'jumps', and $R(x)$ negative jumps.

Thus, the functions $D(x)$ and $R(x)$, obtained as a result of experiments, are discrete.

Indeed, the true distribution curve $D(x)$ or $R(x)$ can be obtained by adjusting the areas of the graph in such a manner that each part of the plane, restricted by the section of the curve on each step, has the area equal to the initial area of a rectangle. The equalisation of the area from the mathematical viewpoint is interpolation, and any interpolation of this type is not unambiguous. Therefore, the distribution curve, shown in Fig. I-2a, represents one of the many possible curves for the examined case. This can be clearly indicated graphically (see Fig.I-2b) showing different layers of the equalisation of the areas

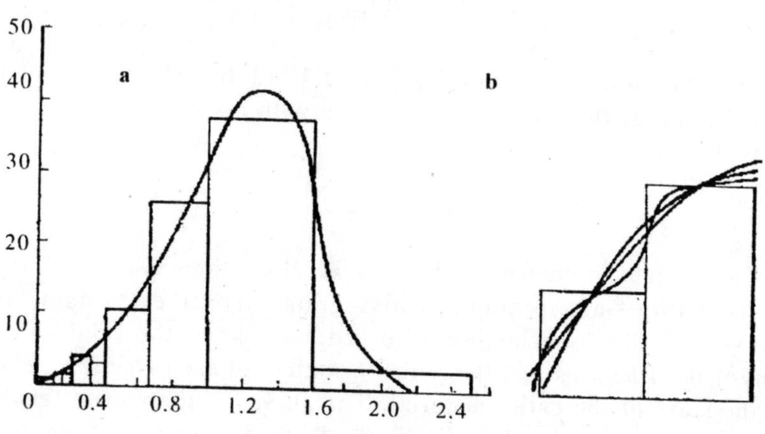

Fig. I-2. Grain size distribution (a). Methods of equalization of areas of the graph (b).

of the rectangle. In this interpretation, quantity x may be regarded as continuous with a sufficient degree of accuracy.

Two comments should be made in this case.

Firstly, the special feature of the determination of the continuous size distribution of grains of the material by means of fraction analysis is that the equalisation of the area is multiple-valued because the size distribution of particles inside each fraction remains unknown. It is therefore necessary to accept that the grain size characteristic, having the linear random quantity x as the abscissa, is itself a two-dimensional random function. This should affect the approximation of this function by different dependences.

Secondly, variable x acquires a continuous set of values. Therefore, the probability of some fixed fraction x_i present in the mixture of particles is equal to zero. At the same time, the sum of the content of particles of all fractions is 1 or 100%.

Here, there are no contradictions, because this is the exact analogy of the claim that the geometrical point has no length, and the section, *i.e.* the population of points, has a non-zero length. Therefore, the content of the particles of some size group in a mixture may be discussed only in relation to the range of sizes (from x to $x+\Delta x$).

Taking these comments into account, it may be assumed that the continuous monotonic distribution function $D(x)$ is differentiated everywhere and has a continuous derivative. This means that there is some function $n(x)$ which can be obtained by differentiating the distribution function $D(x)$ and which is continuous in the range

$$\int_{x_{min}}^{x_{max}} n(x)dx = D(x_{max}) - D(x_{min})$$

The function $n(x)$ is normalised to 100% by the density of the distribution of the mass of the particle with respect to the diameter of particles of partial residues:

$$n(x) = \frac{dD(x)}{dx}$$

The content in percent of the individual fractions is also displayed in the form of a step graph or histogram. Here, the abscissa gives the size of the particles, and the ordinate shows the relative content of the fractions, *i.e.* the percent content of each fraction, related to the mass of the entire material. The curves of the partial residues may be determined using the method of equalisation of the areas of the histograms.

It is believed that the total curves reflect the grain size composition more accurately than the partial ones. However, even for the total curve, unreliability is represented by the possibility of multiple-valued interpolation of 'accurate' points with respect to each other. The examined problem is based on a simple fact indicating that from several quantities it is possible derive an unambiguous mean quantity, but from this mean quantity it is not possible to derive again individual quantities from which the former was formed.

The curves of the partial residues are suitable for the analysis of the processes of refining and classification because they provide clear information on the grain size (fraction) composition of the disperse material. Therefore, they would be preferred in the conclusions and analysis in comparison with cumulative curves. It should be mentioned that the grain size characteristics of the same product, obtained as a result of the experiment, always differ depending on the dispersion analysis method used. This phenomena may ne attributed to the fact that the methods are not absolute. It is likely that this is more due to the fact that the grain size curve is a two-dimensional random function which can be constructed accurately only with a certain probability.

In most cases, the grain size composition is described by the approximation proposed by Rozin and Rummler. This dependence has the form of an exponential power equation of the type

$$\frac{100}{R} = e^{bx^a},$$

where R is the total residue on a sieve with the mesh size equal to x, mm; b and a are coefficients which are constant for a specific material.

In most cases, approximation of the dependence for practical application is carried out in double logarithmic coordinates. The equation of the grain size characteristic of the powder in this case is determined in the form of the equation by a straight line

$$\ln(\ln\frac{100}{R}) = a\ln x + \ln b$$

Regardless of extensive application, this dependence is far from universal. The authors themselves examined four types of grain size composition, with the application of this dependence having a special feature in each type. In processing the experimental material in the best case this dependence in the binary logarithmic coordinates gives a broken line consisting of straight sections. It should be mentioned that the double logarithmic transformation of the parameters conceals large deviations and, consequently, the equalisation of the dependences with respect to the points in these coordinates contains large errors which are several times higher than the value of the parameter. Attempts have been made so simplify this relationship.

S.E. Andreev proposed the following dependence for describing the total curve in respect of passes:

$$D = 100\left(\frac{x}{x_{max}}\right)^k,$$

Here D is the total pass through a sieve with the mesh x; x_{max} is the minimum size of such a sieve (in the direction of increasing size) on which the residue is equal to zero; k is a parameter characterising the curvature of the characteristic. Its value is less than unity on concave curves and higher than unity on convex curves.

V.A. Olevskii proposed the simplified exponential equation

$$\frac{100}{R} = e^{b_0 x}.$$

These relationships describe more or less satisfactorily only finely refined powders. If the grain size of the products increases, these relationships are violated. As indicated by experimental verification, these relationships do not describe the grain size composition of 'truncated' powders produced as a result of carrying out the processes of separation. They also describe unsatisfactorily the grain size characteristic of the powders obtained by methods other than refining, for example, the method of thermal atomisation.

Recently, the dependence proposed by A.N. Kolmogorov has been considered on an increasing scale. He showed that under some assumptions, the normal-logarithmic law of the size distribution of particles in successive refining can be obtained by a purely theoretical method. The main assumption in deriving this relationship was the independence of the probability that each particle splits into a certain number of parts in unit time. In this case, it is assumed that the size distribution of the particles, formed from the initial particle, does not depend on its initial size nor on the number of 'refining' cycles after which the particle formed.

This function has the form

$$D(x) = \int_{-\infty}^{\lg x} e^{-\frac{\lg x - \lg x_{50}}{2\lg x}}\, d\lg x$$

where x_{50} is the size which corresponds to R = 50% in the examined powder.

It should be mentioned that this relationship gives satisfactory results in many cases. In certain cases, it is not justified to use this relationship. Not going into detail, it should be mentioned that the refining process usually takes place with large deviations from the given assumptions. According to A.M. Shteinberg, in examination of ventilated mills, in milling there is always a lower limit in the size of the particles and this is not taken into account by the analytical dependence. The refined materials, separated into fractions, are also not described by the normal-logarithmic law. In addition to the examined relationships, other methods are also used in practice for the approximation of the grain size composition. The best-known methods are those proposed at different times by Martin, Weinig, Goden, Andreazen, Svensson, Griffiths, and others. They also have the shortcomings mentioned above.

Finally, it would be very tempting to find universal mathematical dependences for describing the grain size composition. These attempts are still being made. However, we believe that this task is absolutely hopeless. The point is that the grain size characteristic is a two-dimensional random function which can be constructed only with a specific probability.

All this leads to the problem of the necessity of excluding the mathematical description of the grain size characteristics when examining the processes of fractionation of powders. It is essential to analyse the principle of the fractionation process and optimise the process without preliminary analytical description of the composition of the separation products. Process parameters, invariant to the composition of the initial material, have been found and substantiated for this purpose. The tabulated method of determination of the grain size characteristic, which is the most simple, reliable and accurate method, has been proved to be fully sufficient for this analysis.

References

1. P.C. Reist, Introduction to aerosol science, Macmillan Publishing Company, London and New York (1987).
2. N.A. Fuks, Mechanics of aerosols, Publishing House of the USSR Academy of Sciences, Moscow (1955).
3. L.D. Landau and E.M. Lifshits, Mechanics of solids, GITTL, Moscow (1953).
4. P.A. Kouzov, Fundamental of analysis of the disperse composition of industrial powders and refined materials, Khimiya, Moscow (1974).
5. A.M. Shteinberg, Preparation of coarse dust of brown coal, Metallurgizdat, Moscow (1970).
6. S.E. Andreev, V.A. Zverevich and V.A. Petrov, Fragmentation, refining and screening of natural resources, Nedra, Moscow (1986).

7. M.D. Barsky, V.I. Revnitsev and Yu.V. Sokolkin, Gravitational classification of granular materials, Nedra, Moscow (1974).
8. M.D. Barsky, Optimisation of separation processes of granular materials, Nedra, Moscow (1978).
9. M.D. Barsky, Fractionation of powders, Nedra, Moscow (1980).

Chapter II

METHODS OF OPTIMISATION OF THE SEPARATION OF BINARY MIXTURES

1. DETERMINATION OF THE EFFICIENCY OF SEPARATION

It has been mentioned that the separation processes are used quite widely in different areas of industry. Irrespective of the separation object and equipment in which separation is carried out, the these processes combine general problems associated with the tendency to produce pure products with the maximum utilisation of initial resources taking economic efficiency into account.

A large number of methods have been proposed for evaluating the efficiency of the separation processes. The selection of the optimisation criteria on for these processes is still the subject of constant discussion in special literature. The complexity of the current situation is aggravated by the fact that the results of classification may be characterised by different factors: efficiency, extraction, contamination, concentration, the degree of enrichment, selectivity. ,From the mathematical viewpoint, some parameters are interpreted in different ways, for example, more than 30 equations have been proposed for determining the efficiency of separation. In addition to this, in enrichment, the separation parameters, associated with the separation curve, are used widely. Optimisation on the basis of the separation curve is also characterised by a number of parameters whose physical meaning is often unclear.

Many investigators simply assume that there are no common features between the quantitative methods of evaluation and the separation curves.

Recently, the literature concerned with this problems, has been paying special attention to quantitative parameters and the separation curves have not been used efficiently. This is not always justified and correct.

Therefore, it is important to examine these problems in greater detail.

It should be mentioned that all the parameters of evaluation of the completion of separation, used in practice, have a specific technical meaning and represent a specific information value. Evidently, as the optimisation criteria, it is necessary to select a single indicator, and all other characteristics of the process in relation to this indicator should be regarded as having secondary importance. The definition of the general criterion should be determined by the nature of problems to be solved by separation.

Regardless of the processing line in which the classifier operates, the purpose of the classifier remains unchanged – separation of the initial product in accordance with the given boundary size in such a manner as to obtain the maximum possible fraction difference in the classification products.

An objective method for optimization of the process must be based on the final separation of the feed material into coarse and fine fractions, as described below. The classifier, operating in the ideal fashion, should separate the fine products from the initial material and prevent its passage to the coarse product, *i.e.* it should ensure the maximum possible difference in the fraction composition of the separated target product.

The controlling main feature of the classification process is its mass nature, because the number of particles, taking part in separation, is infinitely large. Therefore, even in the case of a large difference in the properties of the separated particles, there is a certain probability of some of these particles not being included in the product. In a real process, this probability may increase as a result of imperfections in separation equipment.

Thus, in classification, the difference of the fraction compositions of both products differs from maximum. This difference in the operation of real classifiers in comparison with ideal equipment makes it possible to evaluate the degree of completion of separation, *i.e.* the efficiency of separation.

The most extensive and objective representation of the degree of completion of the processes is the main requirement on the method of evaluating the efficiency of separation.

This method should correspond to the following boundary conditions:

1. The numeral indicator should be higher for a process in which, with other conditions being equal, there is a larger difference in the fraction composition of both products.

2. For ideal classification, ensuring complete separation into fractions, the efficiency indicator should have the maximum possible value, which is often regarded as 100% or 1.

3. The efficiency indicator should be equal to zero, if the fraction composition of the classification products does not differ from the initial value, because in this case there is no classification and the material is simply separated into parts. The zero value of efficiency should also correspond to the inclusion of the entire amount of the material in a single product of classification (pneumatic transport regime), because in this case there is also no difference between the fraction composition of the initial mixture and the material of this yield.

4. The quantitative evaluation of efficiency should be unambiguous and should not depend on the type of products used to determine this efficiency, because any classifier, producing two products, carries out a single operation of separation of fine from coarse.

To determine the efficiency indicator corresponding to these requirements, it is important to examine these phenomena taking place during classification in the general form.

Initially, it is necessary to specify the condition of constancy of the initial composition of the binary mixture for all subsequent discussions in this chapter, because if this composition is not constant, additional requirements are imposed on the quality criterion.

A specific separation process will be examined when deriving this indicator. The initial composition will be denoted by the curve ABC (Fig. II-1). It is assumed that, because of technological considerations, this material should be separated on the boundary of the grain size x_0, mm.

The area of the graph, restricted by the curve ABC and the axis of the coordinates, corresponds on a specific scale to the total amount of the initial material. In the ideal case, separation should take place along Bx_0. In a real process, separation takes place with a certain error, *i.e.* part of the fines penetrate into the coarse product, and part of the coarse pieces into the fine product.

It is assumed that the fraction composition of the fine product is expressed by the curve EFK. Thus, in a general case, the are of the graph is divided by the lines of the ideal and real processes into four parts, where D_f is the amount of the fines, extracted into the fine product; D_c is the amount of the fines extracted into the

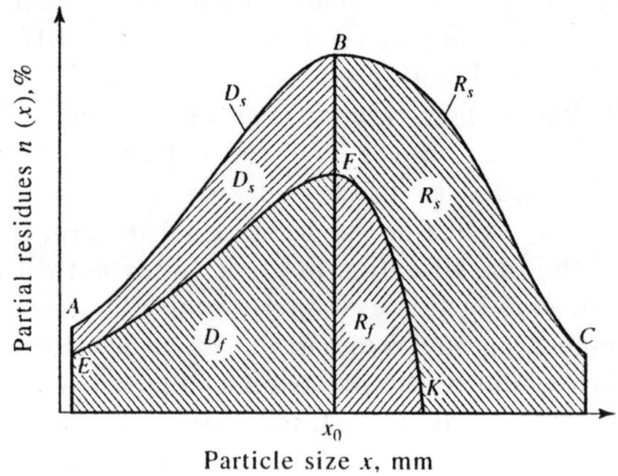

Fig. II-1. Characteristics of initial materials and products of classification in parial residues.

coarse product; R_f is the amount of coarse material included in the fine product; R_c is the amount of coarse material extracted into the coarse product, D_s and R_s is the amount of fine and coarse material, respectively, in the initial product. The dimension of these parameters should be expressed in kilograms, fractions of unity or percent of the total amount of the initial material.

The following equality is follow from Fig. II-1:

$$D_f + D_c = D_s \quad R_f + R_c = R_s$$

$$D_s + R_s = 1 \quad D_f + R_f = \gamma_f$$

$$D_c + R_c = \gamma_c \quad \gamma_c + \gamma_f = 1.$$

On the one hand, the fraction difference of the classification products increases with increasing efficiency of extraction. Extraction is the amount of the product of the given size separated as a result of classification of the product in relation to the amount of the same product in the initial material. Extraction may be determined from the following equations:

for the product of fine classes

$$\varepsilon_f = \frac{D_f}{D_s}$$

for the product of coarse classes

$$\varepsilon_c = \frac{R_c}{R_s}$$

On the other hand, the accuracy of separation decreases as a result of the fact that each product includes particles with the size exceeding the given separation boundary. Each of the separation products is contaminated by another product.

The extraction of the coarse material into fine material (contamination of the fine material) is determined from the expression

$$K_f = \frac{R_f}{R_s}$$

The extraction of the fine material into coarse material (contamination of coarse material) is determined from the equation:

$$K_c = \frac{D_c}{D_s}$$

It is clear that the efficiency of the classification process increases with increasing extraction and decreasing degree of contamination of the products. Neither the value of extraction nor the extent of contamination on their own provide complete information on the changes of the fraction composition of the classification cracks in relation to the initial material because extraction characterises the fraction composition on one side of the given separation boundary, and contamination on the other side. Only the combination of these quantities reflects fully the variation of the fraction composition, achieved in separation. Therefore, it is assumed that the functional relationship of these parameters when calculating the efficiency of classification is determined by the difference between the extraction of one of the products and its contamination. Efficiency is expressed by the following relationships:

for the fine product

$$E_f = \varepsilon_f - K_f \text{ and } E_f = \frac{D_f}{D_s} - \frac{R_f}{R_s} \tag{II-1}$$

for the coarse product

$$E_c = \varepsilon_c - K_c \text{ and } E_c = \frac{R_c}{R_s} - \frac{D_c}{D_s} \tag{II-2}$$

This representation of the relationship of these parameters is proved by the fulfilment of the previously mentioned boundary conditions.

Firstly, the efficiency of the process is determined by the difference in the fraction composition of the classification products, and increases with an increase in the difference in the fraction composition of these products.

Secondly, in the case of ideal separation, the efficiency of classification is 100%, because in this case there is no contamination and D_f = D_s and R_c = R_s.

Thirdly, in separation of the initial material into part without changes of the fraction composition, *i.e.* in the absence of classification, the efficiency of the process is equal to 0. In this case, $D_f = \gamma_f D_s$ and $R_f = \gamma_f R_s$.

Consequently,

$$E_f = (\gamma_f \frac{D_s}{D_s} - \gamma_f \frac{R_s}{R_s}) = \gamma_f - \gamma_f = 0$$

The same result is obtained when the entire material is included in one of the classification products.

Fourthly, the value of the parameter of efficiency is unambiguous and does not depend on the product used to determine this value. In order to prove this, it is sufficient to deduct equation (II-2) from (II-1):

$$E_f - E_c = \frac{D_f}{D_s} - \frac{R_f}{R_s} - \frac{R_c}{R_s} + \frac{D_c}{D_s} = (\frac{D_f}{D_s} + \frac{D_c}{D_s}) - (\frac{R_f}{R_s} + \frac{R_c}{R_s}) = \frac{D_s}{D_s} - \frac{R_s}{R_s} = 1 - 1 = 0$$

i.e. $E_f = E_c$.

Thus, equations (II-1) and (II-2), derived on the basis of analysis of the tasks solved by classification, characterised objectively and equally the efficiency of separation of the initial product and may be used as indicators determining the efficiency of the classification process.

For the continuous process, it is not convenient to determine the efficiency on the basis of the equations (II-1) and (II 2) because in this case it is necessary to use the results of analysis of the samples of initial material and both classification products.

These equations will be converted using the relationships of the content of the coarse and fine materials in the samples.

The following equalities are valid for the continuous product

For the fine product

$$\begin{aligned} \beta + b &= 1; \\ \beta\gamma_f &= D_f; \\ b\gamma_f &= R_f; \end{aligned} \qquad \text{(II-3)}$$

for the coarse product

$$v + f = 1;$$

$$v\gamma_c = D_c; \qquad \text{(II-4)}$$

$$f\gamma_c = R_c;$$

for the initial material

$$\alpha + a = 1;$$

$$\alpha = D_s; \qquad \text{(II-5)}$$

$$a = R_s,$$

In these equations, α is the content of the fines in the sample of the initial material; a is the content of coarse particles in the sample of the initial material; β is the content of the fines in the sample of the fine product; b is the content of the coarse particles in the sample of the fine product; v is the content of the fines in the sample of the coarse product; f is the content of the coarse particles in the sample of the coarse product.

Taking into account equations (II-3)–(II-5), the expressions (II-1) and (II-2) may be determined the following form:

$$E_f = \gamma_f\left(\frac{\beta}{\alpha} - \frac{b}{a}\right),$$

$$E_c = \gamma_c\left(\frac{f}{a} - \frac{v}{\alpha}\right).$$

In the practice of optimisation of the classification processes, it is necessary to use a large number of quality criteria. The analysis of these criteria may be carried out more efficiently by dividing them in advance into individual groups in which it is more efficient to stress their weak and strong aspects.

2. SIMPLIFIED OPTIMISATION INDICATORS

In industry, the efficiency of the classification process is evaluated using different methods. The efficiency of expressing the nature of these tasks, solved by classification, in these methods, will now be analysed.

In chemical and thermal engineering and some other branches of industry, the Rozin–Rummler method is used widely. According to this method, the efficiency of the classification process is characterised by the degree of extraction of the material.

This method, which accurately reflects the work of a dust separation system, is transferred to classification without taking its special features into account. It may be shown that this approach has disadvantages when used for the classification processes.

It is assumed that a certain amount of material with the dispersion characteristic, shown in Fig. II-2 by the curve ABC is to be separated in different systems. The conditions of the processes are selected in such a manner that the extraction of the fines in all cases remains the same. However, these processes may be differ because of the different amount of the large fractions, extracted with the fines.

The example shows that when using this method, the act of separation according to the Rozin–Rummler method, with different quality, will have the same numerical characteristic.

In a classifier, it is possible to create conditions in which the entire amount of the supply of material is included in one of the products of classification without any changes in the fraction composition. In this case, extraction is 100%, although no classification has taken place.

V.P. Romadin has proposed to characterise the classification process by two quantities – the extraction of both classification products.

N.G. Romankov and P.A. Yablonskii evaluate the efficiency of the process by two parameters, but they are related to a specific product of classification, *i.e.* extraction and contamination of the product. These parameters characterise most efficiently the separation process, but the absence of a united criterion complicates comparison of different separation processes. In fact, it is not easy to determine which of the two acts of separation is more efficient, if the results of these acts are as follow:

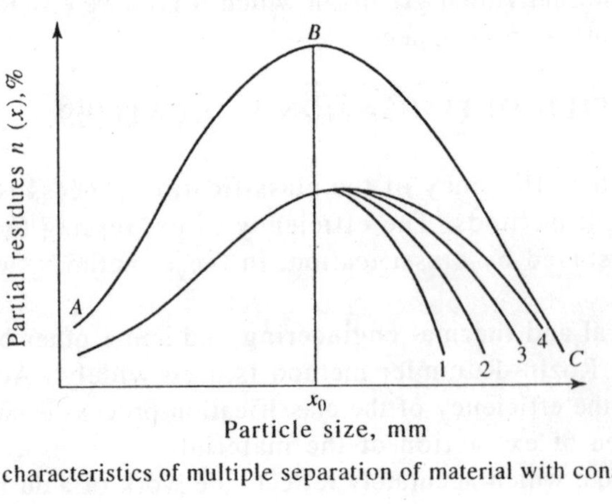

Fig.II-2. The characteristics of multiple separation of material with constant extraction.

$$1) \varepsilon_{f_1} = 95\%, \varepsilon_{c_1} = 71\% \quad \varepsilon_{f_2} = 89\%, \varepsilon_{c_2} = 77\%$$

$$2) \varepsilon_f = 93\%, K_f = 18\% \quad \varepsilon_{f_1} = 86\%, K_{f_1} = 11\%.$$

If the cost of the initial material is not high, the process is sometimes characterised by the quality of the completed product. These evaluations are carried out, for example, in separation of air into components or in the separation of isotopes of heavy hydrogen from water.

At a high degree of extraction, the separation process is sometimes characterised by the losses in the waste of by the yield of the target product.

These parameters are meaningful but it is clear that they reflect incompletely the nature of the tasks sold by separation and, consequently, they can be used as the main parameters of the process. In enrichment, it is often necessary to value the quality of separation on the bases of the degree of enrichment. The degree of enrichment is represented by the relationship:

$$\lambda = \frac{\beta}{\alpha}$$

Where α is the content of the valuable component in the initial material; β is the content of the same component in the enriched product.

In separation on the basis of size, this expression can be related to the same degree to both the fine product and coarse product. For better understanding, it is assumed that in these and in all other subsequent cases, the valuable component is the fine material.

The correspondence of this parameter to the previously formulated boundary conditions will be evaluated:

1. In the case of ideal separation

$$\beta = 1, \alpha = \gamma_f$$

$$\lambda = \frac{D_s}{\gamma_f D_s} = \frac{1}{\gamma_f};$$

2. In separation into parts without any change of the initial and fraction compositions:

$$D_f = \gamma_f D_s,$$

$$\lambda = \frac{\gamma_f D_s}{\gamma_f D_s} = 1;$$

3. For the unambiguous case

$$\lambda_f - \lambda_c = \frac{\beta}{\alpha} - \frac{b}{a} = \frac{D_f}{\gamma_f D_s} - \frac{R_c}{\gamma_c R_s} \neq 0$$

Analysis shows that this indicator is of a limited value and is more suitable for optimising the process in a general case.

It has been attempted two express the efficiency of separation by the ratio of the content of the useful substance in the concentrate on the losses of the substance in the waste:

$$E = \frac{\beta}{v}$$

This is a slightly modified parameter of the degree of enrichment. This equation will now be analysed:

1. In the case of ideal separation

$$v = 0, \beta\gamma_f = D_s$$

This means that in this case $E \to \infty$;

2. In the absence of classification

$$\beta\gamma_f = D_f = \gamma_f D_s$$

$$D_c = (1-\gamma_f)D_s = v(1-\gamma_c)$$

$$E = \frac{D_s}{D_s} = 1$$

which appears to be meaningless;

3. The value of efficiency, calculated from this method, depends on the classification product to which it is related, since

$$E_f - E_c = \frac{D_f \gamma_c}{D_c \gamma_f} - \frac{R_c \gamma_f}{R_f \gamma_c} \neq 0$$

This shows that this method has the same shortcomings as the previous method.

In some cases, the united indicator values for the evaluation of the quality of separation is represented by a criterion describing the product of the extraction by the content of the useful substance in the concentrate:

$$E = \varepsilon\beta$$

This equation may be presented in the following form:

$$E = \frac{D_f D_f}{D_s \gamma_f} = \frac{D_f^2}{D_s \gamma_f}$$

The equation will now be analysed:

1. In the case of ideal separation

$$D_f = D_s = \gamma_f$$

$$E = \frac{D_f^2}{D_f^2} = 1$$

2. In separation without any change of the fraction composition:

$$D_f = \gamma_f D_s$$

$$E = \frac{\gamma_f^2 D_s^2}{D_s \gamma_f} = \gamma_f D_s$$

In this case, the quantitative indicator of the process differs from zero although there are no changes in the fraction composition.

3. This indicator is not unambiguous since

$$E_f - E_c = \frac{D_f}{D_s}\frac{D_f}{\gamma_f} - \frac{R_c}{R_s}\frac{R_c}{\gamma_c} \neq 0$$

Madzhumdar has proposed an identical indicator differing only by the fact that the extraction in this indicator is multiplied by the degree of enrichment

$$E = \varepsilon\lambda$$

This equation may be used to the following form:

$$E = \varepsilon\frac{\beta}{\alpha} = \frac{D_f}{D_s}\frac{D_f}{D_s \gamma_f} = \frac{\varepsilon_f^2}{\gamma_f}$$

The equation will now be analysed:

1. In the case of ideal separation

$$E = \frac{1}{\gamma_f}$$

2. In the absence of separation

$$E=\frac{\gamma_f^2 D_s^2}{D_s^2 \gamma_f}=\gamma_f$$

This indicator is also not unambiguous.

This shows that the Mandzumdar indicator does not completely describes the meaning of the tasks solved by separation processes so that it cannot become the main parameter of optimisation.

Heidenreich has proposed, as the independent indicator of the separation process, to use the following expression

$$\mu=\frac{\alpha-\upsilon}{(\beta-\upsilon)\alpha} \qquad \text{(II-6)}$$

Prior to analysing this equation, it will have to be transformed. For this purpose, the balance of the content of the fines in the initial product and in both classification products will be determined:

$$\alpha=\gamma_f \beta+\gamma_c \upsilon$$

It is now necessary to determine γ_f

$$\alpha=\gamma_f \beta+(1-\gamma_f)\upsilon=\gamma_f \beta+\upsilon-\gamma_f \upsilon=\gamma_f(\beta-\upsilon)+\upsilon;$$

$$\gamma_f=\frac{\alpha-\upsilon}{\beta-\upsilon}$$

Taking this relationship into account, expression (II-6) may be presented in the following form:

$$\mu=\frac{\gamma_f}{\alpha}=\frac{\gamma_f}{D_s}=\frac{D_f+R_f}{D_s}$$

Consequently:

1. In the case of ideal separation

$$D_f=D_s; R_f=0; \mu=1;$$

2. In the absence of classification

$$\mu=\frac{\alpha-\upsilon}{\beta-\upsilon}\frac{1}{\alpha}=\frac{D_s-\frac{D_c}{\gamma_c}}{\frac{D_f}{\gamma_f}-\frac{D_c}{\gamma_c}}\frac{1}{D_s}$$

In this case, the following relationships are valid:

$$D_f = \gamma_f D_s ; D_c = D_s \gamma_c$$

These relationships will be substituted into the resultant equation

$$\mu = \frac{D_s - D_s}{D_s - D_s} \frac{1}{D_s}$$

In the numerator and the denominator there is a zero which leads to an indeterminacy

3. The unambiguity may be verified in the basis of the expression

$$\mu_f - \mu_c = \frac{\gamma_f}{D_s} - \frac{\gamma_c}{R_s}$$

The equality of this expression to zero is possible only in a partial case, at $\gamma_f = D_s \gamma_c = R_s$. In all other cases, there is no unambiguity.

Lincoln has proposed the following equation for determining the efficiency:

$$E = \frac{\beta + \upsilon}{2}$$

This expression does not correspond to the meaning of the tasks solved by classification. In fact, the Lincoln indicator which has no unambiguity is equal to $\beta/2$ in the case of ideal separation, and in separation without change of the fraction composition it is not equal to 0.

All the examine criteria are of interest only from the viewpoint of the characteristic of a specific aspect of the separation process. However, because of these shortcomings, they cannot become the main criteria of quality for the optimisation of the classification processes.

3. UNIQUE INDICATORS OF THE PROCESS

The best known definition of efficiency is that proposed by Hancock and Luyken. Efficiency is the ratio of the actual difference between the extraction of the particles of the given size into the fines and the yield of the fines to the theoretically possible difference (expressed in percent), i.e.

$$E = \frac{\varepsilon_f - \gamma_f}{\varepsilon_{max} - \gamma_{max}} \quad \text{(II-7)}$$

However, the theoretical maximum extraction is equal to 1 and the maximum yield in the conditions of ideal separation is equal to the content of the fines in the initial mixture. Consequently, equation (II-7) in the final form may be written as follows:

$$E = \frac{\varepsilon_f - \lambda_f}{1-\alpha} \tag{II-8}$$

The Hancock–Luyken criterion corresponds to all boundary conditions formulated previously for the analysis of the indicators of the separation process:

1. Ideal separation

$$D_f = D_s = \gamma_f, \varepsilon_f = 1,$$

$$E = \frac{1-\gamma_f}{1-\gamma_f} = 1;$$

2. Separation without any change in the fraction composition

$$D_f = \gamma_f D_s,$$

$$E = \frac{\dfrac{\gamma_f D_s}{D_s} - \gamma_f}{1 - D_s} = 0;$$

In the absence of classification

$$E = \frac{1-1}{R_s} = 0;$$

3. The unambiguity of the criteria may be verified on the basis of the difference:

$$E_f - E_c = \frac{\varepsilon_f - \gamma_f}{1-\alpha} - \frac{\varepsilon_c - \gamma_c}{1-\alpha} = \frac{\varepsilon_f - \gamma_f}{R_s} - \frac{\varepsilon_c - \gamma_c}{D_s} =$$

$$= \frac{\varepsilon_f D_s - \gamma_f D_s - \varepsilon_c R_s + \gamma_c R_s}{R_s D_s} =$$

$$= \frac{D_f - D_f D_s - R_f(1-R_s) - R_c + R_c R_s + D_c(1-D_s)}{R_s D_s} =$$

$$= \frac{D_f - D_f D_s - R_f + R_f R_s - R_c + R_c R_s + D_c - D_c D_s}{R_s D_s} =$$

$$= \frac{D_s - R_s - D_s(D_f + D_c) + R_s(R_f + R_c)}{R_s D_s} =$$

$$= \frac{D_s(1 - D_s) - R_s(1 - R_s)}{R_s D_s} = \frac{D_s R_s - R_s D_s}{R_s D_s} = 1 - 1 = 0$$

$$E_f = E_c$$

The equations proposed by Dean, Madel' and Tyurenkov, reduced to equation (II-8), are also used widely.

Olevskii shows that the equations derived by Luyken and Dean are identical and that the equations derived by Madel' and Tyurenkov can also be reduced to that.

The Chechot equation is used widely in science and practice of enrichment

$$E = \frac{\gamma_f(\beta - \alpha)}{\alpha(1 - \alpha)} \tag{II-9}$$

This equation can be transformed into the following form.

$$E = \frac{\gamma_f \beta - \gamma_f D_s}{D_s R_s} = \frac{D_f}{D_s R_s} - \frac{\gamma_f}{R_s} = \frac{\varepsilon_f}{R_s} - \frac{\gamma_f}{R_s} = \frac{\varepsilon_f - \gamma_f}{R_s} = \frac{\varepsilon_f - \gamma_f}{(1 - \alpha)}$$

The resultant equation indicates that the Chechot equation is also reduced to equation (II-9).

The relatively original physical meaning for understanding efficiency was proposed by G.W. Newton and W.G. Newton. Each of the separated material should be regarded as a mixture of two compositions: initial composition and separated by 100%. Finding the difference in the relative content of these elements, they determined the dependence for the calculation of efficiency:

$$E = \frac{\gamma_f - \gamma_f(\frac{1 - \beta}{1 - \alpha})}{\alpha}$$

After several transformations of this equation:

$$E=\frac{\gamma_f-\frac{\gamma_f-\gamma_f\beta}{1-\alpha}}{\alpha}=\frac{\gamma_f-\frac{\gamma_f-D_f}{1-D_s}}{D_s}=\frac{\gamma_f}{D_s}-\frac{\gamma_f}{D_sR_s}+\frac{D_f}{D_sR_s}=$$

$$=\frac{D_f}{D_sR_s}-\frac{\gamma_f(1-R_s)}{D_sR_s}=\frac{\varepsilon_f}{R_s}-\frac{\gamma_f}{R_s}=\frac{\varepsilon_f-\gamma_f}{1-\alpha}$$

This result indicates that this expression is also reduced to the Hancock–Luyken equation.

Lyashchenko derived an equation for evaluating the quality of the enrichment process, expressed on the basis of the content of the valuable component in the initial material and in the separation products:

$$E=\frac{(\beta-\alpha)(\alpha-\upsilon)}{(1-\alpha)(\beta-\upsilon)}\frac{1}{\alpha}$$

A very similar equation, differing only in the sign of the members, was proposed by Hauser:

$$E=\frac{(\alpha-\beta)(\upsilon-\alpha)}{\alpha(1-\alpha)(\upsilon-\beta)}$$

It may be shown that both these equations are identical with equation (II-8). Previously, it was shown that

$$\gamma_f=\frac{\alpha-\upsilon}{\beta-\upsilon}$$

and, consequently, one can write the following equation

$$E=\frac{\gamma_f}{\alpha}\frac{(\alpha-\beta)}{(1-\alpha)}$$

which is identical with equation (II-9) which is reduced to the Hancock–Luyken equation.

Another equation was derived by Arkhipov:

$$E=\frac{(1-\gamma_f)(\alpha-\upsilon)}{\alpha(1-\alpha)} \tag{II-10}$$

After several transformations of the equation

$$E=\frac{\gamma_cD_s-\gamma_c\upsilon}{D_sR_s}=\frac{D_s-D_s\gamma_f-D_s+D_f}{D_sR_s}=\frac{D_f}{D_sR_s}-\frac{\gamma_fD_s}{D_sR_s}=\frac{\varepsilon_f-\gamma_f}{1-\alpha}$$

which indicates that the equations (II-8) and (II-10) are identical.

Klyachin and Nikitin used the Hancock equation in the form of the ratio of the coefficients of the real and ideal processes:

$$K_c = \frac{\beta - \alpha}{\beta}; K_i = \frac{\beta_i - \alpha}{\beta_i} \qquad \text{(II-11)}$$

The ratio of these values for the case of classification gives

$$E = \varepsilon \frac{(\frac{\beta - \alpha}{\beta})}{(1-\alpha)}$$

It may be shown that this equation can be reduced to equation (II-8)

$$E = \varepsilon_f \frac{(\beta - \alpha)}{(1-\alpha)\beta} = \varepsilon_f \frac{(\beta - \alpha)}{\beta R_s} = \varepsilon_f (\frac{1}{R_s} - \frac{\alpha}{\beta R_s}) =$$

$$= \frac{\varepsilon_f}{R_s} - \frac{\varepsilon D_s \gamma_f}{D_f R_s} = \frac{\varepsilon_f}{R_s} - \frac{\gamma_f}{R_s} = \frac{\varepsilon_f - \gamma_f}{1-\alpha}$$

Other interpretations of the Hancock equation are also available in the literature.

It may be shown that the dependences (II-1) and (II-2), obtained from the analysis of the boundary conditions of classification, are also identical with the expression (II-8)

$$E_f = \frac{D_f}{D_s} - \frac{R_f}{R_s} = \frac{D_f(1-D_s) - R_f D_s}{D_s R_s} = \frac{D_f - D_s(D_f + R_f)}{D_s R_s} =$$

$$= \frac{\varepsilon_f}{R_s} - \frac{\gamma_f}{R_s} = \frac{\varepsilon_f - \gamma_f}{R_s} = \frac{\varepsilon_f - \gamma_f}{1-\alpha}$$

Thus, equations (II-1) and (II-2) do not give any new method of evaluating the efficiency of classification because they are reduced to the previously available equations. However, analysis carried out by means of the proposed boundary conditions shows the principal feasibility of using the Hancock equation (naturally, also all its modifications) for optimising the separation of binary mixtures of a constant initial composition, and also for the determination of weak points of other methods.

PART 4. ANALYSIS OF THE CRITERIA OF QUALITY OF SEPARATION PROCESSES, DIFFERING FROM THE HANCOCK METHOD

In addition to the previously mentioned methods of evaluating the efficiency of separation processes, there are also a number of other methods, claiming the objective representation of the tasks solved by classification. They include the Diamond method. According to Diamond, the efficiency of the enrichment process is the half sum of the extraction of the classification products:

$$E = \frac{\varepsilon_f + \varepsilon_c}{2} \quad \text{(II-12)}$$

The correspondence of this dependence to the boundary conditions will be verified:

1. In the case of ideal separation

$$E = \frac{1+1}{2} = 1$$

2. In separation without any change of the fraction composition

$$\varepsilon_f = \gamma_f; \varepsilon_c = \gamma_{c;}$$

$$E = \frac{\gamma_f + \gamma_c}{2} = \frac{1}{2}$$

Condition 2 shows that in the absence of the fraction difference in the classification products in comparison with the initial material, calculations carried out using the Diamond equation give the results differing from zero.

It is often recommended to use the Fomenko method as a criterion of optimisation of separation processes. The efficiency of the enrichment process according to Fomenko is represented by the product of the extraction of the classification products:

$$E = \varepsilon_f \varepsilon_c \quad \text{(II-13)}$$

Analysis of these equations shows that it has the same shortcoming as the Diamond method. In fact, in separation of the material into parts without changes of the fraction composition, the equation gives results different from zero. This will be indicated on the following example. It is assumed that in the given classification, the material was separated into halves. In this case, ε_f = 0.5 = 50%; ε = 0.5 = 50% and E = 50 × 50:100 % = 25% which contradicts the meaning because there is no classification.

The equations proposed by Tsiperovich has the same shortcoming:

$$E = \sqrt{\varepsilon_f \varepsilon_c} \tag{II-14}$$

Drakely proposed the following dependence for determining the efficiency of the separation process:

$$E = \frac{(\beta - \alpha)}{(1 - \alpha)} \frac{\beta(\alpha - \upsilon)}{\alpha(\beta - \upsilon)} \tag{II-15}$$

This equation will be transformed several times. Taking into account equation (II-6), it may be written that:

$$E = \frac{\beta - \alpha}{1 - \alpha} \frac{\beta}{\alpha} \gamma_f = \frac{\beta\gamma_f - \alpha\gamma_f}{R_s D_s} \beta = (\frac{\varepsilon_f}{R_s} - \frac{\gamma_f}{R_s})\beta = \frac{\varepsilon_f - \gamma_f}{1 - \alpha} \beta$$

This means that the Drakely equation is the Hancock–Luyken equation, additionally multiplied by the content of the valuable component in the concentrate. This complication of the dependence is hardly justifiable.

The Drakely equation distorts the quantitative characteristic of the process because in the case of lower efficiency one can obtain a high content of the valuable component in the concentrate and vice versa. In addition to this, the equation is not unambiguous, and although it has been recommended by a number of authors for determining the efficiency in enrichment of ores, it is likely that it should not be used because of the above shortcoming.

Stevens and Collins proposed a slightly different dependence in the form of the criterion for separation processes:

$$E = \varepsilon \frac{\beta - \alpha}{\alpha} \tag{II-16}$$

This equation can be slightly transformed,

$$E = \varepsilon(\frac{\beta}{\alpha} - 1)$$

The resultant equation will be analysed:

1. In ideal separation

$$\varepsilon = 1; \beta = 1; E = \frac{1}{\alpha} - 1;$$

2. In separation without any change of the fraction composition

$$\beta = \alpha; E = 0;$$

3. This equation is not unambiguous since

$$E_f - E_c \frac{\varepsilon_f^2}{\gamma_f} - \frac{\varepsilon_c^2}{\gamma_c} - \varepsilon_f + \varepsilon_c \neq 0$$

The equation does not correspond the first and third of the boundary conditions.

Lyashchenko proposed a slightly different dependence for the degree of enrichment for evaluating the quality of the process

$$\lambda = \frac{\beta - \alpha}{1 - \alpha} \quad \text{(II-17)}$$

This equation may be in the following form

$$\lambda = \frac{\frac{D_f}{\gamma_f} - D_s}{R_s} = \frac{D_f - \gamma_f D_s}{R_s D_s} = \frac{D_s(\varepsilon_f - \gamma_f)}{\gamma_f (R_s)}$$

The equation will now be analysed

1. In ideal separation

$$E = \frac{\gamma_f - \gamma_f D_s}{\gamma_f R_s} = \frac{1 - D_s}{R_s} = 1;$$

2. In separation without any change of the composition

$$D_f = \gamma_f D_s;$$

$$\lambda = \frac{\gamma_f D_s - \gamma_f D_s}{\gamma_f R_s} = 0;$$

3. The condition of unambiguity may be verified for the dependence

$$\lambda_f - \lambda_c = \frac{D_s(\varepsilon_f - \gamma_f)}{\gamma_f R_s} - \frac{R_s(\varepsilon_c - \gamma_c)}{\gamma_c D_s}$$

The Hancock–Luyken equation shows that

$$\frac{\varepsilon_f - \gamma_f}{R_s} = \frac{\varepsilon_c - \gamma_c}{D_s}$$

Consequently,

$$\lambda_f - \lambda_c = (\frac{\varepsilon_f - \gamma_f}{R_s})(\frac{D_s}{\gamma_f} - \frac{R_s}{\gamma_c})$$

To fulfill the conditions of unambiguity, it is necessary and sufficient that

$$\lambda_f - \lambda_c = 0$$

It is completely clear that

$$\frac{\varepsilon_f - \gamma_f}{R_s} \neq 0,$$

and in the general case

$$\frac{D_s}{\gamma_f} \neq \frac{R_s}{\gamma_c}$$

Thus, the Lyashchenko equation for the determination of the degree of enrichment is quite close to the Hancock–Luyken equation in its properties and differs from it by the fact that the unambiguity condition is not fulfilled.

Trushelevich proposed an equation for the efficiency of the process in the form of the modulus of the difference in the degree of enrichment in enriched and depleted products:

$$E = \left| \frac{\beta}{\alpha} - \frac{\upsilon}{\alpha} \right| \tag{II-18}$$

This dependence will be examined:

1. In ideal separation

$$\beta = 1; \upsilon = 0; E = \frac{1}{\alpha};$$

2. In separation into parts

$$\beta = \alpha; \upsilon = \alpha; E = 0;$$

3. The condition of unambiguity is verified using the expression

$$E_f - E_c = \frac{\beta - \upsilon}{\alpha} - \frac{f - b}{a} = \frac{\frac{D_f}{\gamma_f} - \frac{D_c}{\gamma_c}}{D_s} - \frac{\frac{R_c}{\gamma_c} - \frac{R_f}{\gamma_f}}{R_s} =$$

$$= \frac{\varepsilon_f}{\gamma_f} - \frac{K_c}{\gamma_c} - \frac{\varepsilon_c}{\gamma_c} + \frac{K_f}{\gamma_f} = \frac{\varepsilon_f + K_f}{\gamma_f} - \frac{\varepsilon_c + K_c}{\gamma_c} =$$

$$= \frac{\varepsilon_f + 1 - \varepsilon_c}{\gamma_f} - \frac{\varepsilon_c - \varepsilon_f + 1}{\gamma_c} \neq 0$$

This equation is not equal to zero because the numerators in both fractions are equal to each other with respect of the modulus and the denominators differ. Although the equations proposed by Trushelevich, like Lyashchenko's method, is relatively close to the Hancock method, it also does not satisfy the unambiguity condition. Chapman and Mott proposed to determine the efficiency of enrichment using the equation

$$E = k\frac{\beta}{\upsilon} \qquad \text{(II-19)}$$

where k is a constant quantity which depends of the composition of the initial mixture. This equation characterises very approximately the degree of separation of material into two products, as clearly indicated by its analysis:

1. In the case of ideal separation

$$\beta = 1; \upsilon = 0; E \to \infty$$

2. In separation into parts

$$\beta = \alpha; \upsilon = \alpha; E = k;$$

3. $$E_f - E_c = k(\frac{\beta}{\upsilon} - \frac{f}{b}) = k(\frac{D_f}{D_c}\frac{\gamma_c}{\gamma_f} - \frac{R_c}{R_f}\frac{\gamma_f}{\gamma_c}) \neq 0$$

This indicates the absence of the property of unambiguity for the relationship (II-19).

The identical shortcoming is also characteristic of the Hamilton equation:

$$E = \frac{\alpha(1-\upsilon)}{\beta} \qquad \text{(II-20)}$$

This equation may be presented in the following form

$$E = \frac{D_s(1 - \frac{D_c}{\gamma_c})\gamma_f}{D_f}$$

This equation will be transformed to the following form

$$E = \frac{D_s(\gamma_c - D_c)\gamma_f}{D_f\gamma_c} = \frac{D_s R_c \gamma_f}{D_f \gamma_c}$$

Analysis will be carried out for:

1. Ideal separation

$$D_f = D_s = \gamma_f$$

$$R_c = R_s = \gamma_c$$

In this case

$$E = \frac{D_s R_s D_s}{D_s R_s} = D_s$$

2. In simple separation into parts without any change of the fraction composition

$$D_f = \gamma_f D_s$$

$$R_c = \gamma_c R_s$$

$$E = \frac{D_s \gamma_c R_s \gamma_f}{\gamma_f D_s R_s} = \gamma_c$$

3. This equation does not have an unambiguity.

M.A. Goden proposed the following indicator of efficiency:

$$E = \frac{\beta}{\upsilon}\frac{(1-\upsilon)}{(1-\beta)} \qquad \text{(II-21)}$$

This indicator will be transformed to the following form:

$$E = \frac{\beta}{\upsilon}\frac{f}{b} = \frac{\frac{D_f}{\gamma_f}\frac{R_c}{\gamma_c}}{\frac{D_c}{\gamma_c}\frac{R_f}{\gamma_f}} = \frac{D_f R_c}{D_c R_f}$$

The degree of correspondence of this indicator to the nature of problems to be solved by classification will be expressed:

1. The Goden indicator is characterised by unambiguity since

$$E_f - E_c = \frac{D_f R_c}{D_c R_f} - \frac{R_c D_f}{R_f D_c} = 0;$$

2. In the case of ideal separation

$$\beta = 1; \upsilon = 0; E = \frac{1x1}{0},$$

which has no specific value;

3. In simple separation into parts

$$\beta = \alpha; \upsilon = \alpha; E = \frac{\alpha(1-\alpha)}{\alpha(1-\alpha)} = 1,$$

although no changes have taken place in the fraction composition.

A.M. Rozen slightly improved the equation derived by M.A. Goden and proposed to use this equation in the form of the following dependence:

$$E = \frac{\beta(1-\upsilon)}{\upsilon(1-\beta)} - 1. \qquad \text{(II-22)}$$

The Rozen equation, like the Goden equation, is unambiguous, but it is not suitable for optimisation of the separation processes because in the case of ideal separation and in the case when the entire material is included in the same classification product the calculation of the indicator leads to an indeterminacy.

Similar dependences have been proposed for evaluating the quality of the separation process by K. Cohen in the form:

$$E = \frac{\beta(1-\alpha)}{\alpha(1-\beta)} \qquad \text{(II-23)}$$

or

$$E = \frac{\beta(1-\alpha)}{\alpha(1-\beta)} - 1. \qquad \text{(II-24).}$$

We shall carry out successive analysis of these equations. This dependence may be presented in the following form:

$$E = \frac{D_f(1-D_s)}{\gamma_f D_s(1-\frac{D_f}{\gamma_f})}$$

We shall carry out transformations:

$$E = \frac{D_f R_s \gamma_f}{D_s \gamma_f(\gamma_f - D_f)} = \frac{D_f}{D_s}\frac{R_s}{R_f} = \frac{\varepsilon_f}{K_f};$$

1. For the case of ideal separation

$$\varepsilon_f = 1; K_f = 0; E = \frac{1}{0} \to \infty;$$

2. In separation into parts without any change of the fraction

composition

$$E = \frac{\gamma_f D_s}{D_s} \frac{R_s}{\gamma_f R_s} = 1.$$

3. The unambiguity may be determined from the relationship

$$E_f - E_c = \frac{\varepsilon_f}{K_f} - \frac{\varepsilon_c}{K_c} \neq 0$$

The dependence (II-24) will be examined:

$$E = \frac{\varepsilon_f}{K_f} - 1 = \frac{\varepsilon_f - K_f}{K_f}$$

Equation (II-24) is the Hancock–Luyken indicator, divided by the contamination of the fine product. This addition to the Hancock equation, which reflects accurately the meaning of the separation process, is hardly justified. This is clearly indicated by analysis of the dependence:

1. For ideal separation

$$E = \frac{1-0}{0} \to \infty;$$

2. For separation into parts

$$E = \frac{\gamma_f - \gamma_f}{D_s R_f} = 0;$$

and for the case in which the entire amount of the material is in the same product:

$$\varepsilon_f = 1; K_f = 1; E = 0;$$

3. The unambiguity condition

$$E_f - E_c = \frac{E}{K_f} - \frac{E}{K_c} = E(\frac{1}{K_f} - \frac{1}{K_c}) \neq 0.$$

This equation is not equal to zero because in a general case $K_f \neq K_c$

According to E. Douglas, the efficiency of the separation process is the product of the efficiency of extraction of the useful element into the concentrate, multiplied by the efficiency of removal of rock from the concentrate. Taking into account the factor that in enrichment, the quantitative increase of the amount of the valuable component in the concentrate is

$$\gamma_f(\beta-\alpha)$$

and the theoretically possible increase is

$$\alpha(1-\gamma_f),$$

E. Douglas presented the efficiency of extraction in the following form

$$E_\varepsilon = \frac{\gamma_f(\beta-\alpha)}{\alpha(1-\gamma_f)}$$

The total amount of the material, transferred into the tails, is $\gamma_f(\beta-\alpha)$ and the theoretically possible amount of the material is $\gamma_f(1-\alpha)$. Consequently, the efficiency of removal of rock will be

$$E_c = \frac{\gamma_f(\beta-\alpha)}{\gamma_f(1-\alpha)} = \frac{\beta-\alpha}{1-\alpha}$$

According to Douglas, the total efficiency is

$$E = E_\varepsilon E_k = \frac{\gamma_f(\beta-\alpha)(\beta-\alpha)}{(1-\gamma_f)(1-\alpha)\alpha}$$

Using the relationship $\gamma\beta = \varepsilon\alpha$, E. Douglas finally obtained the following equation:

$$E = \frac{(\varepsilon-\gamma)(\beta-\alpha)}{(1-\gamma)(1-\alpha)} \qquad \text{(II-25)}$$

It may easily be seen that this dependence is the Hancock equation multiplied by the expression $\frac{\beta-\alpha}{1-\alpha}$, i.e.

$$E = E_x \frac{\beta-\alpha}{1-\alpha}$$

This addition to the Hancock equation is hardly justified. Increase in the complexity of the dependence by means of the multiplication, carried out by E. Douglas, removes the advantages of the Hancock equation. This is indicated by analysis of the Douglas equation:

1. For ideal separation

$$E = 1\frac{1-\alpha}{1-\alpha} = 1;$$

2. For separation into parts without any change in the composition

$$E = 0\frac{0}{1-\alpha} = 0.$$

resulting in an indeterminacy;

3. The unambiguity may be determined from the analysis of the equation

$$E_f - E_c = E_x\left[\frac{\beta - D_s}{R_s} - \frac{b - R_s}{D_s}\right] \neq 0.$$

M. Vaga used a very unusual method to evaluate the efficiency of enrichment processes:

$$E = \frac{\beta - \alpha}{1 - \varepsilon_f} \qquad \text{(II-26)}$$

This relationship will now be analysed:

1. For ideal separation

$$\beta = 1; \varepsilon_f = 1; E = \frac{1-\alpha}{0} \to \infty;$$

2. In separation into parts without any change of the composition

$$E = \frac{\alpha - \alpha}{K_c} = 0;$$

3. The unambiguity

$$E_f - E_c = \frac{\beta - D_s}{K_c} - \frac{f - R_s}{K_f} \neq 0.$$

From the viewpoint of analysis, the method proposed by M.Vaga is insufficiently convincing.

Taryan estimates the efficiency of enrichment of coal on the basis of the practical yield of the concentrate and contamination of the concentrate with heavy fractions in the form:

$$E = \frac{\gamma_f(1-\gamma_f) - d_f}{\gamma_f(1-\gamma_f)} \qquad \text{(II-27)}$$

where d is the yield of the heavy fraction in the concentrate.

The insufficient accuracy of this equation is evident because it is not possible to characterise sufficiently he quality of separation only by calculation of the quantities representing the ratio of yields. The values, characterising the distribution between the yields of the useful and non-useful components, are not included in the Taryan equation.

An interesting dependence was proposed by I.A. Veinig

$$E = \frac{\beta - \upsilon}{\beta} \qquad \text{(II-28)}$$

In ideal separation, the calculation using this dependence gives the results equal to 1, because in this case β = 1 and υ = 0. In separation into parts without any change of the fraction composition, the efficiency is equal to 0. However, this dependence does not take into account efficiently the contamination of the concentrate. In fact, when the entire amount of the material is included in the same classification products, *i.e.* when equipment operates in the transport regime, there is no separation and the efficiency of this process should be equal to 0, but according to the equation proposed by Veinig

$$E = \frac{\beta}{\beta} = 1$$

which does not reflect the meaning of the phenomenon taking place.

I.M. Verkhovskii simplified this dependence to the equation:

$$E = \beta - \upsilon \qquad \text{(II-29)}$$

The interesting feature of this equation is that it has an unambiguity. In fact,

$$E_f - E_c = \frac{D_f}{\gamma_f} - \frac{D_c}{\gamma_c} - \frac{R_c}{\gamma_c} + \frac{R_f}{\gamma_f} =$$

$$= \frac{D_f + R_f}{\gamma_f} - \frac{R_c + D_c}{\gamma_c} = 1 - 1 = 0,$$

i.e. $E_f = E_c$.

In ideal separation

$$\beta = 1; \upsilon = 0; E = 1.$$

Again, this indicator does not satisfy the third condition. We shall imagine a separation case in which the entire material supplied into the classification system is included in one of the products of separation. In this case, there was no classification, and the parameter of the process should be equal to 0. According to equation (II-29), in this case $E = \alpha$, which does not reflect the meaning of the process.

The analysis of the methods of evaluating the quality of the separation processes, carried out in this section, has shown that they do not satisfy the set of the boundary conditions, reflecting the problem solved by classification. Consequently, it is not possible to recommended relationships for optimising the classification processes.

Analysis shows that the principle of the classification processes is most efficiently realised by the Hancock–Luyken criterion and that this criterion can be used for optimising separation providing the condition of the constancy of the composition of the initial material is fulfilled.

5. ANALYSIS OF THE APPLICABILITY OF THE HANCOCK DEPENDENCE IN CASES OF CHANGES IN THE COMPOSITION OF THE INITIAL PRODUCT

It is evident that this analysis is essential because crushing–milling systems produce refined materials of different grain composition because of different reasons (the wear of working surfaces, regulating elements of the structure, the variation of the initial characteristic of the material and others, such as moisture content, hardness, concentration and distribution of impurities, etc). As a result of analysis, the Hancock method was selected from all the methods determining the quantitative indicators of the quality of the separation processes. The Hancock method will now be analysed under the condition of changes in the grain size composition of the initial material.

It is assumed that in Fig. II-3, the curve *ABC* shows the initial composition of the product, and the curve *AEF* indicates the composition of the fine product obtained as a result of the realisation of the process. The efficiency of distribution in relation to the size x will be determined:

$$E_x = \frac{D_f}{D_s} - \frac{R_f}{R_s}$$

The initial composition may be change as a result of arbitrary changes in the content of different narrow classes of the product or by changing the ratio of the coarse and fine products in the mixture, i.e. D_s and R_s. The first case for analysis is relatively complicated and complicates the comparison of the results. Therefore, our considerations will be restricted to the second case, and graphically the variation of the relationship of the coarse and fine products may be obtained by simple displacement of the separation boundary, add-

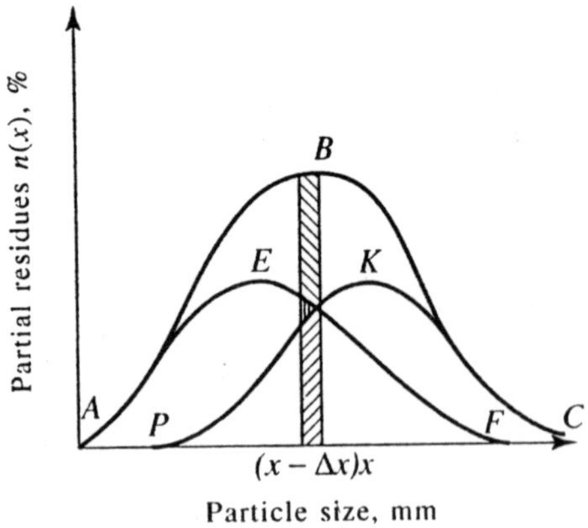

Fig.II-3. Redistribution of products in classification.

ing positive or negative increment to x. The area of the graph, enclosed between the two boundaries, is $\Delta r \Delta x$. Taking this into account, the main parameters of the distribution of the initial mixture are determined from the following equations:

$$D_s(x-\Delta x) = D_s(x) - \Delta R \Delta x = D_s(x) - \Delta R,$$

$$R_s(x-\Delta x) = R_s(x) + \Delta R \Delta x = R_s(x) + \Delta R$$

The part of the shaded section of the area below the curve AEF will be denoted by $-m$, and that above the curve $-l$, i.e. $m + l = \Delta R$. Consequently

$$D_f(x-\Delta x) = D_f(x) - l,$$

$$R_f(x-\Delta x) = R_f(x) + l.$$

The efficiency of classification in relation to a new boundary at separation is:

$$E_{x+\Delta x} = \frac{D_f(x) - l}{D_s(x) - (l+m)} - \frac{R_f(x) + l}{R_s(x) + (l+m)}$$

The resultant equation will now be examined. In this equation, in comparison with the previous equations, the reduction in the numerator and the denominator of the first part of the equation, was added to the numerator and denominator of the second part of the equation, respectively. However, these additions are not proportional to initial or final products of separation and, consequently, the efficiency will change in an arbitrary manner in this case. The efficiency may either increase of decrease, because its change is determined on the one hand by the relationship between R_s and D_s and, on the other hand, by the ratio of these and corresponding parameters of the final products with the addition and decrease.

Thus, even this change of the initial composition leads to an indeterminacy. This is indicated by the experimental examination of this conclusion.

Experiments were carried out in an air classifier with the changes in the composition of the initial material. The initial material was divided in advance into eight narrow classes. From these classes, an initial mixture for experiments was prepared. The content of each class was varied in a wide range: 3.3; 6.7; 10; 12.5; 30; 50; 76.7; 100%.

The content of the remaining classes in each experiment was assumed to be uniformly distributed. In all experiments, both the productivity of the feeding equipment and other technological and design parameters were strictly fixed and assumed to be constant. Figure II-4 shows the dependence of the efficiency of classification (according to Hancock) on the content of the coarse material in the initial mixture. It may be seen that the effect of composition and the efficiency, expressed according to Hancock, is relatively complicated and random. The latter is associated evidently with the fact that the efficiency, calculated from a single boundary size, depends not only on the content of the grains of the boundary size in the mixture but also on the amount and relationship of grains of other classes. Thus, it is concluded

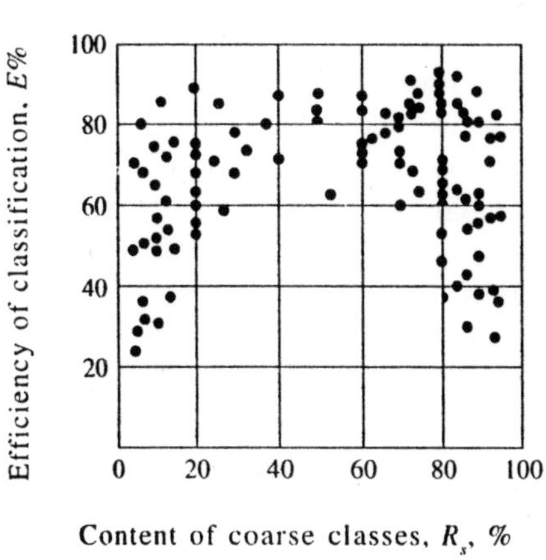

Fig.II-4. Dependence of the efficiency of classification (according to Hancock). on the content of the coarse material in the initial mixture (cascade clasifier $z = 7$, $w = 7.15$ m /s).

that the Hancock equation is also not suitable for optimising separation in a general case. This equation is insufficiently accurate because the results of evaluation are affected by the arbitrary selection of the separation boundary (boundary size, density of separation), and the changing composition of the initial product.

Using this dependence it is not possible to determine, for example, the reason for a decrease in the production indicators: is it the quality of the product because it be affected by less efficient operation of separation equipment, or is it the change of the composition of the initial material.

Firstly, analysis makes it possible to show the insufficient nature of the majority of the currently available methods of optimisation of separation; secondly, it makes it possible to take a critical view regarding the proposed methods of optimisation; thirdly, analysis makes it possible to determine the most objective criterion whose properties are used in further stages in the determination of the quantitative characteristics of the process, invariant to the composition of initial material and the boundary size of separation.

6. METHODS OF DIRECT OPTIMISATION OF SEPARATION PROCESSES

The imperfection of the methods of quantitative evaluation of the completion of the separation processes was the reason for the development of the methods of direct determination of the quality of the process without calculating its indicators. Because of the limited possibilities of measurements of this type, they do not make it possible to fix arbitrary parameters of the process and are designed for determining the conditions of the highest efficiency in which a fraction difference, maximum possible for the given apparatus, is obtained in both classification products.

A number of methods have been developed for the direct determination of the conditions of optimality of the process without calculating its efficiency. They relate the quality of the separation process with the boundary grain size. For any relationship of the regime and design parameters of separation it is always possible to select a class of the size (boundary grain) for which the given process is optimum.

A.Ya. Rubinchik based determination of the boundary grain size on the relationships characteristic of the ideal process, and proposed to determine the optimal efficiency on the basis of the fine class whose content in the initial material is equal to the yield of the fines:

$$D_s = \gamma_f \tag{II-30}$$

To simplify comparison of this condition with other conditions, several transformations will be carried out, since

$$\gamma_f = R_f + D_f \qquad D_s = D_f + D_c$$

From this relationship, the condition according to A.Ya. Rubinchik is reduced to the expression:

$$D_c = R_f$$

F. Bond determines the boundary grain by the content of the fines in the fine product which is equal to the total residue of the coarse grains in the coarse product:

$$\beta_x = f_x \tag{II-31}$$

This relationship may be presented in the following form

$$\frac{D_f}{\gamma_f} = \frac{R_c}{\gamma_c}, \qquad \frac{D_f}{D_f + R_f} = \frac{R_c}{D_c + R_c}$$

Consequently, we obtain

$$\frac{D_f}{R_f} = \frac{R_c}{D_c}$$

A.I. Povarov proposed to determine the boundary grain size as the value of the narrow class of the particles whose content in the initial material and in both classification products is the same:

$$\Delta\alpha_x = \Delta\beta_x = \Delta\upsilon_x \tag{II-32}$$

This condition can be expanded as follows

$$\frac{\Delta D_s}{\gamma_s} = \frac{\Delta D_f}{\gamma_f} = \frac{\Delta D_c}{\gamma_c},$$

And, consequently

$$F_f(x) = \frac{\Delta D_f}{\Delta D_s} = \gamma_f;$$

$$F_c(x) = \frac{\Delta D_c}{\Delta D_s} = \gamma_c$$

Thus, according to A.I. Povarov, the particles of the boundary size in the optimal regime are separated in proportion to the yields of the classification products,

$$F(x) = \frac{F_f(x)}{F_c(x)} = \frac{\gamma_f}{\gamma_c} \qquad \text{(II-33)}$$

Analysis shows that this equation is valid only in a partial case. To confirm this, it will be attempted to find the value of the boundary grain on the basis of analysis of the principle of classification in the most general form.

It is well-known that the optimality condition may be obtained by equating to zero the first derivative of the expression of efficiency with respect to the value of the particle size separation. In the previous chapter, it was determined that the Hancock criterion satisfies more efficiently the principle of the classification tasks. Therefore, analysis will be carried out using equations (II-1) and (II-2). Although it is difficult to expect that this will yield a universal dependence for the optimality conditions (because of the previously mentioned shortcomings of this relationship), the result of this analysis may indicate a path to the formulation of the efficient method.

The optimality condition, determined on the basis of the Hancock criterion, may be expressed as follows

$$E'(x) = 0,$$

and, consequently, the efficiency of classification with respect to the yield of the fines may be expressed by the equation

$$E'_f(x) = \left(\frac{D_f}{D_s} - \frac{R_f}{R_s}\right)' = \frac{D'_f D_s - D'_s D_f}{D_s^2} - \frac{R'_f R_s - R_f}{R_s^2}$$

For efficiency with respect to the yield of the coarse product

$$E'_c(x) = \left(\frac{R_c}{R_s} - \frac{D_c}{D_s}\right)' = \frac{R'_c R_s - R'_s R_c}{R_s^2} - \frac{D'_c D_s - D'_s D_c}{D_s^2} \qquad \text{(II-34)}$$

The efficiency can be expressed as a function of the particle size on the basis of the following considerations. Let the certain amount of the bulk material, whose fraction characteristic is shown in Fig. II-5 by the curve ABC, should be separated into two products on the basis of the boundary size x, mm. In classification of the initial product in some separation equipment it may be assumed that, in a general case, the composition of the fine product is characterised by the curve AFD, and the composition of the coarse product by the curve LMC. In this case, part of the fine particles will be included fully in the fine product, and part of the largest particles in the coarse product. The particles of intermediate size classes from x_L to x_D

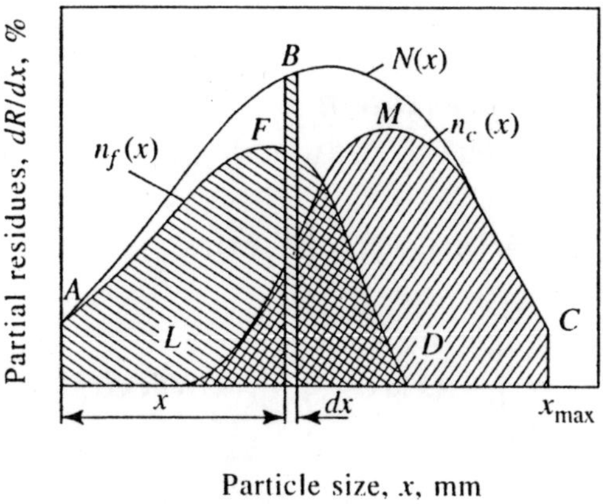

Fig.II-5. Determination of the optimality conditions.

separate between these two products with different degrees of separation. With the known degree of accuracy, the curves $N(x)$; $n_f(x)$; $n_c(x)$, representing the averaged-out fraction characteristics, can be regarded as continuous and differentiable on the basis of the theorem of the calculation of the sums using integrals. If this assumption appears to be insufficient, then to maintain mathematical strictness in the given derivation the relatively continuous integral should be replaced by the sum sign. This replacement has no effect on the final result.

We shall examine a small range of the size dx. The total amount of the material, corresponding to this range is:

In the initial product

$$dG_s = N(x)dx;$$

In the fine product

$$dG_f = n_f(x)dx;$$

In the coarse product

$$dG_c = n_c(x)dx.$$

These relationships show that

$$\frac{dG_f}{dG_s} = \frac{n_f(x)d_x}{N(x)d_x} = F_f(x)$$

$$\frac{dG_c}{dG_s} = \frac{n_c(x)dx}{N(x)dx} = F_c(x)$$

On the basis of these equations, it may be written that

$$D_s = \int_0^x N(x)dx; R_s = \int_x^{x_{max}} N(x)dx;$$

$$D_f = \int_0^x n_f(x)dx = \int_0^x F_f(x)N(x)dx;$$

$$R_f = \int_x^{x_{max}} n_f(x)dx = \int_x^{x_{max}} F_f(x)N(x)dx; \quad \text{(II-35)}$$

$$D_c = \int_0^x n_c(x)dx = \int_0^x F_c(x)N(x)dx;$$

$$R_c = \int_x^{x_{var}} n_c(x)dx = \int_x^{x_{var}} F_c(x)N(x)dx.$$

The optimality condition can be expanded, using the Leibnitz–Newton theorem on the differentiation of the specific integral with a variable upper limit. According to this theorem:

$$\left[\int_0^x f(t)dt\right]' = f(x)$$

Consequently, it may be written that

$$(D_s)' = +N(x);$$

$$(R_s)' = -N(x);$$

$$(D_f)' = +F_f(x)N(x); \quad \text{(II-36)}$$

$$(R_f)' = -F_f(x)N(x);$$

$$(D_c)' = +F_c(x)N(x);$$

$$(R_c)' = -F_c(x)N(x)$$

After substitution of these dependences into equations (II-33) and (II-34), and reducing by the value $N(x)\,dx$, which is not equal to zero, we obtain

$$\frac{F_f(x)D_s - D_f}{D_s^2} - \frac{-F_f(x)R_s + R_f}{R_s^2} = 0;$$

$$\frac{-F_c(x)R_s + R_c}{R_s^2} - \frac{F_c(x)D_s - D_c}{D_s^2} = 0.$$

Consequently, after quite simple algebraic transformations:

$$F_f(x)(R_s + D_s) = \frac{R_f}{R_s} D_s + \frac{D_f}{D_s} R_s; \qquad \text{(II-37)}$$

$$F_c(x) = (R_s + D_s) = \frac{R_c}{R_s} D_s + \frac{D_c}{D_s} R_s. \qquad \text{(II-38)}$$

Finally, we obtain

$$F(x) = \frac{F_f(x)}{F_c(x)} = \frac{R_f D_s^2 + D_f R_s^2}{R_c D_s^2 + D_c R_s^2}. \qquad \text{(II-39)}$$

Evidently, the resultant equation contains information on the boundary grain size because it was derived from the most general dependences characterising the classification.

Several transformations will be carried out. The equations (II-37) and (II-38) will be presented in the following form:

$$F_f(x) = \varepsilon_f R_s + K_f D_s;$$

$$F_c(x) = \varepsilon_c D_s + K_c R_s.$$

Consequently, we shall write

$$F_f(x) = \varepsilon_f R_s + K_f(1 - R_s) =$$

$$= R_s(\varepsilon_f - K_f) + K_f = E_f R_s + K_f;$$

$$F_c(x) = \varepsilon_c D_s + K_c(1 - D_s) =$$

$$= D_s(\varepsilon_c - K_c) + K_c = E_c D_s + K_c.$$

and, therefore,

$$F_f(x) = E_f R_s + 1 - \varepsilon_c;$$

$$F_c(x) = E_c D_s + 1 - \varepsilon_f.$$

Taking into account that $E_f = E_c$ and $F_f(x) = 1 - F_c(x)$, the last equations may be presented in the following form:

$$ER_s = \varepsilon_c - F_c(x)$$

$$ED_s = \varepsilon_f - F_f(x)$$

From this, the optimum efficiency is

$$E_c = \frac{\varepsilon_c - F_c(x)}{R_s};$$

$$E_f = \frac{\varepsilon_f - F_f(x)}{D_s}.$$

According to Hancock, the same efficiency may be written as follows:

$$E_c = \frac{\varepsilon_c - \gamma_c}{D_s};$$

$$E_f = \frac{\varepsilon_f - \gamma_f}{R_s};$$

Taking these relationships into account, the following equation is obtained for the optimality conditions:

$$\frac{\varepsilon_c - F_c(x)}{R_s} = \frac{\varepsilon_c - \gamma_c}{D_s};$$

$$\frac{\varepsilon_f - F_f(x)}{D_s} = \frac{\varepsilon_f - \gamma_f}{R_s}.$$

Consequently, determination of the boundary grain size according to Povarov corresponds to the optimality condition s according to Hancock only in a partial case because the determination is accurate only if the content of the fine and coarse material in the initial mixture is the same ($D_s = R_s$).

The efficiency of determination of the boundary grain size may be indicated by completely different considerations. It was shown that the principle of separation is expressed more sufficiently by the method of determination of the boundary grain size based on the fraction separation curves. In the technical literature, this method is known as the Steinmetser method. According to this method, the boundary grain size is the class of the size divided into half between two separation products:

$$F_f(x) = F_c(x) = 0.5. \qquad \text{(II-40)}$$

For the examined distribution, the size, corresponding to the optimum distribution, is represented by the abscissa of the point of intersection of the curves of the classification products (Fig.II-6). Here, it is denoted by x_0. This size corresponds to the highest efficiency of separation in the examined case, because the total value of contamination of both products (cross-hatched area) in relation to this separation boundary is minimal, as clearly indicated by the displacement of the boundary to the left or right of x_0. In both cases, this leads to an increase of the total value of contamination and a decrease in the difference in the fraction composition of the classification products. It is interesting to note that if the equation proposed by Povarov is a partial case of our dependence, then this dependence is a partial case of the expression of the optimality condition according to Steinmetser. Indeed, at $R_s = D_s$, equation (II-40), may be in the following form

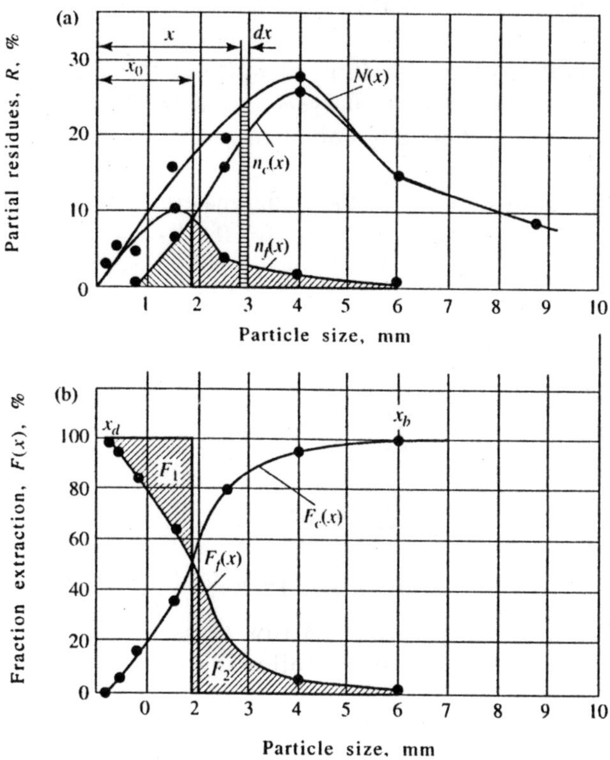

Fig.II-6. Determination of separation curves.

$$F(x)=\frac{F_f(x)}{F_c(x)}=\frac{D_s(\varepsilon_f-K_f)}{D_s(\varepsilon_c-K_c)}=\frac{E_f}{E_c}=1.$$

This indicates the existence of a relationship between the curves of separation and the Hancock quantitative criterion. However, prior to determining this relationship, which is of principal importance in the development of the methods of optimisation of the classification processes, the next chapter will deal with the analysis of the separation curves.

7. FRACTION SEPARATION CURVES

The main properties of the fraction separation curves

It has been assumed that the fraction separation curves were introduced for the first time into the practice of enrichment by Nagel in 1936. The concept of the fraction separation curves is simple. It is based on the determination of the degree of fraction separation of the material of different narrow classes of size in the classification conditions. The fraction separation curves are constructed on the basis of the results of analysis of the separation product and the initial material without any complicated intermediate calculations (Fig. II-6).

The density of distribution of the initial material in respect of narrow classes $N(x)$ will be denoted by $n_f(x)$ for the fine material, and $n_c(x)$ for the coarse material. These distributions are described by the following equalities:

$$\begin{gathered}\sum_n N(x)\Delta x=1\\ \sum_n n_f(x)\Delta x=\gamma_f \qquad \text{(II-41)}\\ \sum_n n_c(x)\Delta x=\gamma_c\\ n_f(x_i)+n_c(x_i)=N(x)\end{gathered}$$

Fraction extraction for each narrow class of the size is determined by the dependences of the following types:

$$F_f(x_i)=\frac{n_f(x_i)}{N(x_i)};$$

$$F_c(x_i) = \frac{n_c(x_i)}{N(x_i).} \qquad \text{(II-42)}$$

Figure II-6 shows the construction of the curve of separation on a real example.

These curves extend to the zone between the points x_a and x_b for which

$$F_f(x_a) = 1; F_f(x_b) = 0; \qquad \text{(II-43)}$$

and vice versa

$$F_f(x_a) = 0; F_c(x_b) = 1.$$

The figure indicates that separation takes place in accordance with the curve $n_f(x)$. The fraction separation curves (see Fig. II-6b) is constructed by the methods of equalisation of the areas of the bases of polyhedrons, determined by calculations. The curve, obtained by the equalisation method, characterises the results of separation processes for adjacent narrow size classes.

Nagel and Tromp already noted on the basis of their investigations that the most important property of these curves is that they are independent (invariant) of the composition of the initial material. In our examination of the separation of bulk grain materials in different pneumatic and hydraulic classifiers it was unambiguously confirmed that this is so. The results presented in Fig. II-4, calculated for the fraction separation curves, are shown in Fig. II-7 for different narrow classes. As indicated by the graph, the fraction separation curves are invariant in relation to the initial composition of the separated material. It may be concluded that these curves are suitable for controlling the operation of separation systems and for developing methods of calculating classification equipment.

Equation (II-43) and the relationship

$$F_f(x_i) + F_c(x_i) = 1$$

indicated the total mirror symmetry of these curves in relation to the horizontal axis, passing through the point of the ordinate with the value of 50%. Consequently, the classification results may be characterised with sufficient reliability by one of the curves and this will determine unambiguously the nature and form of another curve.

It should be mentioned that the fraction separation curve contains complete information on the variation of the grain size composition of the separation products in comparison with the initial material.

The efficiency of separation increases with a decrease in the width of the zone between x_a and x_b. It is clear that in the case of ideal

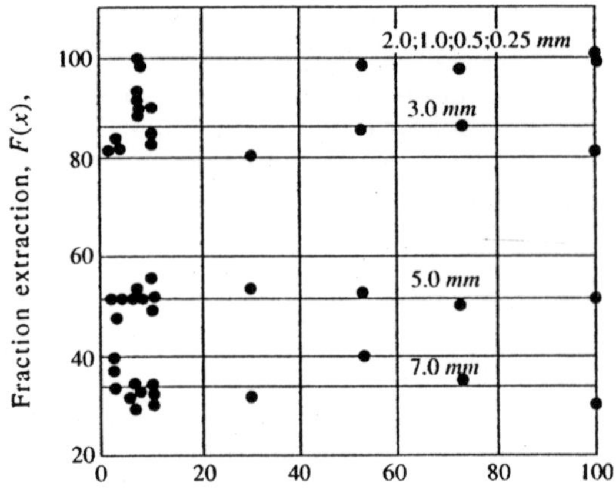

Fig.II-7. Dependence of the degree of fraction extraction of size classes in a fine product on their content in the initial product.

separation $x_a = x_b$. In simple separation into parts without any change of the fraction composition, the functions $F(x)$ are expressed by lines parallel to the abscissa.

It is well-known that the task of classification is to ensure the maximum difference in the fraction composition of the separation products. For the examined case, the size, corresponding to optimum separation, is represented by the abscissa of the point of intersection of the curves. In Fig. II-1, this abscissa is indicated by x_0. The maximum value of efficiency corresponds to this particle size. Owing to the fact that at this point $n_f(x_0) = n_c(x_0)$, the coordinate $F(x_0) = 50\%$ correspond to this point of the fraction separation curve.

It may be shown that the corresponding total area $(F_1 + F_2)$ on the graph is also minimal for this point and increases in transition from this point to either side.

In the practice of enrichment, it has been attempted to investigate separation curves of two types: experimental and calculated. This is carried out on the basis of the fact that incomplete separation appears to be explained by the superposition on the separation the process of the process of simple separation of some part of the material. According to F. Mayer, by excluding this process we can find the true separation curve. This artificial measure distorts the experimental dependence to any previously specified form. The extent of this distortion depends greatly on who carries it out. It is quite difficult to agree with this.

For analysis, it is necessary to use the experimental curves because only they determine the final state of the separation products. In addition to this, it is hardly justifiable to simplify the classification in this manner by introducing assumptions according to which the part of the material is separated without any change of the fraction composition. It is also necessary to investigate in greater detail the physics of the process and, on this basis, describe the separation curves and not carry out analysis using convenient, previously specified relationships.

Indicators of the completeness# of separation

A large number of attempts have been made to find different dependences for the quantitative evaluation of the separation process using fraction separation curves.

In the large majority of cases, taking into account the S-shape of this curve, the calculation dependences are determined from approximation of the curve by the total law of normal distribution. In this case, the quantitative criterion of the process is usually represented by one of the characteristics of the Gauss distribution curves. On the basis of these approximation it is then possible to make far-reaching conclusions, up to the conclusion that the correspondence of the separation curve to the normal distribution is raised to the main general separation law.

However, it is difficult to agree with these assumptions, because of the following considerations. The true curve, reflecting the nature of separation of the material of every narrow class of size and showing its fraction in the total composition of the investigated product, is the function $F_f(x)$ (see Fig. II-6, b). The dependence $F_f(x)$ is obtained as a result of normalisation of the curve $n_f(x)$ to the 100% scale. Its physical meaning is that it is the curve of distribution for the grain size composition of a special type. The curve $F_f(x)$ may show physical separation only in a specific case in which the different size classes have the same content in the initial mixture. Experiments were also carried out in which eight narrow classes were used to produce a mixture from equal fractions containing 12.5% of each product. For these experiments, the curves $n_f(x)$ and $F_f(x)$ completely coincide.

The method of construction of this curve shows that there are no common features between this curve and the normal distribution. Natural differences between these relationships, detected in a large number of experiments, have usually been explained by 'anomalies' of separation curves and are still being related to corrections of these 'anomalies' in the normal distribution.

This approach to the separation process, in which the experimental relationships are 'squeezed' into the previously determined frame, is very harmful because it is pseudophysical.

The similary of the separation curve with the integral Gauss curve of normal distribution has resulted in incorrect interpretation and incorrect coefficient. It would have been better if a large number investigations, carried out in recent years, did not appear, regardless of extensive application of mathematics.

The curve of fraction separation, produced by the equalisation method, is obviously a random curve because the parameters of distribution of the classes of each fraction are not available. If in the case of the grain size composition curves we noted its two-dimensional random state, then for the examined class of the curves it is necessary to accept the three-dimensional or volume random state.

Primarily, the nature of the fraction separation curve is determined by the physical fundamentals of each separation process (screening, gravitation in moving flows, magnetic and electric classification, centrifugal, jigging, etc). They should not be all squeezed into the same frame. The practice of classification shows clearly the principal difference in the form of the separation curves, produced in the realisation of different processes.

The results of our processing of several thousands of experiments with air classification show that the fraction separation curves approaches the normal distribution only in a very small number of cases. Usually, the nature of the function is identical with the dependence shown as an example in Fig. II-6b, i.e. in the general case, it is not symmetric in relation to the vertical axis, passing through x_0. The upper part of the branch is usually considerably steeper than the lower part.

Obviously, in cases in which the identical form of the separation curve and the total curve of normal distribution is justified, it is useful to apply this approximation. However, this fact has not yet provide any substantiation for extensive generalisation.

There are a large number of s-shaped functions which can also be used successfully for approximating the separation curve.

It should be mentioned that in all studies of approximation of the separation curve, attention is given only to the purely formal mathematical (empirical) description of the curve.

It has been assumed that the concept of the 'boundary of separation' is conventional and does not exist in reality because the separation in the real process takes place with respect to the entire range of the sizes from x_a to x_b (see Fig. II-6a) and not with respect to one size. It is not possible to agree completely with the view reflecting the form of the separation curve to some extent, because the concept of the 'separation boundary' has a clear physical meaning. This means that in every specific condition, this boundary corresponds to the optimum condition of separation with respect to some class, and only the change of the technological parameters can result in a change in the value of this class. From this viewpoint, it is fully justified to use the parameter of the 'boundary grain' in enrichment practice.

The best known method of evaluating the efficiency of completion of the process on the basis of the fraction separation curve is the Terr method. This method is based on determination of the value of deviation in the form of the separation curve which is erroneously given the meaning of the mean probability deviation for the normal distribution:

$$E = \frac{x_{75} - x_{25}}{2},$$

here x_{25} is the abscissa of the point at which $F_f(x) = 25\%$; x_{75} is the abscissa of the point at which $F_f(x) = 75\%$.

Terr's deviation can correctly characterise the form of the separation curve when the curve is symmetric because in this case:

$$x_{75} - x_{25} = x_0 - x_{25} = E_T$$

If the curve is non-symmetric, the Terr deviation does not satisfy the equality and cannot be used. Evaluation of the quality of separation of the basis of Terr's deviation has the dimension of the abscissa and provides a value irrespective of where the boundary grain is situated. This indicator does not describe the entire range of the curve, only part of it and, consequently, it does not take into account the differences in the form of the curves in the edge areas. In addition to this, experimental examination of this parameter does not provide a basis for calculating the process results in a general case. These shortcomings are also typical of the indicators proposed by other authors.

For example, F. Mayer proposed to characterise the curve using a more general deviation:

$$E_f = \frac{x_{90} - x_{10}}{2}.$$

In addition to this, in certain cases, the following deviation is used:

$$E = \frac{x_{65} - x_{35}}{2}.$$

On the basis of these indicators it is very difficult to propose a general assumption regarding the quality of the process. It cannot be used for comparison of different acts of classification because of the indeterminacy of the region of the abscissa and the deviation of the form of the curve from the normal or symmetric distribution.

This entire group of the characteristics depends on the scale of the coordinates. The application of these characteristics requires unified diagrams or supplementing the parameters of the process by the appropriate scale of the coordinate.

A more complete characteristic of the shape of the curve of fraction separation is the sum of the areas of the sections of the diagrams to which Drissen refers as the Tromp area (Fig. II-6,b). It is the area enclosed between both branches of the separation curve and the perpendicular line drawn from x_0:

$$F_T = F_1 + F_2.$$

This characteristic embraces the entire course of the separation curve. As the area decreases, the accuracy of separation increases.

Drissen determined the proportionality between the value of this area and E_T on the condition of correspondence of the separation curve to the normal distribution:

$$1.184E_T = E_D.$$

In this case, this indicator can be determined as the ratio

$$E_D = \frac{x_{78.8} - x_{21.2}}{2}$$

The determination of the Tromp area is the basis of determination of the efficiency indicator according to K. Grumbrecht

$$mE_T = \frac{F_T}{(x_b - x_a)}$$

The Tromp area is depicted by Grumbrecht in the form of a rectangle embracing the entire range from x_a to x_b. The height of this rectangle is the mean relative yield.

In addition to this, there is a large number of methods based on the calculation of the static moment of the area in relation to the vertical axis. The determination of the parameters is associated with the construction and careful calculation of the areas. They can be used only if the form of the curves is known, for example, if they correspond to the normal law. If the curves are not governed by the same functional dependence, these parameters should not be used. This can be illustrated quite convincingly on an example of comparison of two separation curves of different shapes having the same Tromp area. Different acts of separation are characterised by the same quantitative value and this does not make sense

T. Eder propósed to characterise the completion of the process of classification by two dimensionless indicators – the coefficient of imperfection (I) and the separation factor (P):

$$I_T = \frac{x_{75} - x_{25}}{x_0}; I_f = \frac{x_{90} - x_{10}}{x_0};$$

$$P_T = \frac{x_{75}}{x_{25}}; P_f = \frac{x_{90}}{x_{10}}.$$

These indicators are derived on the basis of the assumption that the curves of separation and the total curve of the normal law are identical. They can be reduced to the previously examined deviations in the form of the curve:

$$I_T = \frac{2E_T}{x_0};$$

$$I_f = \frac{2E_f}{x_0};$$

$$P_T = \frac{x_0 - E_T}{x_0 + E_T};$$

$$P_f = \frac{x_0 - E_f}{x_0 + E_f};$$

These indicators can be used only in a partial case.

The same shortcomings are typical of the Belyug accuracy parameter used for the separation of coal on the basis of density in hydraulic systems.

$$E_B = \frac{E_T}{x_0 - 1},$$

here $x_0 - 1$ is the density of separation with a correction for the density of water.

Attempts have been made to evaluate the quality of separation as a result of the vector interpretation of the fraction characteristics of the products of both yields. In this case, the indicator is represented by the angle between two vectors in the m-dimensional vector space, where m is the number of different narrow classes in the mixture of the initial material. Undoubtedly, the indicator contains a large amount of information on the substance change of the compositions in comparison with the previously examined indicators, but has a very vague physical meaning. It can hardly be used in practice because the mechanism of comparison of different acts of separation, having m and l classes in its composition, if these classes are situated in the non-intersecting size ranges, is far from clear.

Rumpf and Leschonski evaluated the quality of separation on the basis of the quantitative parameter of the erroneous yield in the following form:

$$\Psi_f = \frac{\int_{x_a}^{x_b} n_c(x)dx}{\int_{x_a}^{x_b} N(x)dx};$$

$$\Psi_c = \frac{\int_{x_0}^{x_a} n_f(x)dx}{\int_{x_0}^{x_a} N(x)dx}.$$

These quantitative relationships, which take into account the relative contamination of both classification products, have the same form as criterial parameters examined previously, i.e. the search range is forcefully closed. These relationships cannot be used for evaluating the quality of the separation processes owing to the fact that they have a number of significant shortcomings which have been examined in detail previously.

Thus, analysis of different methods of evaluation of the efficiency of separation on the basis of the fraction separation curves, used at present, shows that they are clearly insufficiently accurate and not suitable for applications in practice. The main shortcoming of these methods is that they have been derived directly from the separation curve which, because of the previously mentioned reasons, is ambiguous.

Since the separation curves cannot be used for the direct determination of the objective parameters of the process, it will be attempted to use some of their properties for this purpose.

8. THE RELATIONSHIP BETWEEN SEPARATION CURVES AND THE QUANTITATIVE INDICATORS OF THE CLASSIFICATION PROCESS

The development of the methods of optimisation of the separation processes takes place in two different directions, as indicated by our analysis. It has been assumed that these directions greatly differ as a result of the methodology of the approach to understanding the principle of the entire class of the separation processes. This opinion is not correct.

The first direction in the development of the methods of optimisation is historically older and is associated with the names of Lyuken, Hancock, Lyashchenko, Madel, Rozen, Chechot, Rummler, Tyurenkov, and others.

This group of methods is characterised by the derivation of equations for the quantitative evaluation of the efficiency of separation. They have the following advantages:

– The availability of a criterion for the numerical evaluation of the quality of separation which makes it basically possible to compare unambiguously the results of separation in the same classifier, and also in different systems;

– The equations make it possible to determine the optimal obtainable value of the efficiency of operation of any specific system. This characterises its design special features.

Indeed, it is not possible to go above the optimal obtainable efficiency for a fixed separation boundary only by changing the technological parameters of the process. This is possible only with the variation of some design elements, for example, increasing the height of equipment, decreasing its diameter, or by transition to equipment of different design.

However, in addition to these advantages, this group of methods has also important shortcomings:

– The arbitrary selection of the separation boundary (size, distribution density) so that the characteristic of operation of the separation system becomes ambiguous;

– The dependence of the evaluation indicator on the changing composition of the initial product (this circumstance results in an indeterminacy: it is not known whether the decrease in the efficiency of separation, indicated by the criteria, is caused by the characteristic of the design or by the change of the composition of the initial material);

– The optimal technological parameters of separation, determined by these methods, are distorted because they are influenced by the composition of the initial product. This is indicated by the analysis of the boundary grain size determined from the quantitative Hancock criterion. Only in a partial case ($D_s = R_s$), the resultant expression corresponds to the true value of the boundary grain size for which $F_f(x) = F_c(x) = 0.5$.

These shortcomings show that these methods are not suitable for the evaluation of the quality of separation, for its optimisation, nor for comparing the operation of different classification systems which is essential for improving these systems.

The second direction in the development of the optimisation methods is historically younger. It was formed in the middle of the Thirties of the 20th century. It is based on the fraction separation curve.

The main advantages of the separation curves include the following:

– These curves are invariant in relation to the composition of the initial material and this indicates that they are suitable for controlling the operation of classifiers and for comparing the accuracy of their operation;

– Using the separation curve, is it is possible to determine unambiguously and with sufficient accuracy (within the range of the experimental accuracy) the set of the optimum technological parameters of separation.

It should be mentioned that these properties of the separation curves are still underestimated in industrial practice. At the same time, methods based on the application of the separation curves, have a significant shortcomings which complicates the application of these methods: they do not have a reliable quantitative characteristic.

The two examined directions in the development of the optimisation methods appear to supplement each other. However, it is relatively difficult and inconvenient to use the two methods at the same time, although together they may provide true values of the optimum parameters of the process and its quantitative characteristic.

It would be interesting to find a single method having these positive properties of both optimisation directions. We shall attempt to do this on the basis of the fraction separation curves. We shall analyse the nature of these curves from the position of the quantitative criterion.

Optimisation of the process on the basis of the separation curves is carried out as a result of the minimisation of the absolute value of the products of contamination of classification.

To find x_0, it is sufficient to minimise the expression

$$E^I = R_f + D_c$$

or maximise the dependence

$$E^{II} = 1 - E^I = 1 - R_f - D_c.$$

We shall analyse the resultant dependence for the correspondence to the boundary conditions.

1. In ideal separation, $R_f = 0$; $D_c = 0$ and $E^{II} = 1$, which corresponds to the meaning of the process.

2. In simple separation of materials into parts without any change of the fraction composition:

$$R_f = \gamma_f R_s;$$

$$D_c = (1 - \gamma_f) D_s.$$

For this case,

$$E^{II} = 1 - \gamma_f R_s - (1 - \gamma_f) D_s.$$

This equation will be transformed

$$E^{II} = 1 - \gamma_f R_s = D_s + \gamma_f D_s,$$

and

$$E^{II} = R_s - \gamma_f (R_s - D_s),$$

which is not equal to zero in a general case. The dependence is linear in the coordinates of the yield of the product and efficiency (Fig. II-8). Point A corresponds to the value $\gamma_f = 0$; In this case $E^{II} = R_s$; at B $\gamma_f = 1$ and $E^{II} = D_s$. The AB straight line in the examined coordinates has a clear physical meaning: it is the line of zero efficiency in respect of criterion E^{II} in the division of the initial mixture into any parts without changes of the fraction composition.

In the presence of a fraction difference of the products of classification, the curve determined by the dependence E^{II} passes above the straight line AB will and occupies the position of the ACB curve. In this form, it is difficult to use this dependence because its values differ from zero in the absence of separation. This circumstance may result in erroneous conclusions. For this purpose, it is sufficient to subtract in every point of the curve ACB an ordinate equal to the distance from the straight line AB to the abscissa, and this

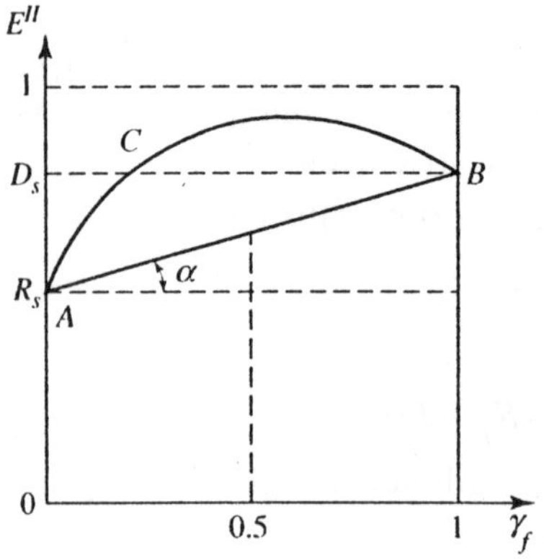

Fig.II-8. Derivation of the relationship between E^{II} and the composition of the initial material.

gives:

$$E^{III} = E^{II} - \left[R_s - \gamma_f (R_s - D_s) \right],$$

Consequently

$$E^{III} = 1 - R_f - D_c - R_s + \gamma_f R_s - \gamma_f D_s =$$

$$= 1 - R_f - D_c - R_s + (D_f + R_f) R_s - (D_f + R_f) D_s =$$

$$= D_f - R_f + D_f R_s + R_f R_s - D_f D_s - R_f D_s.$$

After appropriate transformations:

$$E^{III} = 2(D_f R_s - R_f D_s).$$

This dependence will be analysed:

1. In separation of the material into parts without any change of the fraction composition;

$$E^{III} = 2(\gamma_f D_s R_s - \gamma_f R_s D_s) = 0;$$

2. In ideal separation

$$E^{III} = 2 D_s R_s.$$

This equation indicates that the value of the maximum possible efficiency of separation is determined by the initial composition of the material. This result is very interesting. However, it can be used more sufficiently in normalisation of E^{III} to the 100% scale.

We shall write

$$E^{IV} = \frac{E^{III}}{2 R_s D_s} = \frac{2(D_f R_s - R_f D_s)}{2R2 R_s D_s},$$

and consequently

$$E^{IV} = \frac{D_f}{D_s} - \frac{R_f}{R_s}.$$

The latter is nothing else but the dependence for the classification efficiency, proposed by Hancock.

Thus, analysis of the separation curve results in the well-known quantitative criterion. The results of this conclusion are of great interest. They create suitable conditions for developing a completely new approach to evaluating the separation processes, combining the positive moments of both optimisation directions, accepted in enrichment practice, and show that there are no insurmountable differences between the separation curves and the quantitative parameters.

The main transformations of equation E^{II} resulting in the final analysis in the dependence E^{IV} may be illustrated graphically.

The curve ACB in Fig. II-8 shows the variation of E^{II} in relation to γ_f. The reduction of this dependence to the zero values of the efficiency is possible as a result of subtraction, in each point, of the distance from the abscissa to the straight line AB. This is equal to the transfer of point A to the origin of the coordinates and to the rotation of the curve through an angle equal to

$$\alpha = arctg(D_s - R_s).$$

The reduction to unity as a result of division by $2D_sR_s$ has no effect on the position of the optimum of the dependence and may be taken into account by the appropriate change of the scale of the ordinate. The angle α changes in the range determined by the ratio of the fine and large products in the initial mixture.

At $D_s = 1$, $\alpha = 45°$, at $R_s = 1$; $\alpha = -45$, i.e. $-45° \le \alpha \le 45°$.

Thus, it is possible to transfer from the Hancock criterion to the characteristics of the process invariant in relation to the composition of the initial mixture, i.e. combine the qualitative and quantitative parameters in a single method. In the graphic determination of the optimum of the process, the dependence $E_x = f(\gamma_f)$ must be examined for the extremum in relation to the line inclined under the angle α to the abscissa. This angle varies in the range from $-45°$ to $+45°$, depending on the composition of the initial product.

Analytically, the relationship between the parameters E^{II} and E^x can be determined on the basis of the following considerations.

From the equation for E^{IV} we obtain

$$E_x = \frac{E^{II} - R_s + \gamma_f (R_s - D_s)}{2D_s R_s},$$

and, consequently

$$E^{II} = E_x 2R_s D_s + R_s - \gamma_f (R_s - D_s).$$

This expression makes it possible to determine the separation parameters with the initial composition of the mixture taken into account.

Thus, as a result of the analysis, it was established that the methods of optimisation of the processes of classification using the Hancock method and the separation curves do not contradict each other and are functionally (without correlation) linked together. Taking this functional relationship into account, it is possible to formulate a new method of optimisation of separation. It is based on the methods of evaluation using the Hancock criterion with appropriate corrections for the initial composition of the initial material.

In cases in which it is difficult to compare different separation processes as a result of the low sensitivity of these criteria (in the range of higher values of efficiency), the comparison, described in a number of investigations, must be supplemented by the comparison of the separation curves.

Using the proposed optimisation method it is possible to correct the technological parameters of the process with respect to the initial composition of the initial material. However, it is difficult to use it in the conditions of industrial processes. Using this method, it is difficult to optimise the industrial separation process with a sharp change of the composition of the initial material and during the technological process, because the correction for the composition should always change. Therefore, it is most efficient to solve the problem using the separation parameters invariant in relation to the initial composition. These parameters are determined by the fraction separation curve.

For objective comparison of different classification systems, it is necessary to formulate an objective quantitative criterion on the basis of the fraction separation curves. For this purpose, it is necessary to examine the properties of curves and try to find a method of probable approximation of these curves. In addition to other things, this will make it possible to solve the problem of calculating the process and its results.

9. THE QUANTITATIVE CRITERION OF QUALITY BASED ON SEPARATION CURVES

The proposed method of for the Hancock criterion with corrections for the initial composition makes it possible to determine the true values of the optimum separation parameters. However, the application of the method is not simple, especially in cases in which the initial composition of the initial material changes during a single separation act (experiment). This is most characteristic of the industrial conditions. It is therefore necessary to correct constantly the process as a result of the difficulties in continuous determination of the grain size composition.

There is another approach to the formulation of a qualitative criterion of the quality of the separation process, whose peculiarity is the difference in the parameters forming the basis of the method. Usually, the quantitative criteria for the examined processes are derived on the basis of analysis of the separation of the initial material into two products. Of course, it is then not possible to determine the parameters invariant to the initial composition.

It will be attempted to formulate a quality criterion from a slightly different position. This criterion will be determined on the basis of the curves of fraction separation having the invariance property. It is assumed that parameter x denotes the boundary separation size ($0 \leq x \leq x_{max}$). The quality criterion in relation to the boundary size will be determined not on the basis of the composition of products but on the basis of the separation curve. In this case, for the yield of the fine product

$$E_r = \frac{\sum_0^x F_f(x)\Delta x}{\sum_0^x \Delta x} - \frac{\sum_x^{x_{max}} F_f(x)\Delta x}{\sum_x^{x_{max}} \Delta x},$$

where E_r is the calculated efficiency, determined from the separation curve.

This dependence may be slightly simplified

$$E_r = \frac{\sum_0^x F_f(x)\Delta x}{x} - \frac{\sum_x^{x_{max}} F_f(x)\Delta x}{x_{max} - x}. \qquad \text{(II-44)}$$

The correspondence of criterion E_r to the classification tasks will be analysed.

1. In ideal separation, to the left of the boundary size we have $F_f(x) = 100\%$, to the right $F_f(x) = 0$, and in this case

$$E_r = 100\frac{x}{x} - 0 = 100\%$$

2. Separation without any change in the composition of products (simple division). In this case, the value of $F_f(x)$ to the left and right of the boundary size is the same.

$$E_r = F_f(x)\left[\frac{x}{x} - \frac{x_{max} - x}{x_{max} - x}\right] = 0$$

3. Unambiguity

$$E_{r_f} - E_{r_c} = \frac{\sum_0^x F_f(x)\Delta x}{x} - \frac{\sum_x^{x_{max}} F_f(x)\Delta x}{x_{max} - x} - \frac{\sum_x^{x_{max}} F_c(x)\Delta x}{x_{max} - x} +$$

$$+ \frac{\sum_0^x F_c(x)\Delta x}{x} = \frac{x}{x} - \frac{x_{max} - x}{x_{max} - x} = 0$$

i.e.

$$E_{r_f} = E_{r_c}$$

Thus, this new criterion has all the properties found previously for the Hancock criterion.

We shall examine the 'operation' of this criterion in changes of the initial composition of the material to be classified.

For this purpose, we shall return to Fig. II-4 which shows the dependence of the efficiency, determined by the Hancock method, on the composition of the initial mixture. From this dependence it is not possible to make any conclusion regarding the nature of the effect of composition on the separation indicators. It is natural that using this criterion it is not possible to control the process. This 'non-formalisation' is caused by the fact that, using the Hancock criterion for the analysis of the results of separation in relation to the given boundary size, we *apriori* impose preliminary conditions of the same nature on the effect of each of the products on the process (fine and coarse).

The separation curve shows quite clearly that this is not the case. The classes far away from the boundary separation size, are distributed in the products of classification in completely other proportions than the adjacent ones. Therefore, the variation of the content of the coarse particles in the initial mixture as a result of the variation of different narrow classes leads, on the condition that R_s = constant, to the pattern of the process shown in Fig. II-4.

The same experimental data in processing using equation (II-44) give the dependence shown in Fig. II-9. Comparison of Figs. II-9 and II-4 indicates that the new criterion is characterised by invariance in relation to the changes in the initial composition. Having, at the same time, all positive properties of the Hancock criterion, this criterion may be used widely in the practice of optimisation of separation.

It is interesting to note that the Hancock criterion is a partial case of the determined dependence. We shall examine a binary mixture consisting of two slightly different size classes (narrow classes) at various distances from the boundary size of separation on the condition that this size is between them. Consequently, according to the dependence (II-44) we obtain

$$E_r = \frac{F_1(x)\Delta x_1}{\Delta x_1} - \frac{F_2\Delta x_2}{\Delta x_2} = F_1(x) - F_2(x) = \varepsilon_f - k_f.$$

The last equation is nothing else but the Hancock equation. Thus, the Hancock method is actually valid in the separation of truly binary mixtures. In separation of polydisperse materials into two products, the Hancock method is more difficult to use.

The new criterion will be analysed for the optimality condition.

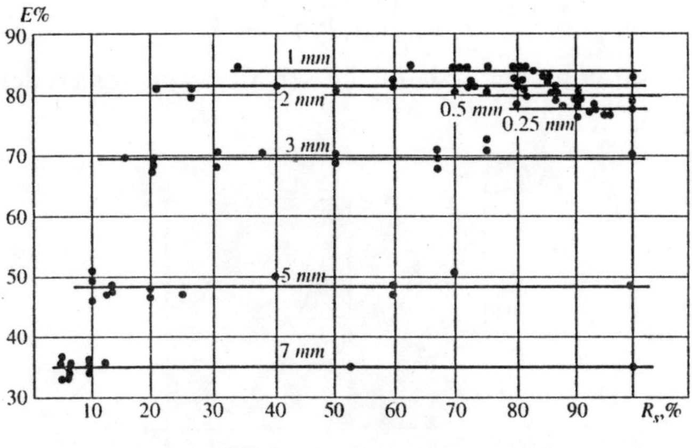

Fig.II-9. Dependence of the efficiency of separation (E_z) on the content of coarse fractions for a cascade classifier at z = 7, w = 7.15 m/s.

To simplify conclusions, the dependence (II-44) will be slightly simplified. Firstly, the sign of the sum will be replaced by the sign of the integral which, as shown previously, does not affect the accuracy of further computations. Secondly, we shall transfer to the concept of the dimensionless size by formalisation of the entire range of the particle sizes to 1. Thus, the dimensionless size parameter will be expressed by the dependence $y = x/x_{max}$. The range of changes for the dimensionless size will be $0 \leq y \leq 1$. In this interpretation, the dependence (II-44) is transformed to the form

$$E_r = \frac{\int_0^y F_f(x)dy}{y} - \frac{\int_y^1 F_f(y)dy}{1-y}. \qquad \text{(II-45)}$$

The separation curve for the dimensionless size is graphically indicated by a line shown in Fig. II-10 which uses the following notations in order to simplify analysis:

$$P_f = \int_0^y F_f(y)dy; I_c = \int_0^y F_c(y)dy; \quad \} \qquad \text{(II-46)}$$

$$I_f = \int_y^1 F_f(y)dy; P_c = \int_y^1 F_c(y)dy. \quad \} \qquad \text{(II-47)}$$

The relationships and the figure show that

$$P_f + I_c = 1; P_f + I_f = F_f; \quad \}$$

$$P_c + I_f = 1; P_c + I_c = F_c.$$

Optimisation on the basis of the separation curves predetermines

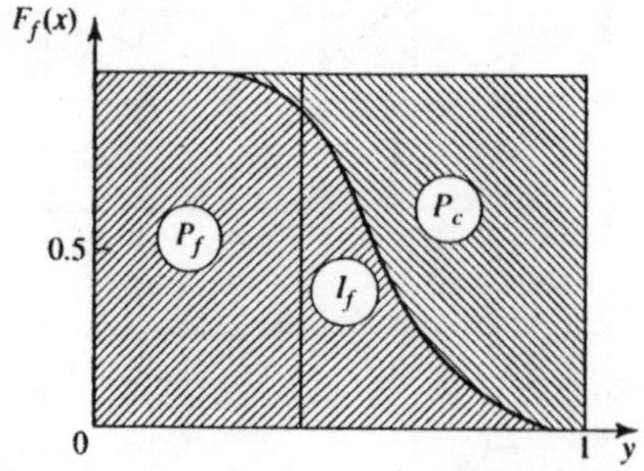

Fig.II-10. Interpretation of the curve of separation to dimensionless size.

the minimisation of the areas indicating contamination, i.e.

$$u^{I} = I_c + I_f$$

or the maximisation of the dependence

$$u^{II} = 1 - (I_c + I_f).$$

In the ideal process $I_c = 0; I_f = 0,$

$$u^{II} = 1.$$

In simple division into parts without changes of the fraction composition, the separation curve is parallel to the abscissa. In this case

$$I_c = F_c(y)y = F_c y;$$

$$I_f = F_f(y)(1-y) = F_f y.$$

Consequently

$$u^{II} = 1 - \{F_f(y) - yF_f(y) + [1 - F_f(y)]\} =$$

$$= 1 - [F_f(y) - 2yF_f(y) + y],$$

or

$$u^{II} = (1-y) - F_f(y)(1-2y),$$

which is not equal to 0 in a general case. The dependence is linear in the coordinates $u^{II} = f\ [F_f(x)]$, and the values of efficiency change in the range from $u^{II} = 1-y$ at $F_f(x) = 0$ to $u^{II} = y$ at $F_f(y) = 1$. This dependence has a clear physical meaning: it is the line of zero efficiency for the new criterion in separation of the initial mixture into any number of parts without changes of the initial composition.

This dependence will be normalised in such a manner that in the absence of separation, the efficiency is equal to 0

$$u^{III} = u^{II} - u_0^{II} = 1 - (I_c + I_f) - (1-y) + F_f(1-2y).$$

After a number of transformations, we obtain

$$u^{III} = 1[P_f(1-y) - I_f y].$$

This corrected dependence is always actually equal to zero in separation into parts without changes of the fraction composition. In fact, after substituting I_c and I_f into this dependence we obtain:

$$u^{III} = 2\left[yF_f(1-y)-(1-y)yF_f\right]=0.$$

However, in ideal separation, using the previously discussed conditions, we obtain

$$u^{III} = 2P_f(1-y) = 2y(1-y)$$

Consequently, for ideal separation, the value of efficiency on the basis of the new criterion u^{III} is determined by the boundary size of separation and not by the relationship of the content of fine and coarse materials in the initial composition, which was characteristic of the Hancock criterion.

This dependence will be normalised for the 100% scale:

$$u^{IV} = \frac{u^{III}}{2y(1-y)} = \frac{2\left[P_f(1-y)-yI_f\right]}{2y(1-y)} = \frac{P_f}{y} - \frac{I_f}{1-y} = E_r. \quad \text{(II-48)}$$

The results of this derivation are very interesting. They show that there is a strong functional relationship between the new criterion and the fraction separation curves. This relationship is close to the one found between the separation curves and the Hancock criterion. However, there is also a large difference in this case. For the determination of the true optimum parameters using a new criterion, correction is carried out in relation to the boundary size of separation. This parameter is more stable than the parameter of the initial composition, or is constant and, therefore, the new criterion may be used for reliable optimisation of the classification processes.

The angle of rotation β of the coordinate axes in relation to the experimental dependence is determined as follows:

$$\beta = arctg(2y-1)$$

Angle β changes in the following ranges: at y = 0, β= –45°, at y = 1, β = 45°, i.e.

$$-45^0 \le \beta \le 45^0.$$

It should be mentioned that β = 1/2 (the boundary size of separation in the centre of the range), the angle β = 0, i.e. only in this case it is not necessary to correct the experimental dependence. Thus, the optimisation of separation by the quantitative criterion is fully justified here. This criterion is characterised by the property of invariance in relation to the initial composition, with the simultaneously determination of the true optimum parameters of the process. This parameter can be used more efficiently and reliably in comparison with the Hancock criterion, because the correction in this case is carried out on the basis of invariable parameter, i.e. the boundary size of separation. This creates suitable conditions for the extensive application of the new dependence for controlling the process. Consequently, it is quite interesting to determine the optimality conditions on the basis of this dependence.

It is well known that the optimum condition corresponds to the value of the first derivative equated to zero, $E_r' = 0$, i.e.

$$\frac{F_f(y)y - P_f}{y^2} - \frac{F_f(y)(1-y) + I_f}{(1-y)^2} = 0.$$

We obtain

$$F_f(y)y(1-y) = P_f(1-2y) + F_f y^2.$$

For the equation for P_f:

$$P_f' = F_f(y)$$

Consequently, the previous equation may be presented in the form of a differential equation

$$P_f' y(1-y) = P_f(1-2y) + F_f y^2,$$

and, consequently

$$\frac{dP_f}{dy} = P_f \frac{(1-2y)}{y(1-y)} + \frac{P_f y}{(1-y)}.$$

We obtain the equation of the first order of the type:

$$y'(x) + P(x)y = Q(x)$$

The solution of this equation must be found in the form of the product of two functions:

$$P(y) = u(y)v(y)$$

The solution of one of these functions is:

$$v(y) = Ce^{-\int \frac{2y-1}{y(1-y)}dy}$$

We determine the expression in the exponent

$$\int \frac{2y-1}{y(1-y)}dy = \int \frac{d(y^2-y)}{y^2-y}dy = \ln(y^2-y),$$

and, consequently

$$v(y) = e^{-\ln(y^2-y)} = y^2 - y;$$

$$u(y) = \int Q(y)e^{\int P(y)dy} = \int \frac{F_f y}{1-y}\frac{dy}{y^2-y} = F_f \int \frac{d(1-y)}{(1-y^2)^2}dy = -\frac{F_f}{1-y}.$$

Consequently,

$$P_f = u(y)v(y) = \frac{(y - y^2)}{1 - y} F_f = yF_f,$$

and the value of the degree of fractional separation is

$$F_f(y) = \frac{\partial}{\partial} \frac{P_f}{y} = F_f$$

It may easily be shown that for the yield of the coarse product

$$F_c(y) = F_c.$$

Thus, the optimality condition on the basis of the new criterion may be written in the following form

$$\frac{F_f(y)}{F_c(y)} = \frac{F_f}{F_c}.$$

Taking into account the fact that the application of new concepts is a very time-consuming process, we would like to draw the attention of the reader to a new criterion, describe its principle and examine it in greater detail. This criterion is several steps closer to true optimisation than the Hancock criterion, which is used widely at the present time, and the new criterion is most general.

The development and justification of the new criterion showed the need for further investigations in the area of optimisation of separation and for formulation of the tasks of these investigations. It should be mentioned that this criterion appeared after solving the problems of the exclusion of the effect of the initial composition. It is now clear that it has not been sufficient to exclude the effect of the initial composition because this resulted in the formation of a previously unknown factor, distorting the evaluation pattern, mainly the boundary size of separation. This means that it is essential to develop an evaluation method which is invariant not only in relation to the initial composition but also the boundary size of separation.

The development of this parameter provides a new solution of the qualitatively new problem. Being invariant in relation to the composition and the boundary size of separation, this criterion will depend unambiguously only on the perfection of the design of the classifier. Thus, it may be necessary to solve the problem of the unambiguous quantitative evaluation of the parameter of 'separation capacity' of specific systems and this will greatly simplify their comparison and selection.

References

1. H. Kirchberg, Enrichments of mineral resources, Gosgortekhizdat. Moscow (1960).
2. V.I. Pavlovich, T.G. Fomenko and E.M. Pogartseva, Determination of enrichment parameters of coal, Nedra, Moscow (1966).
3. A Handbook of ore enrichment, Nedra, Moscow (1972).
4. M.D. Barsky, Fractionation of powders, Nedra, Moscow (1980).
5. Ih. Eder, Probleme der Trennscharfe, Aufbereitungs-Technik, No. 3, 104–106 (1961).
6. R.T. Hancock, Efficiency of classification, Eng. and Min. Journal, No. 210, 237–241 (1920).
7. F.M. Mayer, Algemeine Grundlagen V-Kurven, Autbereitungs-Technik, part I – 1967, No. 8, 429–440, part II – 1967, No. 12, 675–678; part III - 1968, No. 1, 14–23.
8. M.D. Barsky, V.I. Revnitsev and Yu.V. Sokolkin, Gravitational classification, Nedra, Moscow (1974).
9. M.D. Barsky, Optimisation of processes of separation of granular materials, Nedra, Moscow (1978).
10. M.V. Tsiperovich, Enrichment of coal in heavy media, Metallurgizdat, Moscow (1953).
11. N.G. Tyurenkov, United method of evaluating the efficiency of enrichment processes, Metallurgizdat, Moscow (1952).
12. A.F. Taggart, Fundamentals of ore enrichment, Metallurgizdat, Moscow (1958).
13. A.M. Rozen, Theory of separation of isotopes in columns, Atomizdat, Moscow (1960).
14. P.V. Lyashchenko, Gravitational methods of enrichment, Gostoptekhizdat, Moscow (1940).
15. H. Heidenreich, Evaluation of results of industrial enrichment, Gosgortekhizdat, Moscow (1962).

Chapter III

PHYSICAL FUNDAMENTALS OF THE PROCESS OF SEPARATION OF BULK MATERIALS IN MOVING FLOWS

1. THE GENERAL CHARACTERISTIC OF THE CURRENT STATE OF THEORY

The separation of powder materials in moving flows is an exceptionally complicated physical process. This complexity is caused by the movement and mutual effect of a very large number of particles of different sizes in flows of the carrier medium characterised by the maximum heterogeneity in the structure. This movement results in the formation of a wide range of different random factors, with the main factors being: the hydrodynamic and contact interaction of particles in the flows together and with the walls reflecting the flow, the nonuniformity of the fields of velocities and pressures of the flow, the distribution of the solid phase in the flow, the nature of interaction between the phases, and also the effect of the dispersed component on the special features of displacement of the solid medium. Here, it is also necessary to include the main parameters of actual powder materials, characterised by random realisation, such as the size, form, weight of the particles, density, and the distribution of the particles in the size grades.

This completely prevents a pure analytical description of the process, at least on the current level of development of mathematics and physics.

Therefore, in most cases, the examination of the main relationships of this type of processes, and formalisation of the experimental material are carried out on the level of semiempirical theoretical formulations. In the present case, as in the development of any science, knowledge goes from simple to complicated.

The initial examination of the process started with the analysis of behaviour of single particles.

The first publications on this problem appeared in the middle of the previous century. Rittinger examined the relationships governing 19th settling of single spherical particles in an unlimited stationary liquid. In subsequent numerous studies, concerned with examination of this phenomenon, the effect of different factors on these relationships was described: the density of the medium and the material, the final rate of settling of the particles, the drag coefficient of the particles, etc.

The transition in the experiments to real materials with irregular particle shapes has been complicated that studies, concerned with these problems, are still being published widely at the present time.

This indicates the absence of an acceptable model of phenomena taking place even during simple settling of the particles of real materials in a stationary medium.

The attempts for the application, to real processes, of the main relationships, obtained in the examination of the behaviour of single particles in a flow, have not given positive results. In fact, it should be accepted that the rising flow removes from the separation zone completely only the particles whose final settling rate is lower than the velocity of the flow and, conversely, all the particles with a finite settling rate higher than the velocity of the flow will fall out in the direction against the flow.

Everyday experience with fractionation of the powder shows that this is far from the truth.

However, the main results of this scientific direction are not useless; they have not as yet been transferred to the real mass process whose basis they undoubtedly form.

At the same time, it should be mentioned that the examination of the problem of the free settling of single particles of arbitrary shape in the actual separation conditions is slightly one-sided. Usually, the effect on the nature of settling of the moving medium is not examined. In the best case, the region in the immediate vicinity of the particles, the so-called boundary layer, is considered. As regards the medium situated behind this layer, it is assumed that this medium is stationary; this does not correspond to the actual situation. In equipment restricted by solid walls, the settling of particles is always accompanied by the displacement of the medium in the opposite direction. If this phenomenon is actually ignored for a relatively small single particle, then in mass settling in the actual conditions the effect of this phenomenon on the main parameters of the process may become controlling.

The theories, examining the mass separation of the powders in rising flows, are still in the initial stage of development, far from the level required for the solution of practical problems. The most interesting concept is the concept of the mass separation of the mixture in the flow, either along the height of equipment, with respect to the velocity of movement, or with respect to the particle size. It has been shown that the law of actions the masses is valid for this type of separation. Some relationships of the process, detected from the viewpoint of separation, are of considerable interest. However, the construction of theoretical fundamentals of the process from these positions has not as yet resulted in taking into account the specific conditions realised in separation systems.

Recently, many attempts have been made to construct stochastic models of the process. However, it should be mentioned that these constructions are based on the incomplete physical pattern of the process (because of the weakness of the general theory and the exceptionally complicated nature of the separation process) and, consequently, the pure analytical solutions are not as yet capable of providing satisfactory results, which would be at least approximately similar to the experimental data obtained in different systems.

Thus, it should be assumed that, regardless of more than 100 years of development, the scientific direction, concerned with the separation of powders in flows, has not as yet reached the level of development of theory which would be applicable in practice.

The requirements of production are usually satisfied on the level of extensive empirical investigations.

In most cases, the main parameters of the separation process for industrial equipment are investigated on a laboratory or pilot plant model of the process. The parameters, determined in this manner, are transferred in calculations to the operating industrial system. In some cases, this is not successful because of imperfect modelling.

The empirical dependences differ from the theoretical dependences by the fact that they do not form organically from the relationships of the investigated phenomena and reflect them only quantitatively. Any, even the most efficiently constructed experiment, does not make it possible to take into account the entire variety of constant and random factors, both quantitative and qualitative. Here, only an increase in the number of experiments makes it possible to improve the accuracy of determination of the mean effect of the quantitative factors. As regards the qualitative factors, they are more difficult to take into account.

The application of average values greatly reduces the size of the range of application of appropriate generalisations.

The most efficient and justified empirical equation may be used with a satisfactory result only in a limited range, determined by the conditions of derivation of the equation. Extrapolation on the basis of these relationships outside the limits of the experimental range (this is often the case) results in the largest errors. The application of empirical dependences in time is also limited to the same extent because they cannot reflect the development of science and technology over a long period.

In some cases, it has been attempted to improve the accuracy of the available calculation relationships and modify them in relation to the current state of knowledge by introducting different correction coefficients. The application of a large number of coefficients, whose selection is always more or less arbitrary, results in the buildup of errors.

In addition to this, changes in the operating conditions of a machine or equipment often change the entire pattern of the process and may result in a situation in which the appropriate coefficient or a group of these coefficients distort the results of calculations, instead of improving accuracy.

The methods of theoretical calculation in combination with the appropriate laboratory tests make it possible in the majority of cases to solve successfully problems of the development of full-value machines and equipment. In this method, attention is given to solving not only the problem of construction of a specific type of new machine but, in most cases, a substantiated method of calculating and design of a number of machines of the same type or with similar operating principles is developed. Consequently, this method, based on scientific progress, leads to the further development of the appropriate branches of knowledge.

In this connection, it is important to mention the method of formalisation of experimental data using experiment design which has been used extensively in recent years in applied sciences. The application of advanced mathematical apparatus to the processing of experimental data, carried out in accordance with a strictly developed plant, resulting in the calculation dependences in the form of regression equations, produces the vision of scientific significance. In certain studies in recent years, equations of this type have been presented as the basis of process theory. This is one of the most serious mistakes. Without rejecting the usefulness of the method of experimental design, for example, in setting up industrial equipment, it should be accepted that this method is one of the methods of formalisation of the experimental data on a purely empirical level. Ignoring many disadvantages of this method, it is important to mention one of them. The regression equations never help to define the physical fundamentals of the process. It should be mentioned that the extensive application of this method in scientific investigations is not only useless for the development of the theoretical fundamentals of the process but, in many cases, it is greatly harmful because its external scientific appearance generates the impression that extensive theoretical investigations of the physical fundamentals of the phenomenon are not required.

At the same time, the buildup of actual material has a beneficial effect on the development of theory.

The generalisation of a large volume of experimental investigations has made it possible to propose a number of theoretical generalisations for the examined process as a whole, characteristic of all types of equipment without exception. The most significant generalisations include:

1. The necessity for decreasing the random mixing of the solid phase in the flow in organisation of separation as a result of the normalisation of the scale of turbulence pulsations.

The decrease of the large-scale turbulence with simultaneous conservation of the flow energy on the required level is the constant condition of organisation of efficient separation because of two reasons: the degree of mixing, disrupting separation, decreases, and the time, used for separation, increases.

2. The amount of energy supplied by the flow to equipment must be strictly controlled. Excess energy results in the mixing of the separated material, energy shortage does not permit the transport of the removed grain classes. An instrument for the analysis of this phenomenon should be an objective method of evaluating the quality of separation. Unfortunately, this significance of the quality criteria is not always properly understood.

3. The necessity for optimization of the hydrodynamic shape of the chambers of separation equipment. For this purpose, it is necessary to examine the structure of suspended flows, the trajectory of movement of different size grain classes in order to design apparatus in which the separated material is capable of leaving more sufficiently the classification zone through the appropriate exit without mixing. In this respect, the application of working chambers with the flow velocity variable along the height is highly promising. The insertion into the design of apparatus of different mechanical or dynamic decelera-

tion devices, organised in a special manner, also supports the intensification of the separation process.

4. Separation in the gravitational counterflow is not carried out on the basis of the size and shape of particles nor the specific weight of the material. It is carried out on the basis of the generalized hydrodynamic characteristic, which is the velocity of movement of the particles in specific separation circumstances. In specific conditions, the range of the velocity of the particles of different grain classes becomes the highest in the case of non-steady movement conditions (with acceleration and deceleration). Therefore, the transfer of the process to the non-steady conditions of movement of the medium, organised in the appropriate manner, create suitable conditions for improving the quality of separation. In this aspect, the tendency of combining and superposition of fields of another physical nature (vibration, oscillation, sound, electromagnetic, etc) on the gravitational separation, observed in recent years, is highly promising. This results in the intensification of separation without increasing the energy of orderless mixing.

5. The progressiveness realisation of separation in the system of a cascade of elements of the same or different types. Fractionation by standard methods in equilibrium systems is characterised by low efficiency. A simple increase in the height of the systems is usually accompanied by only a small increase in the effect. Therefore, to improve greatly the quality of the process, it is necessary to organise separation on the basis of the cascade principle or even the principle of combined cascades which makes it possible to obtain almost any purity of the separated products. Naturally, this requires appropriate energy expenditure.

Some of the generalisations of the empirical material, accepted universally without objections, are still insufficiently justified and must be examined in detail.

The most important generalisations include:

1. It has been assumed that the velocity of hovering of the particles is actually identical with the velocity of settling of the particles. It is understood that separation takes place in the suspension conditions, but calculations are usually carried out on the basis of the finite velocity of settling of the solid particles in a stationary medium. In this case, it is assumed that the drag coefficients in the hovering and settling particles are identical and do not depend on the turbulisation of the medium which in the flowing flow is different in comparison with the steel medium.

It should be mentioned that the examined parameters are characterised by different physical nature because the hovering velocity relates to the movement of the flow, and the final settling rate to the solid particle.

In addition to this, the flow in medium is always characterised by the establishment of a specific structure (the curve of velocities). The local velocities at different points of the section differ and may greatly differ from the mean velocity which at present is accepted as the controlling parameter of hovering.

2. Usually, in theory it is assumed that the distribution of the solid phase is uniform. However, this does not correspond to the actual situation. The solid particles, placed in the flow, are characterised by a susceptibility for the formation of agglomerates and strands. According to experience, these formations are very stable: they behave as an integral unit in the separation conditions. The breakdown of the resultant aggregates is prevented by the pulling of the surrounding medium into the turbulent wake formed behind the moving particle or a group of particles, and also by the tendency of the other particles to move in the wake of the particles. This is also supported by the decrease of pressure between closely spaced particles. The nonuniformity of concentration is also supported by the distinctive transverse migration of the particles in relation to the flow.

3. In the theory, there is no principal difference for the separation processes carried out in both the liquid and gas medium. Usually, they are described by identical relationships. However, the difference between them is very large because of high density and viscosity of the liquid. This predetermines different rates of organisation of the process, characterised by different conditions of interaction of the phases. Therefore, it is completely justified to assume that the main relationships for wet and dry separation may converge but they may differ greatly. Therefore, a large part of the data, obtained for the flow of the liquid with particles, may be used only to a certain extent for the examination of dry processes, and vice versa.

4. It is assumed the interaction of the particles in the flow and with the wall of equipment may be ignored because of the fact that the effect of this phenomenon on the separation result is very small. The investigations, carried out in recent years, show that this phenomenon has a strong effect on the two-phase flow.

5. Usually it is assumed that the optimum velocity of movement of the flow is equal to the hovering velocity of particles of the boundary size. It is possible that this assumption is justified in the organisation of equilibrium classification. As shown by experience, in cascade systems, the value of the optimum velocity is not only a function

of the particle size of the boundary size but also depends on the area of inlet of the material into the apparatus and the length of apparatus.

All these problems are exceptionally important for the theory and practice of separation and, consequently, they require careful examination. In particular, it should be mentioned that many important aspects of the discussed process have not as yet been solved or even formalised.

The main aspects are as follows:

1. The effect of the initial material, supplied into equipment, on the nature of movement of the two-phase flow is almost completely ignored. It is well known that any flow has a limiting transport capacity, i.e. may carry a certain specific amount of particles of different grain classes. The limiting transport capacity of the flow is associated to some extent with the solid phase: if the amount of the finest fraction is dominant, the capacity increases, and in the case large particles it decreases. Ignorance of the grain size distribution of the initial composition results in insurmountable difficulties in the development of substantiated methods of calculating specific fractionisation systems.

2. The effect of the concentration of the solid phase on the movement of flows in the separation conditions is completely unclear. No answer is available as to the concentration at which it is necessary to consider the interaction of particles in the flow together and the concentration at which this interaction may be ignored, what is the permitted limiting concentration for producing a specific effect, how is this concentration linked with the grain size characteristic of the solid phase, etc.

3. In the currently available theories, there is no formulation of the problem of the relationship between the main factors of the process, the velocity of the flow of the medium, the design of equipment, the composition of the solid phase, on the one hand, and the distribution of different size groups in the separation products, on the other hand. This undoubtedly reflects the insufficient level of theory. The determination of this type of distribution in specific conditions should be the main task in the development of any, even approximate, theory of the process. Without this, it would not be possible to determine the main relationships of separation, developed substantiated methods of calculating equipment and predicting the products obtained in separation.

4. Theory usually does not take into account the effect of the solid phase on the movement of a medium. Both these movements are linked and must be taken into account in the development of

a substantiated model of the process.

These problems and tasks are extremely important for understanding the physical principles of the process. Without understanding and improving their accuracy, it would not be possible to develop an objective theory. Concluding that the pure analytical solution of these problems is not yet possible, the main investigations and constructions will be carried out in future on the semiempirical level.

2. SPECIAL FEATURES OF THE MOVEMENT OF CONTINUOUS FLOWS

A continuous medium is either a droplet liquid or elastic gas. In the conditions of air classification, the gas pressure is usually low. Therefore, the variation of the volume of the gas may be ignored in this case and the gas maybe regarded as an elastic medium. The processes taking place during the movement of continuous media have been the subject of investigations over many years. Initial investigations into the movement of liquid in pipes (Hagen's) were published in 1839. However, systematic investigations of these phenomena became possible only after Reynolds published his outstanding studies.

The general relationships of the movement of continuous media are determined on the basis of the general laws of mechanics by analysis of the behaviour of elementary volumes. If an elementary volumes with the size *dx*, *dy*, *dz* is defined in a moving medium, the following equation may be written for this volume:

$$\frac{\partial w_x}{\partial x}+\frac{\partial w_y}{\partial y}+\frac{\partial w_z}{\partial z}=0 \qquad \text{(III-1)}$$

here w_x, w_y, w_z is the component of the velocity in relation to the coordinates x, y, z.

At every fixed moment of time, the examined moving element is subjected to a set of forces determining the movement of this element. These forces include gravitational force, pressure, internal friction and inertia force.

Taking into account the effect of viscosity forces in a real medium, the equations including all four examined components, form a well-known system referred to as the differential Navier–Stokes equations:

$$\rho\frac{\partial w_x}{\partial t}=-\frac{\partial P}{\partial x}+\mu(\nabla^2 w_x+\frac{1}{3}\frac{\partial \theta}{\partial x})$$

$$\rho\frac{\partial w_y}{\partial t}=-\frac{\partial P}{\partial y}+\mu(\nabla^2 w_y+\frac{1}{3}\frac{\partial\theta}{\partial y}) \quad \text{(III-2)}$$

$$\rho\frac{\partial w_z}{\partial t}=-\rho g-\frac{\partial P}{\partial z}+\mu(\nabla^2 w_z+\frac{1}{3}\frac{\partial\theta}{\partial z})$$

In these equations: $\rho\frac{\partial w_x}{\partial t}$ is the product of mass by acceleration along the appropriate coordinate; ρg is the quantity reflecting the effect of the gravitational force; $\frac{\partial P}{\partial x}$ is the variation of hydrostatic pressure inside the liquid in the direction of the appropriate coordinate axis; $\mu(\nabla^2 w_x+\frac{1}{3}\frac{\partial\theta}{\partial x})$ is a quantity reflecting the effect of internal friction forces and compression and tensile forces, caused by friction; $\nabla^2 w_x=\frac{d^2 w_x}{dx^2}+\frac{d^2 w_y}{dy^2}+\frac{d^2 w_z}{dz^2}$ is the Laplace operator; $\mu\nabla^2 w_x$ is the friction force related to the unit volume of the liquid; $\frac{\partial\theta}{\partial x}$ is the variation of the velocity on the appropriate axes, caused by the effect of compressive and tensile forces formed in the medium during its flow; $\theta=\frac{\partial w_x}{\partial x}+\frac{\partial w_y}{\partial y}+\frac{\partial w_z}{\partial z}$ is the sum referred to as the divergence of the vector of the velocity in the direction of the coordinate axes or divergence of the velocity.

The system of equations (III-1) and (III-2) completely describes the pattern of movement of the elastic viscous medium.

In order to obtain an unambiguous solution of the system, it is necessary to specify the initial values of the velocity fields in space and time and take into account that the velocity should convert to zero on the surface of all solids immersed in the flow and on the walls restricting the flow.

The solution of the system of these equations taking the initial conditions into account has been discussed in investigations of many researchers.

Keller and Fridman showed for the first time that to determine the static moments of any order of hydrodynamic fields of turbulent flows, it is necessary to solve an infinite system of equations, i.e. the system is not closed. The solution of the system becomes possible as a result of introducing various assumptions, idealising the a moving medium. Depending on the assumptions made, the nature of movement changes. Idealised flows according to Newton, Euler, Kutta, Poisell, Hagen, etc. are well known.

Thus, the exact solution of the system of Navier–Stokes equations is not possible. However, the system is interesting owing to the fact that it includes information on the nature of movement of the flows in the form of dimensionless parameters.

These parameters can be determined using the methods of similarity theory.

In all theoretically insolvable cases, the examination and determination of the quantitative characteristics of the process is possible only by means of experiments. In certain conditions, these characteristics become invariant in relation to the scale of the examined phenomenon and, consequently, this makes it possible to apply the results of experiments, obtained on the model, to the entire class of the phenomenon. The principle of this modelling is the determination of requirements (criteria). If these requirements are adhered to, the phenomena taking place in nature and on the model have similar features. The main advantage of the method is that it provides a means of generalisation of the results of even a single experiment.

It has been assumed that similar phenomena also take place in geometrically similar systems if in these systems the ratio of identical quantities is expressed by constant numbers at all identical points. These relationships, referred to as similarity constants, cannot be selected arbitrarily because the values, characterising the phenomena, are not independent of each other and are linked by a specific relationship determined by the laws of nature. In many cases, the relationship may be expressed mathematically. For similar phenomena, the equations should be of the same type.

Because the exact solution of the systems (III-1) and (III-2) is not possible, it is necessary to use approximate methods.

In the numerical solution methods, it is very important to be able to evaluate the order of magnitude of different members of the equations for specific conditions. Consequently, it is possible to ignore the members whose effect is not strong and restrict ourselves to examination of simplified equations.

It has been proved that in the case of steady movement, the members of the equations, containing pressure, are of the same order of magnitude as the non-linear members and their effect is also insignificant.

The flow is stationary if for any point of the flow

$$\frac{dx}{dt}=0$$

where x is the set of physical quantities affecting the movement of the flow.

For this flow it is sufficient to compare only the order of the members describing the forces of internal friction and inertia forces.

If we denote the scale of the velocity by w, the scale of the length along which there are large changes in velocity by l, then the order of the first derivative with respect to velocity is determined by the ratio $\frac{w}{l}$, and the second derivatives $\frac{w}{l^2}$. Therefore, the terms, describing immersion forces, will have the order $\frac{w^2}{l}$ on the terms describing friction forces the order $\frac{\nu w^2}{l^2}$

The relationship of these quantities gives

$$\frac{w^2}{l}:\frac{\nu}{l^2}\frac{w}{}=\frac{wl}{\nu}=\text{Re} \qquad \text{(III-3)}$$

The resultant complex is dimensionless. It was proposed for the first time by O. Reynolds and in his honour it is referred to as the Reynolds number or criterion for the flow. This complex, representing the measure of the ratio of the inertia force to the internal friction force, is the most important characteristic of the flow because the ratio of these forces controls the main properties of the flow of the medium. Inertia forces result in rapprochement of the initially remote volume of the media leading to the formation of flow heterogenities. The viscosity forces, on the other hand, lead to equalisation of the velocity at closely spaced points, i.e. to smoothing of heterogeneities. Therefore, in the case of small Re numbers, when the viscosity forces prevail over the inertia forces, all the fields, characterising the flow, change smoothly and the flow is laminar.

The paths of the individual particles are represented by a straight line, parallel to the axis of the flow, forming a non-intersecting system of curves at bends. Because of internal friction at different points of the cross-section, the elementary small jets have different velocities. The layer, adjacent to the wall of the pipe is characterised by considerably higher friction than the friction between the layers of the liquids. In practice, this boundary layer is stationary. The velocity of its translational movement is equal to 0. The next layer in the direction to the axis of the flow is decelerated as a result of friction on the boundary layer liquid, i.e., is considerably smaller. The second layer moves in the direction of movement of the entire flow with a higher velocity. All subsequent layers of the liquid move parallel in relation to each other with the velocity increasing up to the maximum value on the axis of the pipe (Fig. III-1).

The distribution of the velocities in the cross-section of a circular pipe with radius r_0 may be determined if the forces of internal friction between the layers, moving at different velocities are comparable with the forces of hydrostatic pressure along the length of the examined section (Fig. III-2).

It is assumed that the cylinder moves from left to right. The resultant force of hydrostatic pressure on the cylinder is:

$$P=(P_1-P_2)\pi\ r^2$$

The force of friction along the generating line of the cylinder in accordance with the Newton law is:

$$P_f=\mu F\frac{dw}{dt}=-\mu 2\pi rl\frac{dw}{dr}$$

For steady movement

$$dw=-\frac{P_1-P_2}{2\mu l}rdr$$

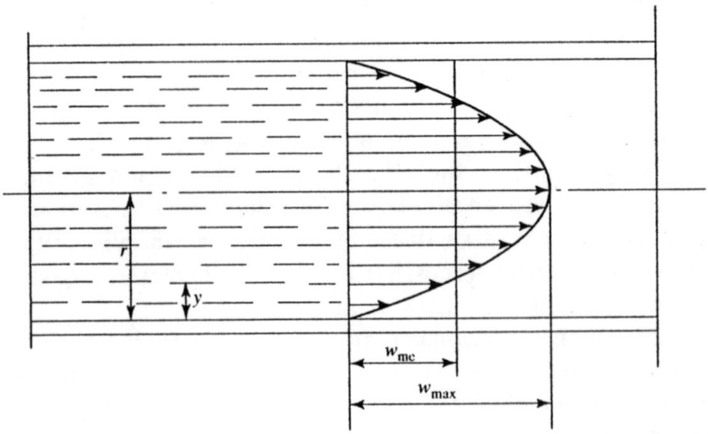

Fig.III-1. Distribution of velocity in the cross section of a circular pipe (laminar flow).

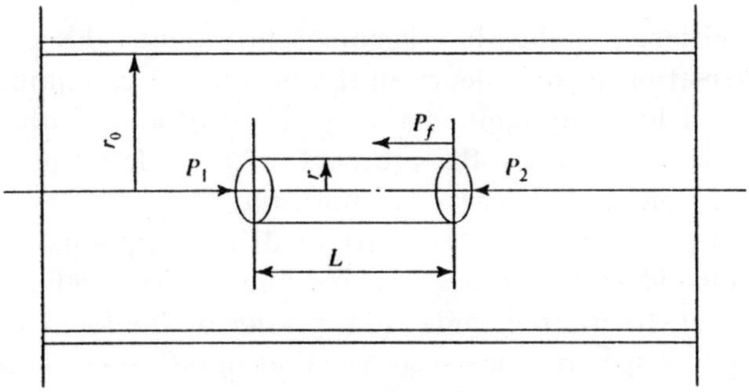

Fig.III-2. Hydrodynamic equilibrium in the flow.

After integration, we obtain

$$w = \frac{P_1 - P_2}{4\mu l}(r_0^2 - r^2) \tag{III-4}$$

Equation (III-4) shows that in the laminar flow, the distribution of velocities in the flow is governed by a parabolic law. The velocity is maximum on the axis of the pipeline at $r = 0$ and is equal to

$$w_{max} = \frac{P_1 - P_2}{4\mu l} r_0^2 \tag{III-5}$$

The elementary volume flow rate through a ring with thickness dr is:

$$dV = wdF = \frac{P_1 - P_2}{4\mu l}(r_0^2 - r^2)2\pi r dr$$

The total flow rate of the medium in integration with respect to radius from the axis of the flow to the wall is

$$V = \frac{\pi(P_1 - P_2)}{8\mu l} r_0^4 \tag{III-6}$$

This equation is used to determine the mean velocity of the flow:

$$w_m = \frac{P_1 - P_2}{8\mu l F} r_0^2 \tag{III-7}$$

Comparison of the equations (III-5) and (III-7) shows that the mean velocity of the laminar flow is half the maximum (axial) velocity:

$$w_m = 0.5 w_{max}$$

At high values of Re, the smoothing effect of viscosity forces is not strong and this results in the formation, in the flow, of local heterogeneities, i.e. the flow becomes turbulent. The main reason for the formation of turbulence in the movement of liquids or gases is the loss of hydrodynamic stability. The turbulent form of movement is characterised by the presence of disordered pulsations of velocity in time and in space (in both the axial and transverse directions). In this case, the pressure at different points of the flow changes in the identical manner. This phenomenon is greatly complicated as a result of disordered, pulsation mixing of the local volumes of gas leading to a specific phenomenon: turbulent diffusion whose intensity is many orders of magnitude higher than that of conventional molecular diffusion. All this predetermines a change in the laws of resistance in this region of flow, accompanied by a large increase in the hydrodynamic losses (quadratic law). This is accompanied by the equalisation of the profile of the curve of average velocity in the centre of the flow and by its rapid drop in the boundary region. Thus, the movement regime of the flow is characterised by the Reynolds number.

Low values of Re corresponding to the stable laminar flow. With increasing Reynolds number, the stability of this flow decreases as a result of a relative increase in the inertia forces.

At some value

$$Re = Re_s$$

the laminar regime loses its stability and movement becomes turbulent. In the case of turbulent movement, the inertia forces greatly exceed the viscosity forces. Consequently, in certain studies it is assumed that it is possible to ignore the viscosity forces when examining turbulent movement.

The acceptance of this assumption simplifies the pattern of the process because it transfers the systems of equations without terms containing viscosity to the equation for the ideal liquid. However, these equations cannot satisfy the boundary conditions of the flow, with the velocity on the walls of the flow changing to zero.

It has been assumed that the initiation of turbulent pulsations takes place in the zone of viscous flow at the walls restricting the flow. As a result of their distance from the centre, these zones generated the periodically repeating ejection of the mass of the medium into the zone with higher velocities. These injections have the form of horseshoe vortices. The scale of vortices of this type is comparable with the scale of the flow, and the frequency of pulsations of velocity and pressure are relatively low. In the case of high Reynolds numbers, the movement of these primary vortices also loses its stability leading to the formation of finer vortices which in turn generate even finer vortices, and so on. It has been established that the vortex formation process takes place along a chain until the finest vortices form (up to Re<1). The movement of the medium inside these formations is of the pure laminar nature, determined by the molecular viscosity of the medium at a specific temperature. It should be mentioned that the direction of the main flow does not affect the orientation of fine-scale of vortices.

In addition to the Reynolds number, the turbulent regimes of the flow are characterised by the following parameters: the average velocity, the intensity and scale of turbulence, and also the frequency of pulsations.

The stabilised profile of the average velocity of the turbulent flow is not established immediately; it forms at a relatively large distance from the inlet into the channel. The medium, arriving in the channel, forms a stagnant layer at the walls. With increase from the inlet, the thickness of this layer increases. This results in the formation of a layer with the laminar flow, and the thickness of this layer is described by the dependence:

$$\Delta = S\sqrt{\frac{v}{w_{max}}l}$$

here S is the height of irregularities on the walls; ν is the kinematic viscosity coefficient; l is the distance from the inlet into the channel; w_{max} is the velocity at the axis of the flow.

Since the flow is continous, deceleration at the walls increases the velocity in the centre and leads to some stretching of the curve. On reaching a specific thickness, the laminar boundary layer loses its stability and becomes turbulent. Therefore, this layer rapidly increases in thickness and overlaps the entire cross-section of the channel, forming a velocity curve characteristic of turbulent motion. In this case, the thin layer at the wall itself remains laminar (viscous sublayer). The thickness of this layer is smaller and, according to different data, varies from 0.1 to 1.8% of the channel diameter.

The thickness of the viscous layer decreases with increasing Reynolds number.

If the boundary layer covers all irregularities on the walls, restricting the flow ($\Delta > S$), its main core will slide on this layer (Fig. III-3a). In this case, the drag coefficient does not depend on the roughness of the walls. With increasing Reynolds number, the thickness of the film decreases ($\Delta < S$), (Fig. III-3,b). The projections of the irregularities of the wall are outside the limits of the boundary layer and are characterised by direct interaction with the turbulent flow, increasing the losses when overcoming friction. In addition to this,

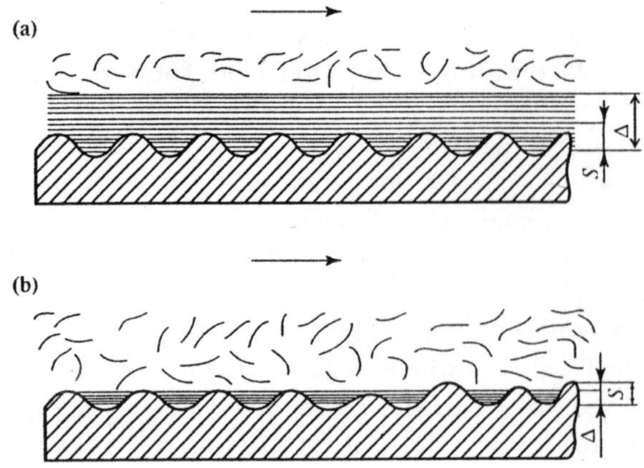

Fig.III-3. Interaction of a flow with a rough wall.

the phenomena taking place in the boundary layer have a strong effect on the intensity of turbulent flows because they introduce additional perturbation factors into these flows. This shows clearly that in the turbulent motion the drag coefficient depends only on the surface roughness of the walls and is independent of the Reynolds number (Fig. III-4).

The defition of relative roughness $K = \frac{S}{r}$ is introduced for ideal pipes. Here S is the mean value of the roughness projections, r is the radius of the pipe.

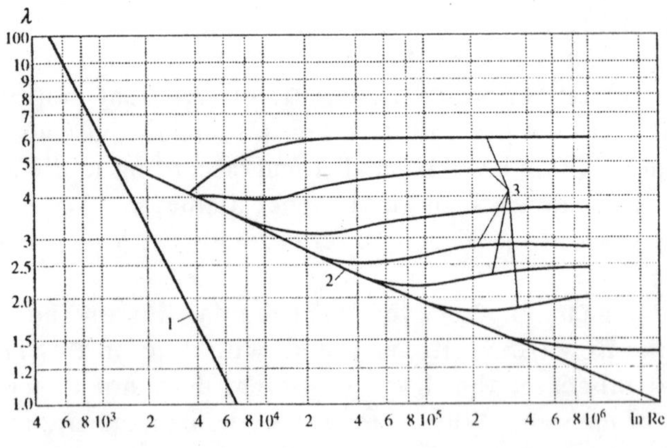

Fig.III-4. Dependence of the drag coefficient of the flow (λ) on the Reynolds number (Re) and roughness of the walls. 1) laminar flow; 2) turbulent flow (smooth walls); 3) turbulent flow (rough walls).

Drag coefficient λ, remaining the dimensionless quantity, becomes a function of two variables, i.e. $\lambda = f\,(\text{Re}, K)$.

The difference in the laminar and turbulent flows is reflected in a number of phenomena which are of great importance for many technical tasks. In the case of turbulent motion, the flow has a considerably stronger effect on the walls or on the solid placed in it and in this case the mixability efficiency of the medium and heat conductivity of the flow greatly increase. Therefore, this shows how important it is to determine the conditions of transition of one type of flow to another.

The value of the critical Reynolds number for different specific cases is determined from experiments.

For the simplest case of movement along a straight circular pipe, it has been determined with a sufficient degree of reliability that of the critical value is

$$\text{Re}_s = 2300$$

However, further investigations show that the values of Re_s, corresponding to the transition from the laminar to turbulent flow may greatly differ in different conditions and are determined mainly by the conditions of entry into the apparatus. For example, experiments have been carried out in which the laminar flow was 'tightened' to the value $\text{Re}_s \approx 20000$ or higher.

These results show that the Reynolds number on its own is not yet an unambiguous criterion of the formation of turbulence. The exact determination of the boundary conditions of formation of turbulence in each specific case should be carried out by experiments on the basis of the ratio of the Reynolds number and the resistance of the flow. Turbulent flow is characterised by a smoother (in comparison with laminar flow) variation of the velocity in the cross-section, with the exception of the boundary layer (Fig. III-5). Because of the complicated nature of the process, the dependence between the mean and the maximum velocity of the flow is not analytically determined in this case.

This relationship is determined by experiments usually in the form of the following exponential equation.

$$\frac{w_y}{w_{max}} = \left(\frac{y}{r}\right)^n,$$

Here y is the distance from the walls; r is the radius of the channel.

According to A.D. Al'tshul', this dependence is expressed more accurately by the relationship

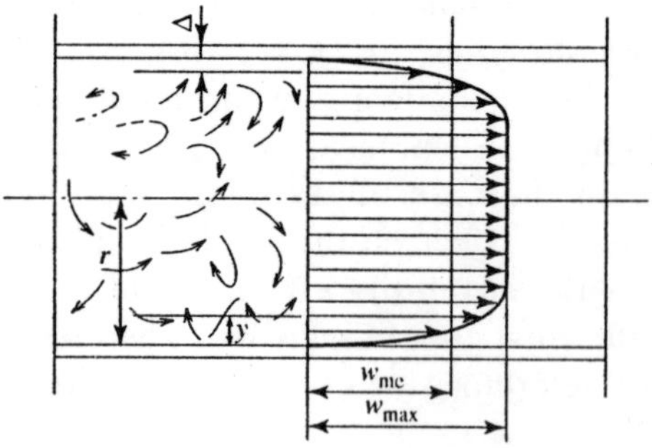

Fig.III-5. Distribution of velocity in the cross section of a circular pipe (turbulent flow).

$$\frac{w_y}{w_{max}} = 1 - 2\frac{\lg\frac{r}{y}}{\frac{0,975}{\sqrt{\lambda}} + 1.35}$$

The ratio of the mean velocity to the maximum velocity, referred to as the quality of the pipe, is determined using the relationship

$$\frac{w_m}{w_{max}} = 1 + 1.35\sqrt{\lambda}$$

and the coefficient, taking into account the nonuniformity of the velocities in the cross-section is determined from the equation

$$\alpha = 1 + 2.65\sqrt{\lambda}$$

Thus, all these characteristics of the turbulent flow are determined only by the drag coefficient and are independent of the Re number. In the turbulent flow, like in the laminar flow, it is possible to determine the layer whose velocity corresponds to the mean velocity of the flow. The distance of this layer from the wall of the pipe is:

$$y_m = 0.232r$$

The calculation of the drag coefficient for the turbulent flow was relatively complicated. Three regions are defined here.

1. **Hydraulically smooth pipes**. In this case, the boundary laminar layer is larger than the absolute surface roughness of the pipe ($\Delta > S$). The thickness of the laminar layer is expressed by the relationship

$$\Delta = \frac{30D}{\mathrm{Re}\sqrt{\lambda}}$$

where D is the diameter of the pipe.

The boundary of the zone of hydraulically smooth pipes is determined from the relationship

$$\mathrm{Re} \leq 27\left(\frac{D}{S}\right)^{1.14}$$

and the friction coefficient

$$\lambda = \frac{1}{(1.8\lg \mathrm{Re} - 1.52)^2}$$

or from the better known equation

$$\lambda = \frac{0.316}{\mathrm{Re}^{0.25}}$$

2. **The turbulent mixed regime**. In this regime, the friction coefficient is determined from the Kolbuk–White interpolation equation:

$$\frac{1}{\sqrt{\lambda}} = 2\lg\left(\frac{K_e}{3.7D} + \frac{2.51}{\mathrm{Re}\sqrt{\lambda}}\right)$$

where K_e is the equivalent or hydraulic roughness.

3. **Self-modelling quadratic turbulent regime**

$$(\mathrm{Re} > 10^5; \Delta < S)$$

Since the thickness of the boundary laminar layer in the general case is not known and, consequently, the type of turbulent regime is also not available, it is most efficient to determine the drag coefficient by the generalised Al'tshul' equation which is valid for the entire range of the turbulent flow

$$\lambda = 0.11\left(\frac{K_e}{D} + \frac{68}{\mathrm{Re}}\right)^{0.25}$$

For rectangular channels, the profile of the average velocity is determined by equations identical to circular sections but with different numerical coefficients. In this case, there is a difference in the profile of the velocity on a narrow and a wide wall.

The scale of turbulence may be characterised by the values of the mean volumes of the gas which, taking part in pulsation movement, retain their integrity for a certain period of time.

In the boundary region of the flow, the length of the mixing path is assumed to be proportional to the distance to the wall restricting the flow.

The intensity of turbulence is determined by the value proportional to the ratio of the mean quadratic velocity of pulsation motion to the mean velocity flow.

The frequency of pulsations characterises the variation of the amplitude values of the pulsation velocity. Its value is determined mainly by the scale of the vortex. Because of the fact that the turbulent flow contains vortices of different scales, there is not one but a whole spectrum of the frequencies of turbulent pulsations.

The pulsation component of the velocity is characterised by a slightly different distribution than the mean velocity. Its highest nonuniformity is detected around the walls. The gradient of this velocity in the centre of the flow is not steep.

In this case, the relative amplitude of the transverse pulsations even in the vicinity of the walls is not determined by the Reynolds number.

Geometrically similar flows at the same value of the Reynolds number are also mechanically similar, i.e. they are characterised by geometrically similar configurations of the flow lines and are described by the same functions. This is the so-called Reynolds similarity law. This law is valid only for steady motion which is not affected greatly by external forces. However, in the case of movements which greatly depend on the external forces, and also for non-steady movements, the similarity law is more complicated. In these cases, for similarity it is necessary to ensure that in addition to the Reynolds number, other dimensionless criteria, which take into account the strong effect of other members of equations (III-1) and (III-2), are also of the same magnitude. From these equations, using the methods identical to the examined method, it is possible to obtain other dimensionless complexes, such as the Strouhal, Froude and Euler criteria. Other similarity criteria, used in hydrodynamics, are the derivatives of these four dimensionless parameters which are regarded as main controlling parameters.

The Strouhal criterion or homochronicity

$$Sh = \frac{wt}{l} \tag{III-8}$$

characterises the forces of inertia in the non-steady flow. This criterion is important where the period of time of changes in the flow is very short and the velocity is relatively high. Therefore, at low values of this parameter, it is possible to use quasi-stationary modelling measures, i.e. replace non-steady movement by steady movement, accepting their instantaneous values as the controlling parameters.

The Froude criterion

$$Fr = \frac{gl}{w^2} \qquad \text{(III-9)}$$

characterises the relative value of the forces of gravity.

The Euler criterion is a measure of the ratio of the force of pressure to the inertia force and is a criterion of the dimensionless pressure:

$$Eu = \frac{\Delta P}{\rho \ w^2} \qquad \text{(III-10)}$$

As already mentioned, the hydraulic quality of the solid walls, restricting the flow, is characterised by the hydraulic drag coefficient. This coefficient is linked by a very simple relationship with the Euler criterion:

$$\lambda = 2Eu \qquad \text{(III-11)}$$

These considerations show that this coefficient is a function of only the Reynolds criterion.

Equation (III-10) shows that

$$\Delta P = \lambda \frac{\rho w^2}{2} \qquad \text{(III-12)}$$

In this dependence, the pressure gradient is proportional to the square of velocity. However, the actual law of relationship between these parameters is far more complicated, because the value of the coefficient of proportionality itself depends on velocity.

The product of the equations for the Euler and Reynolds criteria gives:

$$Eu\,\mathrm{Re} = \frac{\Delta P}{\upsilon w} \qquad \text{(III-13)}$$

It has been established that this dimensionless complex in the range of low values of the Reynolds number is constant, i.e. $Eu\mathrm{Re} \approx$ const.

It is almost impossible to calculate the reliable boundary for the start of the purely turbulent regime of motion in apparatus of any configuration. This must be determined by experiments in every specific case. For this purpose, on the basis of the experimental data it is necessary to express the following dependence for the examined apparatus:

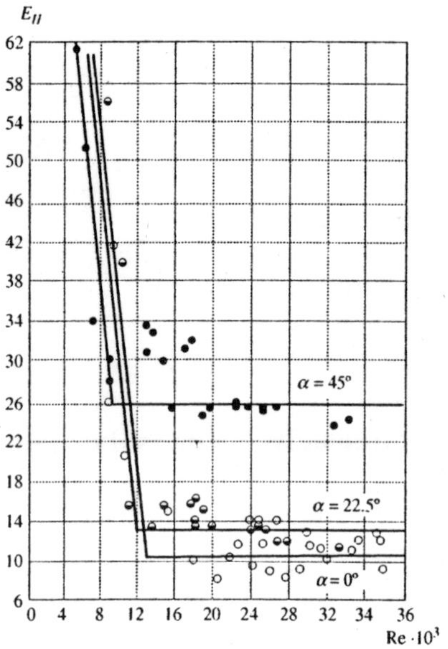

Fig.III-6. Eu = f (Re) dependence for cascade classifiers (z = 4, i^* = 1) with different angles (α) of transfer trays.

$$Eu = f(\mathrm{Re}_s)$$

as indicated for specific classifiers in Fig. III-6.

In the range of purely turbulent motion, the drag coefficient does not depend on the Reynolds number. In the examined relationship, the sought dependence is parallel to the axis of the Reyńolds number. This makes it possible to determine with sufficient reliability both the first and second self-modelling ranges.

All these factors indicate that these phenomena, taking place in moving flows, are very complicated and that there is a large variety of the conditions of the effect of the flows on the restricting walls and on the solids placed in them. This stresses the need for the most detailed consideration of the entire complex of these phenomena when examining the mechanism of the processes of fractionation of powders in moving flows.

3. SETTLING AND HOVERING OF SINGLE PARTICLES

The main relationships governing settling

The movement of individual particles in two-phase flows is affected by so many different factors that the general analysis of this motion is at first sight almost impossible. However, it may be carried out in stages as a result of the transition in analysis from simple phenomena to more complicated ones.

The current views regarding the mechanism of the process of the two-phase flow are based on the phenomena taking place during settling of solid particles in an unlimited still medium.

In examination of these phenomena, all the particles of the regular shape are subjected to usually to strict theoretical analysis. In this case, the settling process is regarded as one-dimensional.

Possible transverse displacement of the particles as a result of the effect of forces, similar to lifting forces, and other factors are not taken into account in this examination.

The rotation of the particles during settling is not considered in the calculations. The movement of particles is regarded only as strictly straight movement in the direction of the gravitational force.

If in some period of time the velocity of the particle is v, then the relative velocity of flow of the medium around the particle

$$w = -v$$

is equal to the absolute velocity of movement of the particle.

At first sight, it may appear that the variation of pressure at different points of the surface of any spherical solid takes place only as a result of overlapping, by the solid, of the part of the space occupied by the medium (Fig. III-7, b).

The simplest possible analytical examination of this phenomenon is based on examining the potential (without viscosity) flow around the sphere.

In transition from point 3 to the points 1 and 2, situated on the middle section, the rate of displacement of the medium by the settling particle increases. According to the Bernoulli equation, this results in a decrease in the pressure in this section.

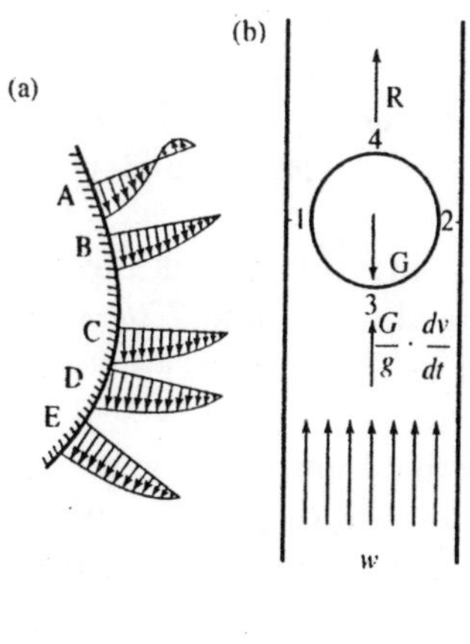

Fig.III-7. Distribution of flow velocities on the surface of particles (a) and the diagram of forces acting on the particle (b).

Behind the middle section, the pressure again starts to increase and reaches the initial value at point 4 resulting in the formation of a field symmetric in relation to the middle section. This shows that the resultant of all forces, applied to the particle, is equal to 0. This consideration results in the physically impossible conclusion on the absence of resistance during movement of the particle. The conclusion, based on these considerations, is well known as the d'Alambert paradox. This paradox is the result of the excessive schematisation of the investigated process. From this, firstly, it is clear that it is necessary to take into account the viscosity of the medium and, secondly, it is necessary to examine the phenomena taking place at the phase boundary.

As a result of the presence of viscosity forces in the liquid or gas medium the so-called boundary layer forms at the interface. The phenomena taking place in this layer greatly differ from the phenomena examined previously because the layer is characterised by the dissipation of energy as a result of a large change of the velocity of displacement of the medium in the vicinity of the particle.

On the surface of the particle, the effect of bonding forces results in the formation of an elementary layer which moves together with the particle. The velocity is transferred from this layer to neighbouring elementary masses of the medium as a result of the viscosity forces. This results in a monotonic decrease in the velocity in the boundary layer along the normal to the surface from v to 0. In this case, the curve of distribution of velocity has the characteristic form which also reflects its continuous decrease, initial from the surface, and the smooth transition to the stationary medium (Fig. III-7a). This distribution of velocity in relation to the normal corresponds to the negative derivative:

$$\frac{dw}{dy} < 0$$

which gradually decreases with increase of the distance from the surface and tends to zero on approaching the external boundary of the layer.

This pattern forms in the frontal part of the solid or on the entire surface of the solid during continuous flow-around. The situation formed in movement with separation of the boundary layer is different.

In the region of increasing pressure, the medium is inhibited not only by internal friction but also by an increase of pressure along the surface of the particle resulting in movement of the medium from areas with higher pressure to areas with lower pressure, i.e. against the direction of flow-around. In this case, some part of the medium on the surface of the boundary layer moves in the opposite direction. Consequently, The ratio $\frac{dw}{dy}$ at the boundary of the layer in some range becomes positive.

The shape of the velocity distribution curve in this case is completely different (position A) in comparison with the conditions of continuous flow and acquires the usual formal only on approach to the surface of the solid. In this section, separating the zones of the continuous flow and the flow with breaks, the profile of the velocity representing a curve limiting for curves of both types (position B) should be established. In this curve, the region of reverse flow decreases to a point situated on the surface of the layer. At this point

$$\left.\frac{dw}{dy}\right|_{y=\Delta} = 0$$

the position of separation of the boundary layer is linked with it.

In this area, the boundary layer swells up and separates from the particle surface.

The hydrodynamic pattern of the process becomes completely asymmetric. Consequently, this results in the formation of the equivalent pressure force, determining the resistance of the settling solid in the stationary medium. If the boundary layer is laminar prior to separation from the surface, then after separation it behaves as a free stream in the submerged space and rapidly becomes turbulent. The interfacial surface, being the surface of tangential discontinuity of the velocity, becomes unstable and rapidly folds up into one or several vortices.

According to S.L. Soo, these phenomena start to take place when the Reynolds number for a sphere is $Re \approx 51$. Other investigators indicated that vortex formation starts at $Re = 17$.

G. Shlichting published interesting dependences for the distribution of velocity in a laminar boundary layer on a sphere in the reverse flow. The sphere was suspended in a magnetic field with an incident horizontal flow flowing around it. He showed that the effect of the settling solid, even of a regular shape, extends to large distances into the solid medium.

The presence of extensive dissipation of energy in the entire volume of the turbulent wake and also the formation of the interface at separation from the boundary layer lead to a high resistance to the settling of the particle. In this case, the resistance decreases with a decrease in the width of the turbulent wake, i.e. with an increase of the distance of the separation point on the particle surface. Thus, the viscosity forces are the primary reason for the dynamic interaction of the solid and the medium of the dual type. Firstly, these forces are manifested in the form of a friction resistance at relative displacement of the particle and the medium. Secondly, the viscosity of the medium determines the formation of dynamic counterpressure forces.

The nature of settling of the particle is determined by a system of forces consisting of the weight of the particle in the investigated the medium and the resistance of this medium (Fig. III-7). The weight of the particle may be expressed by the equation:

$$G = mg_0$$

where m is the mass of the particle; g_0 is the freefall acceleration in the moving medium.

It is well known that

$$g_0 = g\frac{\rho - \rho_0}{\rho_0}$$

where g is gravitational acceleration; ρ, ρ_0 is the density of the solid and the medium, respectively.

It should be mentioned that the value of the static lifting force for the air is three orders of magnitude lower than the weight of the particle, since

$$\frac{\rho}{\rho_0} > 1000$$

Consequently, for the gas medium it may be assumed that

$$g \approx g_0$$

without decreasing the accuracy of calculations.

On the other hand, for liquid media, it is important to take this correction into account because the order of magnitude of the specific weight of the liquid is the same as that of the solid particles.

In a general case, the resistance of the particle is determined by the dependence

$$R = \lambda F \frac{v^2}{2} \rho_0$$

where λ is the resistance coefficient of the particle; F is the middle section of the particle, m^2; ρ_0 is the density of the medium, kg/m^3; v is the velocity of the particle, m/s.

The resistance coefficient is an important characteristic and determines the total effect of the friction forces and dynamic pressure.

Thus, the general equation of movement of the particle with strictly vertical settling of the particle in any stationary medium may be expressed by the equation:

$$m\frac{dv}{dt} = -mg_0 + \frac{1}{2}\lambda\ Fv^2\rho_0 \qquad \text{(III-14)}$$

The general solution of this equation has the form:

$$v = -\sqrt{\frac{g_0}{K}}th\left(t\sqrt{g_0K}\right) \qquad \text{(III-15)}$$

where

$$K = \frac{\lambda F\rho_0}{2m}$$

The hyperbolic tangent is characterised by a limit equal to 1 to which it tends asymptotically. Theoretically, this limit is obtained at infinity. However, it may be assumed with the accuracy sufficient for practice, that this function reaches the limiting value at an argument of 2.5. Consequently, the duration of the transition process may be determined from the following equation:

$$t\sqrt{g_0K} = 2.5 \qquad \text{(III-16)}$$

After this time, the particle moves at a steady velocity referred to as the finite settling velocity or incidence velocity.

The dependence (III-16) shows that:

$$v_0 = \sqrt{\frac{g_0}{K}} = \sqrt{\frac{2mg(\rho-\rho_0)}{\lambda F\rho_0^2}} \qquad \text{(III-17)}$$

For a circular particle, the finite settling velocity in an unlimited medium is:

$$v_0 = \sqrt{\frac{4gd(\rho-\rho_0)}{3\lambda\rho_0}} \qquad \text{(III-18)}$$

The determination of the value of the finite settling velocity using this dependence is possible after determining only the resistance coefficient, i.e. all other quantities are determined unambiguously.

It has been established that all quantitative special features of the settling process, determining the resistance coefficient: the thickness of the boundary layer is a function of the liftoff angle of the moving medium, the position of the liftoff point, the profile of velocity in the boundary layer and the nature of its variation, depend on the Reynolds number calculated for the particle, i.e.

$$\mathrm{Re} = \frac{vd}{\upsilon}$$

Here v is the velocity of the particle; d is the particle diameter; v is the kinematic coefficient of viscosity of the medium.

The dependence of the resistance coefficient on the Reynolds number for the sphere, determined by experiments, is shown in Fig. III-8.

The range of very low values of the Reynolds number, i.e. the region of continuous (break-free) laminar flow-around, is indicated by the straight line.

Within this region, self-modelling is evident and the law of inverse proportionality operates, i.e. $\lambda\mathrm{Re}$ = const. This region is defined by very low Reynolds numbers. The viscosity forces are controlling in this region. The coefficient of resistance for this region was determined by Stokes as

$$\lambda = \frac{24}{\mathrm{Re}} \tag{III-19}$$

This dependence hold for the Reynolds numbers of up to 0.2, but is often used in the range up to 2.

The Stokes law was derived assuming the medium behaves as a continuum, i.e. as a fluid. In settling in a gas, the pattern of the process changes. Here, the main dependences have a lower limit and are applicable only when the Knudsen number for the particle is considerably lower than 1, i.e.

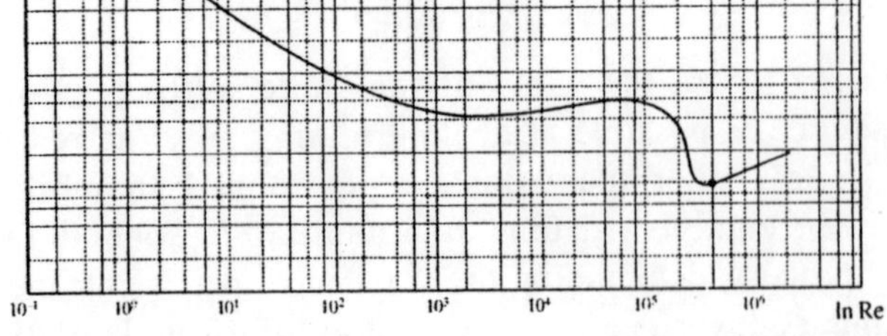

Fig. III-8. $\lambda = f$ (Re) dependence for a shperical particle with single settling in a stationary medium.

$$Kn = \frac{l}{d} << 1 \tag{III-20}$$

where l is the mean free path of the gas molecules. The value of this parameter may be expressed by the relationship:

$$l = 4.03 \frac{\mu}{\rho_0 w} \tag{III-21}$$

where μ is the dynamic viscosity of the gas; ρ_0 is the density of the gas; w is the velocity of the gas.

When the length of the free path of the gas molecules becomes of the same order of magnitude as the particle diameter, the velocity of the particle becomes higher than that indicated by the Stokes law. This phenomenon results in movement of the particle over large distances. The theory of this Brownian motion has been developed quite sufficiently.

Analysis shows that the Brownian effects are only very small in actual technological processes because the particles in these processes are relatively large for these effects to operate. Practice shows that the influence of this effect on particles with a size of 3-5 μm can be ignored. However, the Brownian effect has a strong influence on the coalescence of these or even larger particles.

From the moment of initiation of separation of the boundary layer, self-modelling is disrupted. The pattern of flow of the medium around the particle becomes considerably more complicated.

The resistance coefficient in this regime cannot be determined analytically because the physical pattern of flow-around is not completely clear. This region is referred to as the transition region and includes the range of the Reynolds numbers from 2 to 100 and, according to some data, 2 to 500.

The coefficient of resistance in this region is expressed by different empirical approximating dependences, for example:

$$\lambda = \frac{13}{\sqrt{\text{Re}}}$$

or

$$\lambda = \frac{18.5}{\text{Re}^{0.6}}, \qquad \text{and so on}$$

A shortcoming of equations of this type is that they are limited by the range of the Reynolds numbers and that they are not sufficiently exact. It is assumed that compound equations, for example

$$\lambda = \frac{24}{\sqrt{\text{Re}}} + \frac{4}{\sqrt[3]{\text{Re}}} \qquad \text{(III-22)}$$

or

$$\lambda = \pi \, (0.128 + \frac{12.8}{Re}) \qquad \text{(III-23)}$$

are more exact. At the Reynolds numbers Re > 1000, the rearrangement of the flow-around regime is interrupted. A specific form of the interaction of the particle with the medium is established. The stability of this form is determined by the constancy of the angle of liftoff of the boundary layer (82°). The value of this angle is stable in relation to the changes of the Reynolds number, and this determines the constancy of the resistance coefficient and the presence of self-modelling during movement of the medium in pipes.

The resistance coefficient in this range, corresponding to $1000 < \text{Re} < 1 \times 10^5$, is stabilised and, according to various sources, has the values from 0.42 to 0.50.

The boundary layer in this settling remains laminar. For settling of the particles, the nature of variation of the resistance coefficient is considerably more complicated than in the movement of the medium in the form of a flow with solid walls. In the conditions of the external problem, the range of constancy of the resistance coefficient is considerably more complicated than for the movement of the medium by the flow with the solid walls. In the conditions of the internal problem, the region of constancy of the coefficient of hydraulic resistance is not limited at the top and extends a long way on the scale of the values of Re. In Fig. III-8 the range of the second self-modelling region is replaced by the range of a very large decrease in coefficient λ.

This unexpected effect should be explained on the basis of the nature of interaction between the boundary layer and the external medium at their boundary.

In the second self-modelling region, the stabilisation of the boundary layer takes place after separation of the layer from the particle surface.

With increasing Reynolds number, the point of the start of turbulisation approaches gradually the liftoff point. At a specific value, the start of turbulisation overlaps the point of separation of the boundary layer, i.e. some part of the boundary becomes turbulent. The size of this part increases with increasing Reynolds number. This results in the rearrangement of the flow-around pattern, the angle of liftoff increases to 120° and, the width of the turbulent wake behind the particle rapidly decreases and the resistance rapidly decreases. This phenomenon is referred to as the crisis of resistance.

The crisis of resistance is explained by the increase of the intensity of exchange of the amount of motion between the boundary layer and the rest of the medium. The time to the start of this phenomenon decreases with increase of the degree of perturbation of the flow-around, i.e. with a decrease in the critical value of the Reynolds number for transition to the to the turbulent regime in the boundary layer.

As indicated by the graph (Fig. II-8), the transition from laminar resistance (Stokes law) to turbulent resistance (Newton law) does not occur as in the case of a hollow pipe but it takes place gradually in the range of relatively high values of the Reynolds number. This is explained by the relatively larger (in comparison with the particle size) thickness of the laminar layer at their surface. In pipelines, the thickness of the boundary layer is negligible in comparison with the flow diameter and, consequently, the transition from laminar to turbulent resistance is more rapid.

Finite settling velocity

The particle, falling in a viscous medium under the effect of gravitational force, starts to move finally at a constant velocity (v_0) at which the forces of gravity are balanced by hydrodynamic forces. The value of this finite velocity is regarded as the most important parameter of organisation of various processes with disperse materials: classification, pneumatic transport, fluidised bed, etc. The determination of this parameter has been the subject of a large number of theoretical and experimental studies.

Expression (III-18) determines the finite settling velocity of an individual isolated circular particle in an unlimited medium.

For the region of action of the Stokes law according to (III-19)

$$v_0 = \sqrt{\frac{4gd(\rho - \rho_0)\mathrm{Re}}{72\rho_0}}$$

and, consequently

$$v_0 = \frac{d^2(\rho - \rho_0)}{18\mu g}$$

For the region of validity of the Newton law λ = const and the finite velocity is determined from the relationship (III-18). The settling velocity in the transition region is determined by the methods of successive approximation (trial calculations) because the required velocity v_0 is included as an argument also in the Reynolds number and the resistance coefficient. The solution of this problem is greatly simplified by calculating the dependence:

$$\mu \mathrm{Re}^2 = f(\mathrm{Re})$$

or the dependence:

$$\frac{\mathrm{Re}}{\lambda} = f(\mathrm{Re})$$

The interesting feature of these relationships is that they do not depend on the relative velocity of the particle and the medium:

$$\lambda \mathrm{Re}^2 = \frac{4g}{3} \cdot \frac{d^3}{\upsilon^2} \cdot \frac{\rho - \rho_0}{\rho_0} \tag{III-24}$$

$$\frac{\mathrm{Re}}{\lambda} = \frac{3}{4g} \cdot \frac{v_0}{\upsilon} \cdot \frac{\rho_0}{\rho - \rho_0} \tag{III-25}$$

In this case, it is necessary to use the dependences

$$v_0 = \frac{\upsilon \mathrm{Re}}{d} = \frac{\mu \mathrm{Re}}{\rho_0 d} \tag{III-26}$$

$$d = \frac{\upsilon \mathrm{Re}}{v_0} = \frac{\mu \mathrm{Re}}{\rho_0 v_0} \tag{III-27}$$

For the transition region, the resistance coefficient is expressed in certain cases by the approximate independence

$$\lambda = \frac{A}{\mathrm{Re}^n}$$

Consequently, for the fall velocity in the transition region of the following relationship will be valid

$$v_0 = \sqrt{\frac{4g}{3A} \cdot \mathrm{Re}\, d \frac{\rho - \rho_0}{\rho_0}}$$

and after substitution

$$\mathrm{Re} = \frac{v_0 d \rho_0}{\mu}$$

We obtain

$$v_0^{2-n} = \frac{4}{5} g^{1-n} \frac{d^{1+n}}{A\mu^n} \cdot \frac{\rho - \rho_0}{\rho_0^{1-n}}$$

In some cases, the same solution is presented in the criterial form by the dependence of the type

$$\text{Re} = f(Ar)$$

For steady motion

$$\frac{\pi}{6} \frac{d^3}{}(\rho - \rho_0) g = \lambda \frac{\pi}{4} \frac{d^2}{} \cdot \frac{v^2}{2} \rho_0$$

After carrying out appropriate transformations, we obtain

$$\frac{gd^2}{\upsilon^2 \rho_0} \cdot \frac{d(\rho - \rho_0)}{3} = \frac{\lambda}{4} v^2 \frac{d^2}{\upsilon^2} \qquad \text{(III-28)}$$

The dimensionless complex

$$\frac{gd^3(\rho - \rho_0)}{\rho_0 \upsilon^2} = Ar$$

is the Archimedes criterion. Taking this into account, (III-28) can be transformed to the following form

$$\frac{Ar}{3} = \frac{\lambda \text{Re}^2}{4} \qquad \text{(III-29)}$$

Substituting, into equation (III-29), the value of λ, corresponding to different flow-around conditions, one obtains

For the laminar regime

$$\text{Re} = \frac{1}{18} Ar;$$

for the turbulent regime

$$\text{Re} = 1.742 Ar^{0.5};$$

for the transition regime

$$\text{Re} = 0.153 Ar^{0.714}$$

A universal equation, linking these two criteria for all flow-around conditions, has been proposed

$$\text{Re} = \frac{Ar}{18 + 0.6\sqrt{Ar}}$$

The settling of single particles in a medium, limited by solid walls, has a number of special features. The solid particle overlaps a part of space and, consequently, the medium, displaced by the particle, moves in the opposite direction. The curve of distribution of velocity in the vicinity of the wall is shown in Fig. III-9. The general nature of the interaction of the particle with the wall depends on the shape and size of the particle, its position and orientation, and also on wall and geometry.

Thus, a circulation movement of the medium forms around the

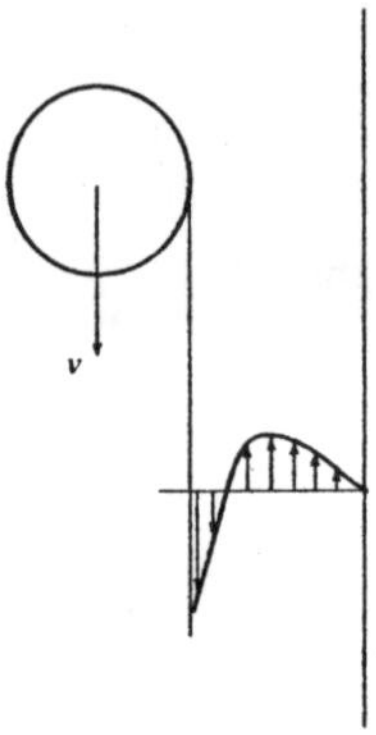

Fig. III-9. Distribution of velocities in settling of a particle in the vicinity of a wall.

particle, settling in a limited space. The volumes in the immediate vicinity of particles move together with the particle, and the volume situated at some distance outside the limits of the boundary layer moves in the opposite direction. This increases the resistance and decreases the settling velocity.

In turbulent movement, the effect of the wall on the settling of the particle becomes slightly weaker but it cannot be completely ignored because, especially in mass settling where this effect greatly corrects the actual velocity of settling of every individual particle in comparison with the calculated values. There are a number of empirical relationships which make it possible to consider the constraint in the settling of particles of this type. The following equations are used most widely. The equation derived by V.A. Uspenskii

$$v = v_0\left[1-\left(\frac{d}{D}\right)^2\right] \qquad \text{(III-30)}$$

where D is the channel diameter, m.

The equation derived by R.B. Rozenbaum:
for the laminar movement regime

$$v = v_0\left(1-0.6\frac{d}{D}\right)\left(1-\frac{d}{D}\right)^2 \qquad \text{(III-31)}$$

for turbulent movement

$$v = v_0\left[1+2.1\frac{d}{D}\left(1-\frac{d}{D}\right)\right]\left(1-\frac{d}{D}\right)^2 \qquad \text{(III-32)}$$

The equation derived by A.S. Kemmer

$$v = v_0\left[1-\left(\frac{d}{D}\right)^2\right]^{1.5} \qquad \text{(III-33)}$$

and a number of other equations.

In certain conditions, the constraint determines the values associated with both the migration of particles in the direction normal to the settling direction, and with rotation of the particle.

The behaviour of spherical and irregular shape particles differs. In laminar flow-around, the particles of the regular shape fall in the majority of cases in the position in which they are introduced into the medium. The velocity of these particle depends on the initial orientation and, consequently, identical particles may have different settling velocities. In the transition region, the orientation of the particles becomes unstable and is accompanied by oscillations whose amplitude increases with increasing Reynolds number. In the turbulent region, the particles, irrespective of the initial orientation, fall or hover in the position resulting in the maximum value of their resistance coefficient. Particles of irregular shapes are characterised by a more distinctive tendency for rotation and transverse migration during settling. Thus, the process of simple settling of even individual particles in a stationary medium is characterised by a highly complicated nature and indeterminacy. Naturally, the behaviour of the particle in moving media is characterised by the phenomena which are far more complicated and indeterminate.

Special features of the interaction of the particle with the moving medium

In comparison with settling, in this case external turbulence results in a large decrease of the resistance coefficient of the particle as a result of a decrease of the upper critical value of the Reynolds number with simultaneous displacement into the tail part of the circular line of liftoff of the flow from the sphere. This displacement is already observed at Re>400. At lower values of the Reynolds number this effect weakens: there is no displacement of the separation point but the resistance slightly increases as a result of an increase in the dissipation of energy in the region of the wake.

If the particle is small in comparison with the smallest scale of turbulence, the particle reacts to all pulsations typical of turbulent motion. This feature may be a basis for the first determination of the difference between the behaviour of large and small particles in a flow. Coarse particles take part mainly in the linear movement of the medium, the small particles follow the turbulent vortices. The resistance of the small particles to movement is determined by the viscous nature of the surrounding medium. Since the velocities of the particle and the medium differ by the value of slip, the presence of the particles in the flow increases the intensity of dissipation. This predetermines the exceptionally complicated nature of movement of the particles which consequently does not fit the framework of the cellular model proposed by Chen, according to which the particle moves together with some volume of the deformed medium. This approach with a large number of stipulations may be used efficiently only for laminar flow conditions.

Understanding of the mechanism of the investigated phenomenon is greatly complicated by the formation of secondary motions – oscillation and rotation of the particles which has a strong effect on the resistance coefficient. According to S. Sow, the oscillations are not detected at $Re<80$ and always occur at $Re>300$. The rotation of the particles in the liquid flow may be caused by different reasons. If a particle is in a gas with a velocity gradient, the particle starts to rotate. Although the velocity of shift in turbulent vortices may be high, this effect is self-compensated as a result of the random nature of turbulence and its influence on the rotation of the particles is not strong. An exception is the flow in the vicinity of the wall. In this layer, during movement of the liquid or gas, the mass of the medium is attached to the rotating particle and this increases the velocity of flow on one side of the particle and reduces the velocity on the other side. The phenomenon, known as the Magnus effect, forces the particle to move into a region with a higher velocity (to the axis of the flow). However, in accordance with accurate experimental data, the particles concentrate in the ring-shaped layer whose distance from the axis of the pipe is equal to approximately half the pipe radius. The transverse effect of the flow may also form as a result of the displacement of the point of separation of the boundary layer during rotation of the particle.

It should be accepted that the general practice of this phenomenon is exceptionally complicated, and only idealised cases have been studied more extensively. The results of these investigations are useful because they indicate the comparative importance of different factors. According to R. Boothroyd, in the laminar flow (or a laminar boundary layer) the ratio of the transverse force to the force determined in accordance with the Stokes law is:

$$\frac{F_n}{F_c} = \frac{0{,}121d}{D} \mathrm{Re}_n f \qquad \text{(III-34)}$$

where f is the friction coefficient during movement of the gas.

Analysis of the relationship shows that the tendency for the movement of particles in the direction normal to wall is quite strong. The phenomena of this type are often used in practical applications when the layer of the flowing liquid is not large, for example, in enrichment on gates, and tables, etc. The force, transverse in relation to movement, has a strong effect on the nature and results of classification. It has been reported that in some cases at Re ≈ 10 the particle moves towards the axis of the flow, and at 16 < Re < 120 the particle moves towards the wall. It may be seen that the problem of the interaction of the particle with the moving medium is far from solved.

However, in practice, it is necessary to examine systems containing large numbers of particles. Until recently, the investigations of this type into two-phase flows were carried out on the level of determination of the velocity of hovering of the particles in a rising flow. This determination was carried out without taking into account rotation of particles, transverse migration, the absence of collisions between the particles and the wall, i.e. the complex process was reduced to a linear unidimensional problem. In this case, the velocity of the flow of the medium is assumed to be determined and identical in the entire cross-section of the channel. In these extremely idealised conditions, the relative velocity of flow around the particle is determined by the dependence:

$$w_b = v - w$$

where v is the velocity of movement of the solid particle; w is the velocity of the rising flow.

The general equation of motion of a spherical particle in these conditions may be presented in the following form:

$$m\frac{dv}{dt} = -g_0 + \frac{1}{2}\lambda$$

This equation was transformed to the following form:

$$\frac{dv}{dt} = -g + K(v - w)^2$$

where

$$K = \frac{\lambda F \rho_0}{2m}$$

The dependence in this form is the Riccati equation which is reduced to the differential equation of the second order.

The solution of this equation gives

$$v = w - \sqrt{\frac{g_0}{K}} th\left(t\sqrt{g_0 K}\right) \qquad \text{(III-35)}$$

Comparison of (III-15) and (III-35) shows that at any compared moment of time, the velocity of movement of the particle in the counterflow appears to be equal to the velocity of the particle during is settling in a stationary medium + the velocity of the flow itself.

The second multiplier of equation (III-35) is the hyperbolic tangent asymptotically approaching its limit. After some time, the velocity of the particle becomes almost constant and in subsequent stages it is no longer dependent on time and is determined by the relationship:

$$v = w - \sqrt{\frac{g_0}{K}}$$

The value of the velocity is referred to as the steady velocity of movement of the particle and is determined only by the velocity of movement of the flow. Of greatest interest is the limiting case in which the steady velocity is equal to zero. The velocity of the flow of the medium, fulfilling this condition, is referred to as the hovering velocity and is determined from the relationship:

$$w_0 - \sqrt{\frac{g}{K_0}} = 0$$

consequently,

$$w_0 = \sqrt{\frac{4gd(\rho - \rho_0)}{3\lambda\rho_0}} \qquad \text{(III-36)}$$

The results and the dependence (III-18) are usually used for concluding that the finite velocity of settling and the hovering velocity in the counterflow for a spherical particle are completely identical. The dependence (III-36) is a consequence of the equation obtained from the extreme idealisation of the phenomenon. In this case, it is accepted without discussion that the coefficient of resistance of the settling and free-falling particles are the same and do not depend on the Reynolds number, i.e. they do not depend on the turbulence of the medium whose value in the flow is different in comparison with that in a still medium.

The problem of the relationship of the hovering and settling velocities of the same particles has not been studied in detail. In practice, there are no reliable data on the possibility of determination of one characteristic parameter on the basis of another parameter for all ranges of variation of resistance. Therefore, the value of the hovering velocity for each specific case is determined by experiments, mainly in the sections of stabilised movement. This is carried out using different methods and experimental procedures: visual, photoelectronic, marked particles, high-velocity filming, instantaneous sectioning of parts of the channel, etc.

It is assumed that the coefficient of resistance of the particles increases with increasing acceleration, and the effect of acceleration on the value this coefficient may be very strong. It has been determined that for gas media, when $\frac{\rho}{\rho_0} > 1000$, the resistance coefficient does not depend on the sign of acceleration and is equal to the value of this parameter for the sphere with a constant flow around it.

The examination of the pattern of interaction of the particle and the flow is greatly complicated in transition to particles of irregular shapes. In most cases, the behaviour of these particles in a flow is not steady, with a distinctive tendency for rotational movement and migration. In the majority of cases of industrial powders, the shape of particles is such that they have no axis of symmetry and, consequently, the effect of the flow on them results in a moment whose value is unstable. This causes the formation of higher Magnus forces.

The problem of the behaviour of particles of irregular shapes has been studied insufficiently. The problem of the shape factor of these particles has not as yet been determined and is still the subject of discussions.

Urban divides all particles of irregular shapes into two groups.

The first group includes particles where the separation of the boundary layer is unambiguously determined by the presence of an angle. The distinguishing feature of these bodies or particles is that the cross-section of the body either increases or remains constant and the angle greatly changes. These bodies have a resistance coefficient independent of the Reynold number in the entire range of variation of Re. These are rectangular sheets, small cylinders, cones, hemispheres, oriented in the appropriate direction in relation to the flow.

The second group includes round solids with poor flow-around, with no sharp edges, and the cross-section of these solids does not decrease suddenly in the direction of the flow. Here, the area of liftoff of the boundary layer is determined by the nature of flow-around and, consequently, the resistance coefficient depends on the Reynolds number.

The resistance coefficient of the particles of irregular shapes is reduced to the appropriate characteristics of the equivalent shere by different methods.

The shape factor is represented by the ratio of the coefficient of resistance of the solid to the coefficient of resistance of the equivalent sphere:

$$K_g = \left(\frac{\lambda}{\lambda_0}\right)_{d_0=idm;\mathrm{Re}=idm}$$

As the difference between the shape of the particle and the shape of the sphere increases and as the roughness of the surface of the particle increases, the coefficient of resistance increases and the hovering velocity of the particle decreases. It has been established that the coefficient of resistance of particles of irregular shapes depends not only on the geometry of the particles but also on the Reynolds number, i.e.:

$$\lambda = K_g \lambda_0 = f(K_g;\mathrm{Re})$$

It is evident that the dynamic and geometrical coefficients of the shape are linked by the following relationship:

$$K_g = f(K\ ;\mathrm{Re})$$

For particles of irregular shapes, this dependence has been studied insufficiently and the question of determination of the coefficient of resistance of this type of particles is still the subject of discussions.

It is only known that the dynamic shape factor in the transition the region increases with increasing Re. This indicates a strong dependence of the resistance coefficient on the Reynolds number for irregular particles in comparison with the sphere.

The region of self-modelling for particles of irregular shapes (λ = const) starts at lower Reynolds numbers. The displacement in this case increases with an increase of the geometrical shape factor. This circumstance indicates that turbulisation at the surface of nonspherical particles started earlier than at the surface of the sphere.

According to V.A. Uspenskii, the particles of irregular shapes are oriented in the flow in such a direction that the resistance becomes maximum possible. This results in early turbulisation of the medium in the tail part. Z.R. Gorbis confirmed this by experiments using aluminium cylinders.

The surface roughness of the particles also affects the resistance coefficient. The particles with a rough surface, with other conditions being equal, are characterised by a lower hovering velocity. The effect of the surface roughness of the particles, especially particles of irregular shapes, has been studied in sufficiently and it is almost impossible to take this factor into account in theoretical calculations, especially for fine particles.

If the values of w_0 and λ are determined by experiments, the application of these parameters in analytical calculations makes it possible to take into account efficiently all secondary effects. Therefore, the following equation can be used for the hovering velocity, determined by experiments:

$$mg_0 = \lambda\ F\frac{w_0^2}{2}\rho_0$$

i.e. the resistance of the particle is equal to its weight in the medium and, at the same time, the velocity of the particle $v = 0$, i.e.

$$\frac{\lambda\ F}{2}\rho_0 = \frac{mg_0}{w_0^2}$$

In a general case, the resistance of the particle is:

$$R = \lambda F\frac{(w-v)^2}{2}\rho_0$$

Taking this into account, we obtain an interesting relationship containing all characteristics of the particle in the form of parameters which can be determined quite easily by experiments and which does not include the resistance coefficient in the explicit form:

$$R = \frac{mg_0(\ w-v\ )^2}{w_0^2} \qquad \text{(III-37)}$$

Experimental examination of the relationship of the hovering and settling velocities of particles

The indeterminacy, existing in the relationship of these parameters, and also the importance of these parameters for the resultant level of the development of the theory of the process, has predetermined the need for carrying out special experimental investigations.

It should be mentioned that in the currently available literature, the hovering velocity is, according to the majority of the authors, the mean velocity of the rising flow at which the particle is suspended, i.e. the velocity of the particle is $v = 0$ in relation to the walls of equipment. In this case, no attention is given to the question of the plane of the cross-section in which the particle is suspended, i.e. the result stemming from from assumptions on the uniform profile of the curve of the velocity of the continuous medium.

Experiments were carried out in water using special equipment, whose diagram in shown in Fig. III.10. The main element of equipment is the vertical transparent cylindrical pipe 1, with a height of 3 m, diameter 100 mm. Water is supplied into the lower part of the pipe using pump 2, the flow velocity through equipment is regulated using the valves 3 and 4 and measured with the flowrate meter 5.

At entry into the pipe there is a chamber 6 and a stabilising insert 7. Water is discharged from the pipe through the sleeve 8 and the container 9. In order to prevent the displacement of the solid particles, from both sides of the pipe, partitions 10 and 11 were installed on both sides of the pipe.

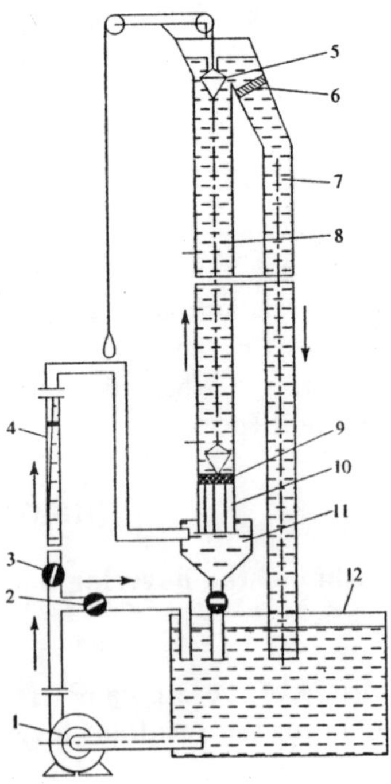

Fig. III-10. Experimental equipment for determining the hovering and settling velocities of particles.

The material was supplied to the pipe in the upper part. The settling velocity for the same particle was determined many times in the experiments. Special device 12 was used for these measurements, which made it possible to return the particle to the upper initial position.

The settling velocity was measured in a section 1500 mm long situated at a distance of 1000 mm from the upper edge of the pipe.

All the determinations of the velocity were carried out at a constant temperature of water.

The investigations were carried out on five types of materials with different density.

From each narrow fraction into which the initial mixture was divided, 10 grains were taken in a random manner and weighed on an analytical balance. The diameter of the particle was determined as the mean arithmetic value of the measured values.

The settling velocity was determined by the measurement of the time of passage of the solid particle through the reference section of the path during free settling of the particle.

The hovering velocity of the particle was determined by measuring the velocity of movement of water during weighing of the particle in the flow. By light tapping of the pipe in the experiments it was possible to ensure free fall the particle strictly in the centre of the flow and the flow rate of water was measured in this position.

It should be mentioned that the settling and hovering velocities in the experiments were determined in succession and many times for the same particle (7–8) and average values were subsequently determined.

After these experiments, it was possible to find the stable difference between the settling and hovering velocities for the investigated materials, and the settling velocity was always higher than the hovering velocity by some value.

For the measured hovering and settling velocities of the particles, the value of the Reynolds number was calculated:

$$\mathrm{Re} = \frac{v_0 d}{\upsilon}$$

$$\mathrm{Re}_0 = \frac{w_0 d}{\upsilon}$$

where Re is the value of the Reynolds number, determined in settling of the particle; Re_0 is the Reynolds number determined in the hovering of the particle; v_0 is the settling velocity of the particle (experimental value); w_0 is the hovering velocity of the particle (experimental value); d is the equivalent diameter of the particle, calculated in respect of the volume, υ is the kinematic coefficient of viscosity. The dependence of the relative difference of the Reynolds numbers $\left(\dfrac{\mathrm{Re}-\mathrm{Re}_0}{\mathrm{Re}}\right)$ for the same particles on the value of the Archimedes number (Ar) is presented in Fig. III-11.

According to this experimental dependence, the difference in the settling velocity of the small and light particles reaches very high values. With increase of the size of the particles and of the specific weight of the material this difference monotonically decreases. This shows that, in a general case, the mean velocity of the flow of the medium, ensuring hovering of the solid particles, is not equal to the final velocity of settling of the particles, and this difference must be taken into account in the determination of the optimum conditions of movement of the medium for organising the processes of gravitational enrichment.

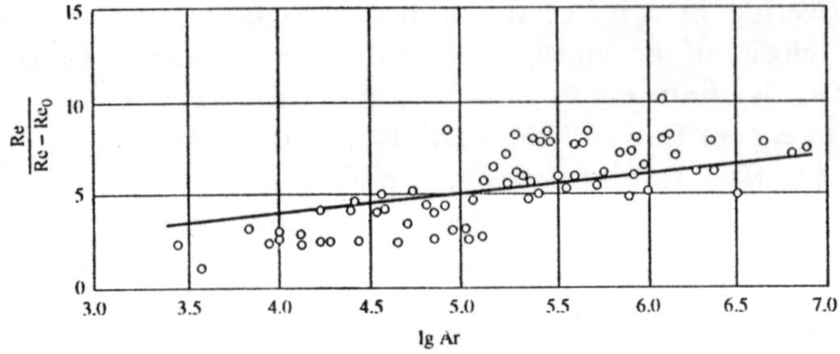

Fig. III-11. Dependence of $Ar = f\left(\frac{Re - Re_0}{Re}\right)$ for settling and hovering particles.

It should be mentioned that according to the experimental data, the settling velocity is always higher than the hovering velocity and this difference increases with increase in the degree of ordering of the flow conditions of the medium, i.e. with a decrease of the value of the Reynolds criterion in the flow. The settling velocity does not exceed the hovering velocity by more than a factor of two. This result of the experiments is slightly unexpected from the position of the investigated theoretical fundamentals of the process. These data can be used to obtain the relationships between the velocity of hovering and settling in the following form:

$$\frac{w_0}{v} = 1 - \frac{1}{1.16 \log Ar}$$

It will be attempted to find what determines this situation: random factors of the process or, possibly, some other factors. It is interesting to compare the experimental velocity of settling with the local velocity of the flow v_0 which suspends the particle on the axis of the pipe, i.e. examine this relationship taking the structure of the flow into account. For this purpose, the experimental data will be processed by the following procedure.

Initially, the Reynolds criterion of the medium is determined from the mean velocity of the flow w_0 in relation to the walls of the pipe:

$$\mathrm{Re}_n = \frac{w_0 D}{\upsilon}$$

This is followed by the calculation of the 'quality' of the pipe (the ratio of the mean velocity of the flow to the velocity of the axis of the flow). In the investigations, the flow regime of the medium is varied from laminar to turbulent. It was taken into account that the maximum values of the Reynolds criterion in the experiments did not exceed 4×10^4 and, taking into account the absolute equivalent surface roughness of the pipe produced fron organic glass, s = 0.01 mm, we obtain the product of the co-factors:

$$\mathrm{Re}_{max} \cdot \frac{s}{D_p} = 4 \cdot 10^4 \cdot \frac{0{,}01}{100} < 10$$

This inequality shows that none of the experiments extends outside limits of the region after the smooth turbulent regime.

The ratio $\frac{w_0}{v_0}$ in the range of the Reynolds number up to 4000 was determined from the experimental graphical dependence, valid for hydraulic originals pipes, presented by A.D. Al'tshul.

In the remaining range of the values of the Reynolds criterion we use the expression for the 'quality' of the pipe in the range of a smooth turbulent flow:

$$\frac{w_0}{v_0} = 1 - \frac{0.835}{\log \mathrm{Re}}$$

This is followed by the determination of the velocity at the axis of the flow w_0, equal to the true velocity of suspension of the particles.

Comparison of the local hovering velocity with the settling velocity is shown in the graphs in Fig. III-12. The graph shows that the settling velocity of particles of irregular shapes and different size and density is in fact identical with the local hovering velocity. The difference in the values of the line averaging the position of all points with the straight line $w_0 = f(v)$ does not exceed 5%. This difference may be explained by both the actual discrepancy in the velocity of hovering and the settling velocity of the particles of irregular shape and also by the presence of a systematic error in the experiments which could have affected the experimental results, obtained in particular for the coarse particles with a large specific weight. On the whole, with the acceptable degree of accuracy, the difference in these parameters can be ignored.

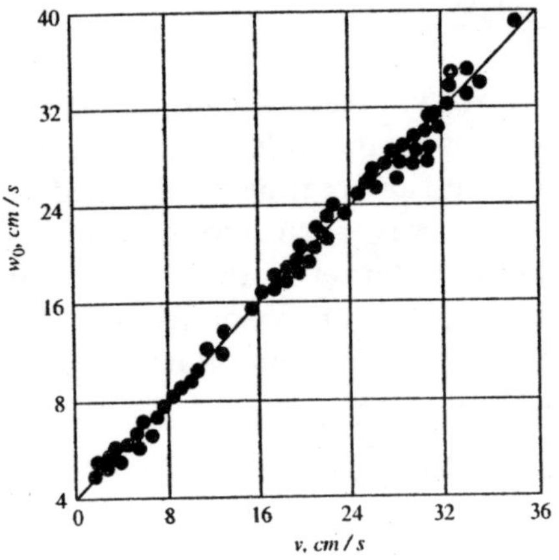

Fig. III-12. Ratio of the settling velocity of the particle to its local hovering velocity.

Thus, it has been established that the settling velocity of the individual particles in water is similar to the local velocity of hovering but not to the mean velocity of the flow, as mentioned previously.

This result can be used to draw three very important conclusions for the development of the theory and practice of gravitational processes.

Firstly, the hovering velocity of the solid particles in the flow should be determined taking into account the curve of the velocity (structure) of the flow. This velocity differs greatly from the mean velocity (calculated by conventional methods) in relation to the entire cross-section of equipment.

Secondly, the indisputable condition of organisation of highly efficient separation is the tendency to equalise the curves of the flow velocity in the cross-section as a result of the appropriate composition of equipment for classification. This requirement may be regarded as one of the main principles of rational organisation of gravitational separation.

Thirdly, these systems must be designed taking into account the need to ensure, in all separation zones, the maximum constancy of the force effect of the flow on the particles of the separated material. It should be mentioned that the determined ratio is valid for laminar and turbulent conditions of movement of the medium in relation to the particles.

Fourthly, in the development of mathematical models of the process it is important to take into account the curves of the velocity of the flow in the cross-section of equipment, and not the mean value of the velocity.

PART 4. SPECIAL FEATURES OF THE FORMATION OF THE TWO-PHASE FLOW IN THE SEPARATION CONDITIONS

1. Mass settling of particles

Special features of the mass settling of particles were investigated in the development of siphon hydraulic classifiers for the classification of metallic concentrates. Ignoring purely quantitative relationships, observed in these investigations which are of practical interest, attention will be given to the qualitative pattern of the observed phenomena.

The experiments were carried out in equipment produced from organic glass, with a height of 3 m and a diameter of 100 mm. The united of settling of metallic concentrates of several narrow size classes equal to (0–0.14); (0.14–0.28); (0.28–0.56) and (0.56–1.0) mm, was investigated. The density of the powder was in the range from 3400 to 4100 kg/m^3. The experiments consisted of the examination of the simultaneous settling of 100 g of the powder of each of the given fractions. The observed pattern may be described as follows. Initially, the powder particles move as a packed group. Subsequently, single particles begin to lag behind the main group and fill uniformly the entire cross-section of the vertical pipe. The velocity of the packed group of the particles moving together is greatly higher than the velocity of movement of the single particles. During descent, new and new portions of particles separate from the main group, but over a distance of 3 m the core does not completely break up. The pattern of this settling is shown in Fig. III-13.

The lower part of the packed group is cup-shaped. This phenomenon may obviously be explained as follows. The joint settling of the particles is accompanied by the displacement of a large amount of the medium resulting in intensive turbulent movement of the liquid in the front part, with a large number of lagging particles pulled into this part. In this zone, these particles may move even upwards.

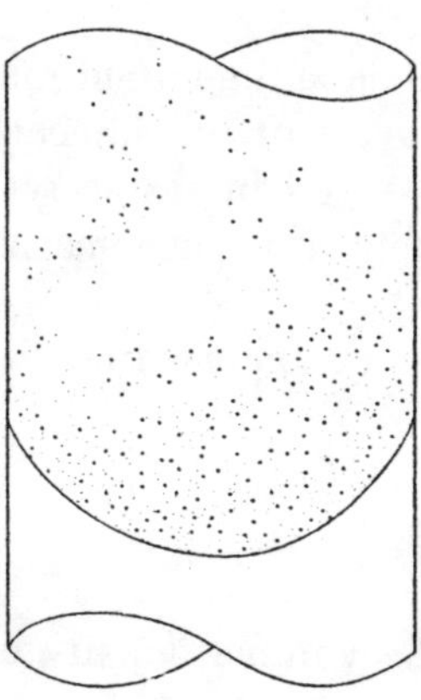

Fig. III-13. Mass settling of solid particles in water.

It may be expected that in the flows of gas suspensions, in which the clusters and agglomerations of the particles form much more easily, this phenomenon will be more distinctive. This type of settling is accompanied by the formation of high local concentrations in the core and by the nonuniform concentration in the remaining parts. Finally, the core of the particle should be disrupted by erosion as a result of the resistance of the medium, and the hovering velocity should decrease in accordance with a decrease of the size of the core.

The core becomes cup-shaped owing to the fact that the particles, distributed in the core above the phase boundary, move more rapidly than the particles situated in the lower part. This core may catch up with a small cloud of the particles and absorb it and, consequently, its size increases.

This phenomenon formulates very important questions regarding the organisation of input of the material into equipment for separation. Evidently, the material should be introduced in the maximally airated form and periodically with a cycle ensuring the separation of the previous portion prior to supplying the next portion. If the material is supplied into the system in the packed condition, it is important to use, firstly, a mechanical device for disrupting the core of the particles and the empty height of apparatus for erosion of the remaining parts by the flow of the medium. The phenomenon of the formation of the core of the particle was not detected for larger particles which, evidently, can be regarded as independent.

This phenomenon could be avoided at extremely low concentrations of the fine particles. In the region of disruption of the core, the concentration in the cross-section of the solid particles is equalised. In this case, the resistance force, acting on the particle, is high for a single particle because of the two reasons:

1. The velocity gradient in the vicinity of particles increases because of the proximity of the particles.

2. In the examined case, the medium is displaced upwards and this increase the velocity gradient even further.

At a relatively high concentration of the particle an important role is played by the interaction of the boundary layers of the adjacent particles. The traces behind the particles, observed in single settling of the particles, disappeared almost completely. This phenomenon result in a change in the resistance coefficient.

According to (III-18), the following equation can be used for the core of the flow:

$$v_a = \sqrt{\frac{4gD(\rho_c - \rho_0)}{3\lambda_a \rho_0}}$$

where D is the diameter of the pipe, ρ_c is the volume density of the particles in the core, λ_a is the coefficient of resistance in settling of the suspension. For a particle inside the core:

$$v = \sqrt{\frac{4gd(\rho - \rho_0)}{3\lambda\rho_0}}$$

At $v_a > v$ a core forms, at $v_a \approx v$ the particles move independently.

D. Happel and H. Brenner described interesting experiments simulating the settling of particles with uniform concentration.

Two identical particles, with parallel settling, rotate against each other. In this case, the nature of settling is determined by the conditions shown in Fig. III-14. The crosshatched area shows the distribution of the vertical velocity of the medium along the lines of the centres O_1O_2. If two identical particles settle on the vertical line one after the other, the rear particle acquires a high settling velocity and catches up with the front particle. Thus, the doublet formed in this manner increases its velocity. In settling of three spheres, when one of the spheres is situated in the vertical plane passing in the centre between the two other spheres, and all three particles have the same size, the external spheres move apart allowing the rear sphere to pass between them and then they again come together behind the third particle. If all these spheres fall along a single vertical axis (Fig. III-15) a 'doublet' forms A and B and they travel faster than C (position I). At some moment of time from the start of movement, the spheres A and B catch up with sphere C and the distance between all three spheres becomes the same (position II). This 'triplet' is, however, not stable because the central sphere starts to move towards the sphere C forming a doublet with the latter, with the doublet moving away from A (position III). A relatively complicated situation is produced in this case.

The constricted settling in uniformly dispersed systems has been

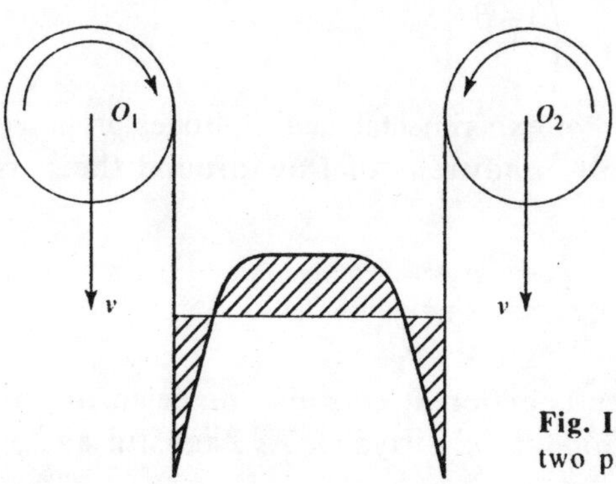

Fig. III-14. Combined movement of two particles.

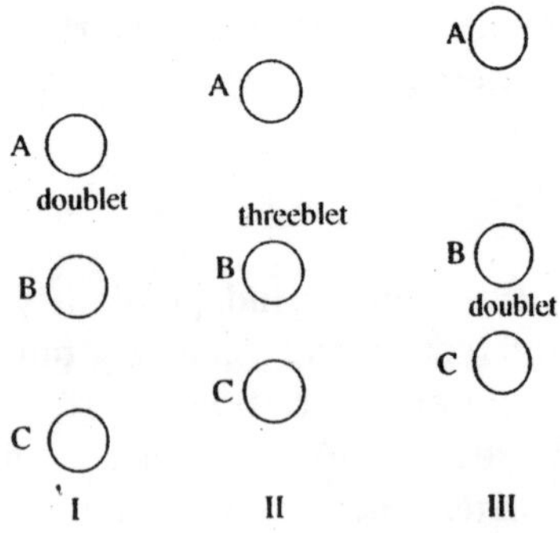

Fig. III-15. Schematic representation of movement of three spheres.

studied in a relatively large number of investigations. A number of empirical relationships including the volume fraction, occupied by the particles, have been proposed. The following are the best known. The Hirst–Lyashchenko correction coefficient, taking into account the effect of concentration on the settling velocity, is expressed by the empirical dependence:

$$\varphi = (1-\beta)^n$$

where β is the volume concentration of the solid-phase; n is the experimental parameter.

According to Lyashchenko, $n = 3$ and Hancock defined this parameter as $n = 2$, Finney $n = 1$.

On the basis of experiments carried out with gravel and sand, Mintz and Schubert found that the value of this parameter changes in the range 2.25–4.6. They showed convincingly that the value of this parameter cannot be constant and depends on the conditions of flow around the particles.

I. Kachan determines the hovering velocity in the constricted conditions by the parameter:

$$\varphi = \left(\frac{1-\beta}{\beta}\right)$$

On the basis of generalisation of experimental data, O. Todes proposed a dependence, common for all conditions of flow around the particles:

$$\mathrm{Re} = \frac{Ar(1-\beta)^{4.75}}{18 + 0.6\sqrt{Ar(1-\beta)^{4.75}}}$$

A unique interpretation of the coefficient φ is also made in the relationships proposed by A. Goden, D. Liflyand, A. Zagustin and V.

Kizeval'ter. The common moment, reported by different authors, is the distinctive dependence of the rate of suspension on the concentration of the material in the laminar flows and a less marked dependence in the turbulent flows. This difference becomes greater with increasing volume concentration of the particles. The value of the coefficient, which depends on concentration, is greatly affected by the shape of the particles, and this effect differs in the free and constricted conditions.

On the basis of analysis of the publications, it may be concluded that at a low concentration $\mu \leq (1-1.5)$ kg/m^3, the value of the parameter φ can be assumed to be equal to unity, with the accuracy of up to 5%. However, the experiments show that the effect of concentration at these values is not reflected in the interaction of the uniformly dispersed particles with each other directly or through the boundary layers, and it is reflected in the formation of the core of the jointly settling particles. This phenomenon is also observed at considerably lower consumption concentrations and must be taken into account.

2. Mass suspension of the particles in the flow

The processes of gravitational fractioning of the powders are organised at the velocity of movement of the medium ensuring the free transport of relatively fine classes and settling against the flow of the relatively coarse particles.

Until recently, the examination of the mechanics of two-phase flows in a large majority of cases was carried out only for the monodisperse composition of the solid component.

However, the relationships detected as a result of this examination can not be used for obtaining qualitative and quantitative dependences with special reference to the separation process in which the solid-phase represents a usually polydisperse material with a relatively wide range of the size. This composition of the solid-phase greatly changes the hydrodynamic circumstances of the process in connection with the formation of new phenomena which do not take place in the flows with the single fraction material. The two phase flows with the polydisperse composition of the solid component have been studied in a very small number of investigations, although recently this problem has been given special attention.

In analysis of the relationships of the rising flow with a polydisperse material, examination showed new important aspects of the flow. The main of these aspects is the collision of the particles in the flow and the formation of agglomerates, moving as an integral unit. These mechanisms are linked together and to a large extent are the consequence of the polydisperse nature of the particles and the turbulence of the two-phase flows. The formation of aggregates is most distinctive for gas suspensions. This phenomenon forms as a result of the constant 'pulling' of the surrounding medium into the turbulent wake, formed behind the moving particle. The particles move more rapidly in the direction of the hydrodynamic wake of a result of the formation of a local pressure gradient. This results in the formation of a conglomerate consisting of two or more particles. The formation of conglomerates is also supported by local nonuniformities of the pressure which are most distinctive in the case of turbulent flow. In the approach of two or more particles in the flow, the velocity of their mutually directed flow increases as a result of the instantaneous reduction of the distance between them.

The interaction of the solid polydisperse particles in the flow with each other is a very complicated physical process. The colliding particles may agglomerate, may simply separate, exchanging pulses, if prior to collision the particles were aggregates, then after the collision the aggregates may be completely or partially disrupted or, on the other hand, they may grow. A collision takes place mainly as a result of different velocities of movement of the solid phase.

The relative velocity of these particles (and of clusters of these particles) may be a consequence of different reasons: the size, configuration of the aggregates, the nature of local turbulent vortices, etc.

This is explained by the complicated nature of the phenomenon which, evidently, cannot be investigated directly by experiments because any contact device in the flow cannot influence the condition and behaviour of the particles of the aggregates. In some cases, the effect of agglomeration is not strong, for example, gas suspensions with coarse and granulated particles belong these systems. In the flows with the particle size smaller than 60 μm this phenomenon is on the other hand extremely pronounced and its intensity increases with a decrease in the size of the particles.

The aggregation phenomenon has a negative effect on the efficiency of the separation process. Therefore, in organising separation it is important to investigate special measures for the continuous or periodic disruption of the aggregates.

In some cases, aggregation is used with a positive effect, for example, when trapping dust. It is well-known that coarse, rapidly falling particles are capable of displacing smaller particles from the suspension. This phenomenon was referred by N. Fuchs as 'kinematic coalescence'.

The occurrence of a mechanical interaction of the particles in the flow has been confirmed by the simplest experiment. In equipment, whose principal scheme is shown in Fig. III-16, the selected velocity of airflow was such that in the conical part of equipment it was possible to develop a suspended layer of spheres with a diameter of 12–15 mm and $\rho = 6000$ kg/m^3. In the experiments, the layer was stabilised along the entire height of the cone. Subsequently, fine-dispersion coal dust ($d < 0.25$ mm) was supplied into the airflow in quantities which were so small that the transparency of the flow was not impaired (< 0.1 kg/m^3). Under the effect of this dust, the heavy and thick spheres, suspended in the column, were ejected into the cylindrical part of equipment to a height of up to 400 mm from the edge of the column. This experiment clearly demonstrates the nature of the effect of interaction of the particles.

At the same regime parameters, the frequency of interaction of the particles depends greatly on the physical properties of the material and, primarily, on the elasticity of the particles. The number of interactions between the particles of different sizes increases with increase of the concentration of the material in the flow. Evidently, the nature of movement of any of the fractions of the separated material in the flow is closely linked with the distribution of the size of particles of other classes.

Exchange of pulses takes place during the collisions of particles and aggregates. The coarse particles are accelerated in the direction of the flow, the small particles are inhibited in their movement. This leads to a conclusion according to which in the case of high concentrations all particles of the diffraction mixture assume approximately the same velocity of movement of it is detected, in, for example the during vertical pneumatic transport. However, there is one large difference in the behaviour of suspensions with coarse and fine particles.

When examining the resistance of the two-phase flow to movement in pipes, it was assumed that the introduction of the solid component increases the pressure losses.

For a moderate ratio of the flow rates of the solid and gas phases, Gasterstadt introduced the relationship:

$$\Delta P = \Delta P_0(1 + K\mu)$$

where ΔP is the resistance of the two-phase flow; ΔP_0

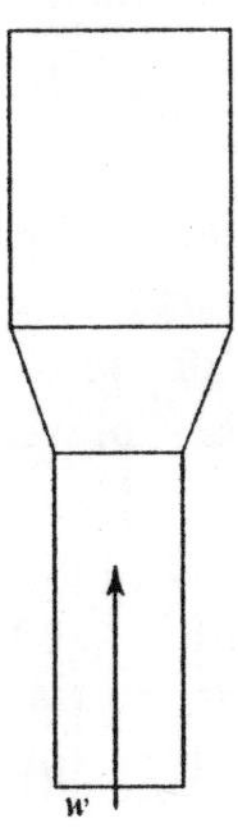

Fig. III-16. Equipment for suspending heavy and coarse particles.

is the resistance of pure air; μ is the concentration of the solid phase; K is a coefficient.

For many years, all investigations were carried out not doubting the validity of this assumption.

Ya. Urban confirmed this assumption by the fact that the introduced the principle of additivity to the pressure losses from the pure flow and the solid phase separately.

However, the authors of this book have noted that, in certain conditions, when solid particles are added, the pressure losses during flow through a pipe decrease to a lower level than even in the case of the pure flow. This phenomenon is characteristic only of the small particles and does not occur in the flows with coarse particles. It has been established that at the concentration of the solid particles causing this effect, the profile of the velocity of the gas medium is almost constant because of the presence of the solid particles in the flow. In this case, the mean concentration of the particles is such that the distance between them is 10 or more times greater than their diameter. So far, this effect has not been unambiguously explained.

The experimental results show the dual effect of the particles both in generation and in suppression of turbulence. In the studies carried out by P. Ribender and M. Reiner it is shown that if the particles cannot follow the movement of the vortices, they will stabilise the flow and create suitable conditions for the laminar flow.

Here, we can specify the second definition of the coarse and fine particles on the basis of their behaviour in the flow. The coarse particles are those whose introduction into the turbulent flow increases the resistance of the flow, and the fine ones are those which reduce the resistance in the specific conditions.

Thus, in the flow of suspensions with fine particles one can expect a large increase in the thickness of the viscous boundary layer with low turbulence which is not subjected to any disruptions, with the exception of large random vortices. The thickness of the viscous layer increases several times in this case.

The particles are not capable of following the reduction of the velocity of the medium in the boundary layer. In long apparatuses, this results in a large increase of the concentration of the particles at the wall. The transfer of the particles to this layer takes place mainly as a result of turbulent diffusion in the core of the flow. Here, examination also shows the operation of the mechanism of slipping of the particles past the region of low turbulence as a result of the exit of the particles from the surrounding vortex due to their inertia.

Consequently, the rising gas flow may be characterised by the

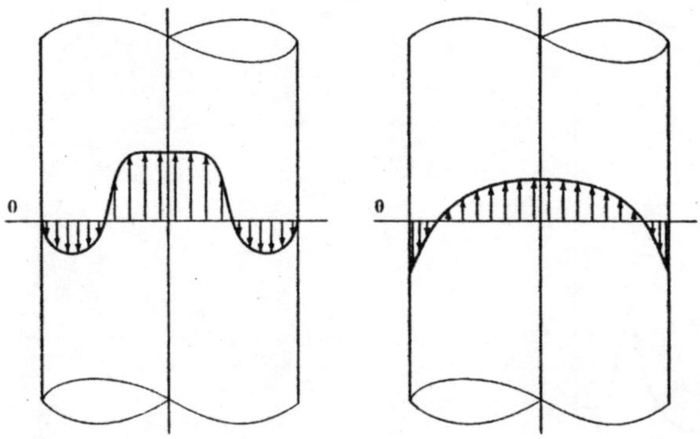

Fig. III-17. Different profiles of velocities of particles of a narrow class in a flow.

establishment of different profiles of the velocity for the solid phase, with the typical profiles shown in Fig. III-17.

All this determines the conditions in which at the mean velocity of the rising flow, sufficient for the displacement of the fine particles, some of these particles move downwards, against the direction of the flow, and part of the coarse particles, whose hovering velocity is considerably higher than the mean velocity of the medium, are displaced upwards into the fine product.

This results in the formation of an effect supporting the constant displacement of the material to the walls of the channel. It has been reported that, penetrating into the region in the vicinity of the wall, the particle may start longitudinal displacement along the walls without leaving this region. It has also been established that this movement is quite short during the rising movement of the particles and relatively long during their downward movement.

Because of the migration of the particles in opposite directions (the reasons for this have been examined in detail), the maximum concentration is usually not formed at the wall but somewhere in the middle of the distance between the axis of the flow and the wall (Fig.III-18). This has a negative effect on the resultant gravitational classification in hollow systems because the increase of the concentration in the peripheral part of the flow impairs the separation conditions, and the downward movement of the particles in the region in the vicinity of the wall results in the heavy penetration of fine particles into the yield of the coarse products.

To prevent this phenomenon from taking place, it is necessary to provide for a constant or variable removal of the material from the walls into the centre of the flow. It is clear that in this case

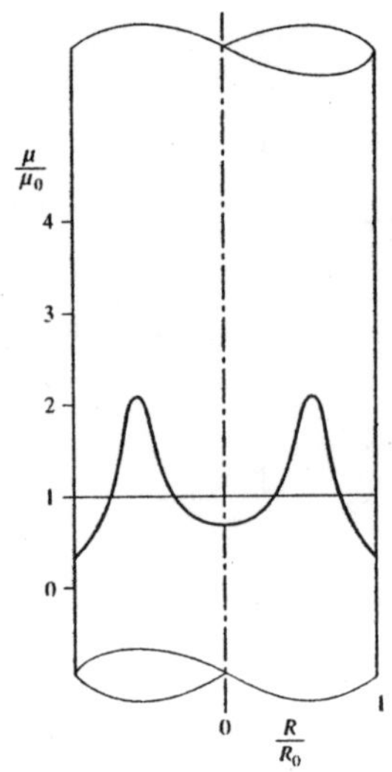

Fig. III-18. Distribution of concentrations of narrow class particles in a flow. μ_0 is the concentration at the axis of the flow, R_0 is the radius of the channel (pipe).

the efficiency of separation will increase.

Sometimes, in order to understand and examine the phenomena taking place in the two-phase flows, they are regarded as a single-phase pseudo-homogeneous medium with high viscosity and density.

This approach is insufficiently effective for the efficient description of the main phenomenon of the disperse flows in the separation conditions because the approach is basically pseudophysical or reduces the flow mechanism of the two-phase flow to the flow of the single-phase medium. If simplifications of this type can useful to some extent in examination of the properties of continuous transport flows, then for the case examined here it is not possible to accept assumptions of this type.

Therefore, we are facing a completely new problem of modelling the separation process. Unfortunately, in all significant studies of the two-phase flow the regimes ensuring separation of the flow are not even mentioned.

An efficient approach to the phenomenon of modelling for the pneumatic transport regimes has been used by V. Bart.

Taking into account the fact that in a general case it is not possible to satisfy all similarity conditions, Bart emphasized the most important parameters for modelling. He stressed the following three parameters:

1. The Froude number for the flow

$$Fr = \frac{gD}{w^2}$$

2. The Froude number calculated from the finite velocity of settling of the particles

$$Fr = \frac{gd}{v_0^2}$$

3. According to Bart, for the similarity of the movement of the gas and the particles, the force of interaction of the particles with the medium G at of the sliding velocity Δv should be linked with the weight of the particles Q by the following relationship:

$$\frac{G}{Q} = \left(\frac{\Delta v}{w}\right)^{2-K}$$

where K is a coefficient changing in the range from 0 to 1.

For the flows with coarse particles $K = 0$.

Analysis of these parameters shows that they are not suitable for the separation process in two-phase flows. The first parameter characterises the flow to some extent, the second parameter in the separation conditions has one value for the particles of any boundary size. The third relationship contains parameters which are almost impossible to measure.

One can agree with Bart's conclusion according to which the Reynolds number plays a secondary role for the two-phase flow.

For this type of flow, in addition to the finite settling velocity and the hovering velocity, it is also necessary to determine another parameter: the minimum velocity of transfer. This velocity is the mean velocity of the flow at which there is no obstacle to the flow. It can be easily shown that this velocity is slightly higher than the hovering velocity of the appropriate particles.

The calculations of the processes of classification in the flows are efficient only if we find the conditions for determination of the regimes ensuring any particle displacement of the particles from the apparatus and, as a partial case, the minimum velocity of transfer.

Thus, the formulated problems are very important for examining and understanding the mechanism of the separation process. To solve these problems, it is necessary to carry out extensive experimental and theoretical investigations.

2. Carrying capacity of two-phase flows

The aim of these investigations was to explain the force effect of the two-phase flow on a fixed sensor in the conditions of the process of classification of the bulk material and examination of the profile of the force effect of the two-phase flow in the cross-section of apparatus and the profile of the curves of the force effect of the continuous medium in the two-phase flow. A pipe with a circular cross-section was selected for the investigations.

The measuring system is shown in Fig. III-19. The circular pipe contained four holes with a diameter of 6 mm at a distance of 50, 250, 450 and 750 mm from the lower edge. In these holes, the sensor 2 (glass pipe, diameter 4.7 mm), capable of moving along its axis, was installed at a fixed distance of X = 100, 95, 85, 75, 55, 55, 45 and 35 mm. The force effect of the solid medium, received by the sensor, is transferred to the lever device 3 by means of the steel needle 4 to the balance with the measurement range from 0 to 500 g with the scale divided in 0.1 g divisions. This balance records continuously the reaction of the measuring system to any perturbation. The moving table 6 is used for centring the sensor during its displacement in relation to the axis of the hole.

All experiments were carried out on periclase (ρ_m = 3600 kg/m^3). The flow concentration of the material was maintained on the level μ = 1.5 kg/m^3. The charge of the material for the experiments was 2.5–3 kg. Grain size analysis was carried out on a set of sieves with the mesh size of 0.75; 0.5; 0.3; 0.2; 0.14 mm. The holes for the supply of material into the apparatus were made at a height of 350 mm from the lower edge of the pipe. The total height of the pipe was 1200 mm. The air was supplied into the pipe from the bottom and its velocity was 2.86; 3.96; 4.92; 5.66; 6.3 m/s.

The resultant profiles of the carrying capacity of the two-phase flow were identical. Figure III-20 shows the results for a flow velocity of w = 3.96 m/s at a flow concentration of the material of μ = 1.5 kg/m^3. Figure III-21 shows that the carrying capacity of the two-

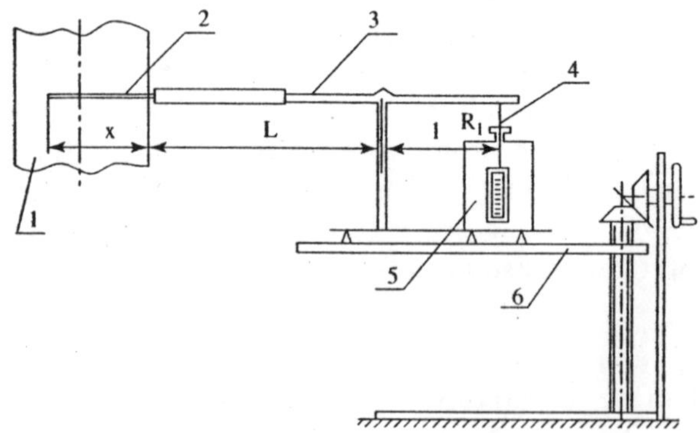

Fig. III-19. Measuring system. 1) examined section of the channel; 2) sensor; 3) lever devices; 4) needle; 5) balance; 6) moving table.

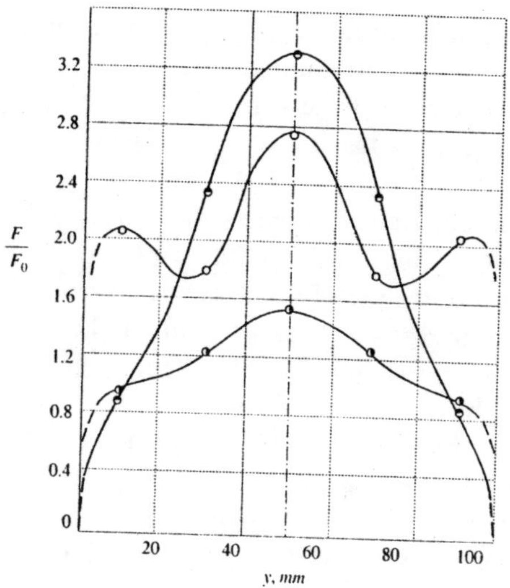

Fig. III-20. Profiles of the curve of the carrying capacity of the two-phase flow at w = 3.96 m/s and μ = 1.5 kg/m^3. The distance from the measurement point to the lower edge of the pipe: ◗) 100 mm; ○) 300 mm; ◓) 700 mm.

phase flow greatly differs from that of pure air. For example, at the axis, the carrying capacity may exceed the effect of pure air by a factor of 3 or more. The profile of the curves of the carrying capacity of the two-phase flow forms depending on the measurement point. A sharper profile and the maximum carrying capacity at the axis are characteristic for the upper part of apparatus. At the lower positions of the section in the appratus the carrying capacity of the flow on the flow axis is lower. In particular, it is important to note the lower curve produced at the point situated close to the area of introduction of the material into apparatus. This profile is characterised by the extremely high nonuniformity resulting from high

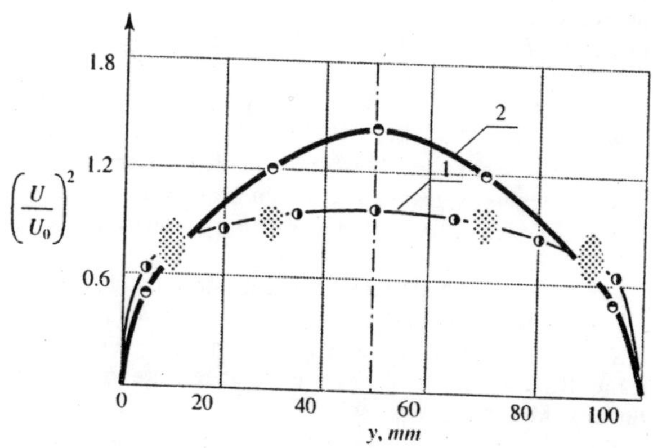

Fig. III-21. Profiles of carrying capacity: 1) pure air; 2) two-phase flow.

local concentrations of the material and the nonuniform distribution of the material in the cross-section of the apparatus.

The curves of the carrying capacity of the solid-phase in the two phase flow is greatly deformed in the classification conditions, their peaks become sharper.

The analysis of the mechanism of the process and the state of the problem from the viewpoint of the development of physical fundamentals of the problem shows quite clearly that the conditions are not yet suitable for the purely analytical examination of the problem. Therefore, the main relationships in the integral representation are usually determined purely by experiments. This method makes it possible to establish relationships of this type in purposeful examination of the phenomenon. Unfortunately, the currently available large amount of empirical material, has usually been obtained in the examination of different systems from the viewpoint of their efficiency without efficient elaboration of the problem of the investigation for the special features of the physical formulation of the process, and does not contain elements of generalising relationships.

References

1. S.L. Soo, Fluid dynamics of multi-phase systems, Blaisdell Publishing Co, 1971.
2. N.A. Fuks, Mechanics of aerosols, Publishing House of the Academy of Sciences of the USSR, 1987.
3. P.C. Peist, Introduction to aerosol science, Macmillan Publishing Co, 1987.
4. A.D. Al'tshul' and P.A. Kiselev, Hydrodynamics and aerodynamics, Stroiizdat, Moscow, 1975.
5. G.L. Babukha and A.A. Shraiber, Interaction of particles of polydispersed material in two-phase flows, Naukova dumka, Kiev, 1972.
6. Z.R. Gorbis, Heat exchange and hydrodynamics of disperse continuous flows, Energiya, Moscow, 1970.
7. L.D. Landau and E.M. Lifshits, Mechanics of solids, Gosgortekhizdat, Moscow, 1953.
8. A.S. Monin and A.M. Yaglom, Statistical hydromechanics, volume 1, 1965, volume 2, 1967, Nauka, Moscow.
9. N. Urban, Pneumatic transport, Mashinostroenie, Moscow, 1967.
10. V.A. Uspenskii, Pneumatic transport, Mashinostroenie, Moscow, 1983.
11. E.P. Mednikov, Turbulent transport and settling of aerosols, Nauka, Moscow, 1984.
12. M. Barsky, Fractionation of powders, Nedra, Moscow, 1980.
13. M. Barsky and E. Barsky, General trend of gravity separation, in: Proceedings of the XXI International Mineral Processing Congress, Rome, 2000.
14. G. Happel and H. Brenner, Hydrodynamics at low Reynolds numbers [Russian translation], Mir, Moscow, 1976.
15. B.V. Kizeval'ter, Theoretical fundamentals of gravitational enrichment processes, Nedra, Moscow, 1979.

Chapter IV

STATISTICAL FUNDAMENTALS OF THE PROCESS

1. JUSTIFICATION OF THE STATISTICAL APPROACH

As indicated by the previous considerations, the two-phase flow is an extremely complicated physical phenomenon. Evidently, the approach used for the construction of the theory of such flows is insufficient. As soon as it is necessary to examine the problems of mass transfer in the flows of this type, the inefficiency of the existing theories becomes evident, regardless of the extensive application, especially in recent years, of various methods of mathematical modelling.

The problem of the separation of the solid phase in the two-phase flows organised in the separation conditions, has not been sufficiently studied yet.

This is undoubtedly one of the most complicated and confusing problems of the theory and, in most cases, it is attempted either to bypass this problem or restrict its examination to empirical relationships.

The explanation of several aspects of the process and the definition of the most general relationships governing the process, are possible only with the application of statistical approaches.

The principal distinctive feature of the statistical approach is that this approach is based on the definition of the state of the entire system and not individual objects, as at the application of analytical methods. Although in this approach it is necessary to avoid using a large number of partial factors, the approach is nevertheless quite fruitful because it makes it possible to explain the general pattern of the process.

It is well-known that the behaviour of a population of solid particles forming a two-phase flow together with the continuous medium, is also described, strictly speaking, on the basis of classical mechanics. In principle, the behaviour of the entire continuum can be specified by the behaviour of each individual particles and, consequently, the following equation may be written for these particles:

$$\frac{dv}{dt} = P_i; \qquad \frac{dX_i}{dt} = v_i;$$

or

$$\frac{d^2X}{dt^2} = P_i$$

where P_i is the force acting on the i-th particle in relation to unit mass; X_i is the radius vector of the i-th particle, v_i is the vector of the velocity of the i-th particle. In the general case, P_i consists of gravitational forces, the forces of the flow, and also the interaction of the i-th particle with other particles and the walls restricting the flow. In order to determine completely the behaviour of the system from the viewpoint of this approach, it is necessary to solve $6N$ (N is the number of particles in the flow) differential equations of the first order with $6N$ unknown quantities. It is also necessary to specify $6N$ initial values of all parameters. It is completely clear that this problem cannot be solved even using high-speed computers, not only because of the large number of the particles, but also owing to the fact that all these equations are linked together because the force of the specific particle is, at every moment of time, the function of the position of all remaining particles of the system, i.e.

$$P_i = f(X_j); \qquad (j = 1; 2; 3......N)$$

Even if it is assumed that, after time-consuming examination and expensive experiments, it is possible to solve this problem to a certain extent, the resultant information will be completely useless since using the large amounts of the data, determining the magnitude and direction of each particle at different moment of time, it is hardly possible to make any specific conclusions. It should be mentioned that the number of the particles in the system in the conditions of fractionation of the powders has the order $N \approx 10^{10}$.

It is evident that the general behaviour of the entire system is associated in some manner with the behaviour of the set of the parts forming the system. Examination of the bonds of this type is the subject of statistical mechanics. Since the investigated system contains a large number of the particles, it is necessary to determine a method for the description of the 'mean' behaviour of the particles and, subsequently, link the behaviour with the experimental results.

We examine a certain number of solid particles moving in any direction together with a moving uprising flow through a limited volume of the space. This volume may be regarded as the natural space of the entire separating system or of its part. In the flow of this type we are interested only in the mass distribution of the initial powder more accurately in the fractional extraction of different particle sizes to the upper and lower products. Therefore, we shall not examine

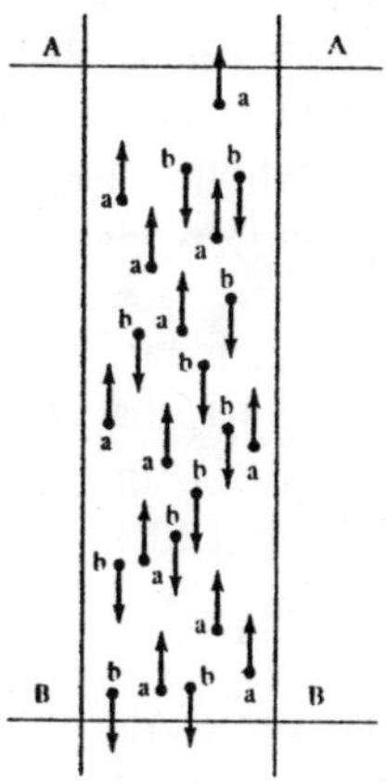

Fig. IV-1. The statistical model of the process.

the true velocity of the particles, and examine, for each particle, the projection of the velocity to the vertical axis (Fig. IV-1).

The direction of projection of the velocity of every particle may be oriented only in two methods: upwards or downwards. It should be mentioned that the probability of this orientation for each particle does not depend on the orientation of the other particles. So far, we are not interested in any other parameters of the process: neither the true direction of the velocity, nor the interaction of the particles with each other and with the walls restricting the flow, nor in the local nonuniformities of the concentration and other characteristics of the flow, only in the instantaneous projection of the velocity of the particle on the vertical axis. In particular, it should be stressed that in the statistical examination we initially ignore the very value of the projection. Taking into account only the direction of the projection, we denote, in accordance with the presented graph, the value of the probability of the direction upwards by *a* and downwards by *b* (it should be mentioned that *a* and *b* are not necessarily numbers, only symbols).

In principle, for different two-phase flows the main axis can also be placed in a different position: for example, for horizontal or centrifugal flows in the horizontal position, for inclined flows in the inclined position. For the examined case of gravitational separation it is natural that the axis should be made vertical.

The system (apparatus) will represent the set of all particles, passing in both directions through the limited space of the flow in the direction of height, for example, in Fig. IV-1, the system is restricted by the lines *A* and *B*.

The object of the present examination is not any system, it is only a system with a steady process. Since the separated space is not characterised by the constant buildup of material, because the total yield of both products of separation in the steady process is always equal to the initial feed, it may be assumed that the number of the particles in the separated volume is approximately constant (see chapter VII-7).

It is interesting that the total number of the particles, located in the volume of the apparatus at some fixed moment of time, may be very large. In actual conditions, in fractionation of powders in real systems, the number of particles at d = 0.1 mm only at 100 kg of the product is $N = 5 \times 10^{10}$. For smaller particles, the number of the particles will be considerably greater.

The set of this type of particles, passing through the examined space, is regarded in further examination as a statistical system. We use only one concept of the statistical mechanics, namely, the concept of the stationary state of the system of the particles. Globally, this concept means that the probability of detecting the particle in any element of the volume is independent of time, i.e. all the investigated physical properties are explicitly independent of time. This means that the stationary states of the systems examined here can usually be counted, although the number may be very large in this case. The possibility of certain fluctuations in the statistical system will be assumed.

From the mathematical viewpoint, the disorder in the system is determined by the number of different methods by which the specific set of the objects can be divided. As the number of this objects increases, the probability of these objects being distributed in a random manner increases, and the objects will not be in any ordered state. Since these concepts assume a determining importance in further examination, we explain them using a suitable example. As a system, we shall use a pack consisting of 36 cards.

In the normal conditions, the pack is in the condition of the random distribution of the cards. The probability of the cards in the pack being grouped in any order is small (of course is this is not done intentionally), for example, the suits can hardly be distributed in any specific order. The number of different methods of the distribution of the cards in the pack is evidently equal to 36!, since there are 36 possibilities of the selection of the first card, 35 possibilities of selecting the second card, 34 possibilities for the third card, etc. There is another important comment which should be made. If it is assumed that all 36 cards are identical, for example, aces of hearts, then there is only one method of the distribution: they would always be presented in the completely ordered form.

Below, we try to count the number of different methods which can be used to distribute the particles in any two-phase flow in order to satisfy specific restrictions imposed on the system. For this purpose, it is necessary to explain initially the parameters which can be used to separate one particle from another.

2. NUMERICAL EVALUATION OF THE STATE OF THE STATISTICAL SYSTEM

Initially, we assume that the examined system in the stationary state consists of N identical particles, and the particles are placed in the flow, resulting in division of the particles into both exits. It is clear that in the conditions, similar to hovering, the system, consisting of such a particle, is characterised by two different stationary states, one state with the velocity directed upwards, and the other one with the velocity directed downwards. The system of two particles is characterised by four states (*aa*; *ab*; *ba*; *bb*), the system consisting of three particles is characterised by eight states (*aaa*; *aab*; *aba*; *baa*; abb; *bab*; *bba*; *bbb*), and so on. Consequently, the total number of all possible states of the system, consisting of N particles, is written in the form 2^N. It should be stressed again that each particle can be oriented by two methods, regardless of the orientation of the remaining particles. From the position of the process examined in this case, the given process represents the potential possibility of separating the particles in each of the states. Potential extraction refers to the number of all particles in the stationary system, with their velocity oriented upwards. If the number of particles in the system is N, the potential extraction changes for different stationary states of the system in the range $0 \leq \varepsilon \leq N$.

In the examination of the dynamic characteristics of the system, it is necessary to use parameter N and also two other parameters for the upper and lower orientation. This is inconvenient. Therefore, we introduce another parameter reflecting the value of the potential extraction in connection with the number of the particles in the following manner:

$$\varepsilon_{nf} = \frac{N}{2} + z \qquad \text{(IV-1)}$$

$$\varepsilon_{nc} = \frac{N}{2} - z$$

The new parameter z will be referred to as the separation factor. It is suitable owing to the fact that its value characterises unambiguously the separation and it is not necessary to use two parameters, ε_{nf} and ε_{nc}. In relative units, if both parts of the dependence (IV-1) are divided by N, we obtain:

$$F_f = \frac{1}{2} + K \qquad \text{(IV-2)}$$

$$F_c = \frac{1}{2} - K$$

In this case, it is clear that F_c and F_f differ only by a constant, and their derivatives will be identical as regards the modulus.

In the physical plan, the separation factor is equal to the number of particles by which F_c deviates on departure from the optimum regime for some class. It is clear that in the optimum regime for this class $z = 0$. In this case, $F_c = F_s = \frac{1}{2}$.

We have shown that the number of states of the system is 2^N. It is interesting to note that the magnitude of the possible values of potential extraction in this case is $(N+1)$. In our example, we can obtain three values of the separation factor for two particles:

1) *aa* – both particles are oriented upwards ($z = +2$);
2) *bb* – both particles are oriented downwards ($z = -2$);
3) *ab* and *ba* – particles have different orientation ($z = 0$).

It should be noted that the latter values of the system are self-similar.

Thus, the number of states is larger than the number of possible values of the potential extraction. For example, at $N = 10$, there are $2^{10} = 1024$ states for only 11 different values of the potential extraction. It is quite easy to find an analytical expression for the number of states with $\left(\frac{N}{2}+m\right)$ particles with the velocity oriented upwards, and $\left(\frac{N}{2}-m\right)$ particles with the velocity oriented downwards. It is convenient to regard N as an even number. We are interested in cases in which the value of N is very high, and in this situation it is not important whether N is even or odd.

The difference

$$\left(\frac{N}{2}+z\right)-\left(\frac{N}{2}-z\right)=2z$$

Of course, at any given moment of time, each particle may acquire only one value of the contribution to the general separation factor. We shall examine a system consisting of N particles at the moments of time following each other t_1; t_2; t_3... t_m, and the number of such examinations is high and equal to m. It is assumed that in each examination the system was in one of its states. The value $n(i)$ is the number of cases in which the system was in the condition i (i.e. in the self-similar condition i). Consequently, the probability of this state is:

$$P(i)=\frac{n(i)}{m}$$

With an increase of the number of examinations m, the value $P(i)$ will tend to some limit. It should be mentioned that from the definition of the probability:

$$\sum_i P(i)=1$$

In other words, the probability of the system being in any state is equal to one. Here, it is necessary to determine the mean value of any physical quantity for the investigated systems.

If in the condition i the relevant physical quantity has the value $A(i)$ then its mathematical expectation is

$$<A>=\sum_i A(i)P(i)=\frac{1}{m}\sum A(i)n(i), \qquad \text{(IV-3)}$$

where $P(i)$ is the probability of the system being in the state i; $n(i)$ is the number showing how many times in the series of m examinations the system will be detected in the condition i.

This is the natural determination of the mean value of A. It is evident that for the system consisting of N particles there are $N!$ methods of their distribution. However, it may be assumed that amongst them there are n_i particles ensuring that the value of z_1 is obtained. From this viewpoint, the particles situated in the group n_1 are indistinguishable from each other. Another concept must be introduced here.

The number of the stationary states of the system or of its part, characterised by the same separation factor, or by its value situated in a narrow range, is referred to as the self-similar number. States of the system, self-similar in relation to each other, will be those which ensure the same separation factor, and their number must be taken into account in the determination of the total number of states φ. If the separation factor z_i can be realised by different methods y_i, it will be assumed that the state z_i is y_1-multiple of self-similar ones.

We stress two principal moments in the determination of self-similarity.

Firstly, this definition is applicable not to the states of the system which differ greatly, but only to the value of the separation factor.

Secondly, the practical determination of self-similarity in the conditions of the real process is determined to a large degree by the efficiency of the experimental procedure. When using a more accurate procedure, it is possible to find a difference in the extraction where it would appear that there is no such difference, if the particles are divided into smaller classes.

When the number of particles is restricted, it is quite easy to find the self-similar states. We have shown that if the total number of states for N particles is 2^N, then the number of separation factor values is only $(N+1)$.

If the specific configuration is selected randomly, the probability of finding this configuration is $\frac{1}{2^N}$. If this configuration has C self-similar states, its probability is $\frac{C}{2^N}$.

We make two further comments and then carry out calculations.

First, without examining the details of the process, it will be assumed that any of the states of the system, self-similar in relation to each other on the basis of the separation factor, are equally probable.

Secondly, there may be states of the system in which the statistical properties of the system from the viewpoint of the examined process are no longer interesting, i.e. the probability is vanishingly small.

To decribe any single state of the system, we can use a suitable image, as in Fig. IV-1, or a symbolic form:

$$a_1 b_2 a_3 a_4 b_5 a_6 b_7 \ldots\ldots\ldots a_i b_j \ldots b_N \qquad \text{(IV-4)}$$

This equation shows the state of the system with the fixation of the direction of projection of the velocity of each specific individual particle.

The product N of the co-factors in (IV-4) can be written without taking into account the order number of the particles, i.e. it is not important from the position of the results of the examined process. Since the projection of the velocity of every particle has only two orientatios, the total number of the states of the system consisting of N particles is:

$$(a+b)^N \tag{IV-5}$$

For a general case, this dependence can be developed using a Newton's binomial theorem:

$$(a+b)^N = a^N + Na^{N-1}b + \frac{1}{2}N(N-1)a^{N-2}b^2 + ... + b^N$$

The equation can be written in the more compact form:

$$F = (a+b)^N = \sum_{K=0}^{N} \frac{N!}{(N-K)!K!} a^{N-K} b^K$$

where K is the current number of the term.

It is more convenient to carry out selection of the states in other ranges, namely in the range of variation of the separation factor from $-\frac{N}{2}$ to $+\frac{N}{2}$. In this case:

$$F = \sum_{-\frac{N}{2}}^{+\frac{N}{2}} \frac{N!}{(\frac{1}{2}N+z)!(\frac{1}{2}N-z)!} a^{\frac{N}{2}+z} b^{\frac{N}{2}-z}$$

The expression $a^{\frac{N}{2}+z} b^{\frac{N}{2}-z}$ enumerates all possible factors of separation in the range $-\frac{N}{2} \le z \le +\frac{N}{2}$,, and the binomial coefficients indicate the number of self-similar states of the system with the fixed number of the particles, oriented upwards or downwards. We carry out calculations on the condition that $N >> 1$ and $z \le \left|\frac{N}{2}\right|$:

$$\varphi(N;z)=\frac{N!}{\left(\frac{1}{2}N+z\right)!\left(\frac{1}{2}N-z\right)!}. \qquad \text{(IV-6)}$$

Taking logarithm of the left and right parts, we obtain the equation:

$$\ln\varphi=\ln N!-\ln\left[\left(\frac{N}{2}+z\right)!\right]-\ln\left[\left(\frac{1}{2}N-z\right)!\right] \qquad \text{(IV-7)}$$

We examine individual parts of this expression:

$$\ln\left[\left(\frac{N}{2}+z\right)!\right]=\ln\left(\frac{N}{2}\right)!+\sum_{k=1}^{z}\ln\left(\frac{N}{2}+k\right)$$

$$\ln\left[\left(\frac{N}{2}-z\right)!\right]=\ln\left(\frac{N}{2}\right)!-\sum_{k=1}^{z}\ln\left(\frac{N}{2}-k+1\right)$$

Taking this into account, we can write the sum of the expressions:

$$\left[\left(\frac{N}{2}+z\right)!\right]+\ln\left[\left(\frac{N}{2}-z\right)!\right]=2\ln\left(\frac{N}{2}\right)!+\sum_{k=1}^{z}\ln\frac{\left(\frac{N}{2}+k\right)}{\left(\frac{N}{2}-k\right)} \qquad \text{(IV-8)}$$

assuming that $\frac{N}{2}-k+1$ is approximately equal to $\frac{N}{2}-k$, and the second term in (IV-8) is:

$$\sum_{k=1}^{z}\ln\frac{\frac{N}{2}+k}{\frac{N}{2}-k}=\sum_{k=1}^{z}\ln\frac{1+\frac{2k}{N}}{1-\frac{2k}{N}}=\sum_{k=1}^{z}\frac{1+x}{1-x} \qquad \text{(IV9)}$$

where $x=\frac{2k}{N}$

It is clear that always $x \ll 1$.

We carry out additional calculations in order to reveal the content of (IV-9):

$$e^x = 1 + x + x^2 + ...$$

Taking into account that $x \ll 1$, we can confine ourselves to the first two terms, consequently:

$$e^x \approx 1 + x; \text{ i.e. } \quad x \approx \ln(1+x)$$

$$e^{-x} \approx 1 - x; \text{ i.e. } -x \approx \ln(1-x)$$

This means that:

$$\ln\frac{1+x}{1-x} = 2x; \qquad \ln\frac{1+\frac{2k}{N}}{1-\frac{2k}{N}} = \frac{4k}{N},$$

and the dependence (IV-9) has the following form:

$$\sum_{k=1}^{z} \ln\frac{1+x}{1-x} = \frac{4}{N}\sum_{k=1}^{z} k = \frac{4(z+1)z}{2N} \approx \frac{2z^2}{N}$$

(IV-10)

Thus, the dependence (IV-6) is converted to the form:

$$\varphi(N;z) \approx \frac{N!}{\left(\frac{N}{2}\right)!\left(\frac{N}{2}\right)!} e^{\frac{-2z^2}{N}}$$

(IV-11)

The results can be presented in the following form:

$$\varphi(N;z) = \varphi(N;0)e^{\frac{-2z^2}{N}}$$

i.e., this equation should be treated as showing that the number of states of the examined system for any value of the separation factor is equal to the number of equilibrium states (z = 0) multiplied by the exponent.

The value of the coefficient at the exponent can be determined using the Stirling equation:

$$\frac{N!}{\left(\frac{N}{2}\right)!\left(\frac{N}{2}\right)!}=\frac{\sqrt{2\pi N}N^{N}e^{-N}}{\pi N\frac{N!}{2^{N}}e^{-N}}=2^{N}\sqrt{\frac{2}{\pi N}}$$

Taking this equation into account, (IV-11) is transformed to:

$$\varphi=2^{N}\sqrt{\frac{2}{\pi N}}e^{\frac{-2z^{2}}{N}}$$

This dependence will be analysed at different ratios of the probabilities of the yield of the class into fine and coarse products. Initially, we examine the simplest case for the conditions of equilibrium distribution of the narrow class.

3. MAIN STATISTICAL CHARACTERISTICS OF THE SEPARATION FACTOR

We analyse the resultant dependence (IV-11). The validity of this equation can be verified if the equation is summed up with respect to all values from $-\frac{N}{2}$ to $+\frac{N}{2}$:

$$\sum_{z=-\frac{N}{2}}^{z=+\frac{N}{2}}2^{N}\sqrt{\frac{2}{\pi N}}e^{\frac{-2z^{2}}{N}}$$

Summation with respect to all resultant values of z gives the integral:

$$\int_{-\infty}^{+\infty}2^{N}\sqrt{\frac{2}{\pi N}}e^{\frac{-2z^{2}}{N}}dz=2^{N}\sqrt{\frac{2}{\pi N}}\int_{-\infty}^{+\infty}e^{\frac{-2z^{2}}{N}}dz$$

We introduce a new variable

$$\frac{2z^{2}}{N}=y^{2}$$

Consequently

$$\sqrt{\frac{2}{N}}dz=dy;\quad\text{and}\quad dz=\sqrt{\frac{N}{2}}dy$$

Talking this into account, the examined dependence is transformed to the form

$$2^N\sqrt{\frac{2}{\pi N}}\int_{-\infty}^{+\infty} e^{-y^2}\sqrt{\frac{N}{2}}dy = \frac{2^N}{\sqrt{\pi}}\int_{-\infty}^{+\infty} e^{-y^2}dy$$

Also, according to the handbook data we obtain

$$\int_{-\infty}^{+\infty} e^{-y^2}dy = \frac{\sqrt{\pi}}{2}\int_{-\infty}^{+\infty} 2^N\sqrt{\frac{2}{\pi N}}e^{\frac{-2z^2}{N}} = 2^N,$$

which accurately corresponds to the total number of states of the system.

The distribution determined by the right-hand side of the dependence (IV-11) is the Gauss distribution. It has a maximum with a centre at $z = 0$.

For this type of curves, the measure of relative width of the distribution is the rms deviation. The value of this deviation is $\sigma = \sqrt{N}$

The ratio of the rms deviation to the maximum value is: $\frac{\sqrt{N}}{N} = \frac{1}{\sqrt{N}}$.

If the total number of particles forming the the system is, for example, $N \approx 10^{10}$, the relative width of the distribution will be on the order of 10^{-5}. This means that in the present case we obtain a sharp maximum at the mean value $z = 0$.

The physical meaning of this is that the potential degree of fraction separation, obtained in reality in a specific system, is not in principle the only possible, but is most probable of all variants.

Here, it was shown that the probability of this state is so much greater than any other apparent distribution that it can be regarded as the only possible for the given condition, i.e. determinate. This is unambiguously confirmed by all available experimental material.

We introduce another concept for the investigated system. The main parameter of distribution in this system is the velocity of the rising flow. The value of the separation factor ($2z$) is functionally linked for a narrow class of the particles with the velocity of the flow which is the kinetic parameter of the distribution. It will be attempted to formulate the potential parameter of the distribution whose importance is evident.

On the one hand, the distribution in this system is formed under the effect of the gravitational field (g), and on the other hand under the effect of the rising flow (w).

The potential parameter will be referred to as the 'lifting factor' and denoted by the symbol I:

$$I = 2zgd \tag{IV-12}$$

It should be mentioned that it has the dimension of the square of velocity.

At first sight, this parameter differs by a constant from the separation factor. This is valid only for the gravitational field. As regards centrifugal fields of separation or fields of any other nature, it is evident that for these fields it will be of considerable generalising value.

Summing up, the total distribution pattern is determined by the gravitational field, the mean velocity of the rising flow and the particle size of the narrow size class.

4. DETERMINATION OF ENTROPY FOR THE TWO-PHASE FLOW IN THE SEPARATION REGIME

We examine a system consisting of two narrow classes of particles selected in such a manner that they are close from both sides to the same dimension.

In other words, these particles have only slightly different aerodynamic properties and, primarily, the hovering velocity of these particles is similar.

We examine their behaviour in the system using the following procedure. Initially, the examined particles are of only one class, for example, those which are close to the left to the separation boundary. When the number of states of this type of system is determined, we shall start to examine the second particles. Subsequently, we examine the characteristics of the system formed by the particles of both classes. It is assumed that the realisation of the united system in this case is determined by some fixed values of the separation factor z_1 and z_2, different from zero. The number of permissible self-similar states of the first system will be denoted by φ_1. Each of these states may be realised with any of φ_2, permissible states of the second system.

Thus, the total number of states in the joint system is:

$$\varphi = \varphi_1 \varphi_2 \tag{IV-13}$$

This relationship is supplemented by the following relationship

$$z = z_1 + z_2$$

i.e.

$$z_2 = z - z_1$$

The number of particles in the system is constant

$$N = N_1 + N_2 = \text{const}$$

The realisation of the joint system can be characterised quite efficiently by one of the separation factors, we assume that this is z_1. To obtain the total number of all permissible states, it is sufficient to sum up the product (IV-13) with respect to all possible values of z_1, i.e.

$$\varphi = \sum_{z_1} \varphi_1 \varphi_2 \tag{IV-14}$$

It is well known that this product has a sharp maximum at some value $z_1 - z_{m1}$ and this value of the parameter determines the most probable realisation of the joint system. Consequently, the number of states in the most probable configuration is:

$$\varphi_1(N_1 z_{m_1})\varphi_2(N_2; z - z_{m_1}) \tag{IV-15}$$

It is obvious that if the number of particles at least in one of the systems is very large, then in relation to the changes of z_1 this maximum is exceptionally sharp. The presence of a sharp maximum indicates that the statistical properties of the joint system are determined by a relatively small number of configurations.

This results in an important conclusion according to which for the distribution with a sharp maximum the average properties of the system are precisely determined by their most probable configuration. This means that the previously determined mean value of the physical quantity with respect to all permissible configurations may be replaced by their mean value for only one most probable configuration. It will be shown what this approximation means for (IV-15). Taking into account the previously made conclusion, it may be written that

$$\varphi = \varphi_1(N_1; z_1)\varphi_2(N_2; z_2) = \varphi_1(N; 0)\varphi_2(N_2; 0)e^{\left(-\frac{2z_1^2}{N_1} - \frac{2z_2^2}{N_2}\right)}$$

We examine this dependence in the function of z_1. Consequently (IV-15) may be written in the following new form

$$\varphi = Ae^{-\left[\frac{2z_1^2}{N_1} + \frac{2(z-z_1)^2}{N_2}\right]} \tag{IV-16}$$

It should be mentioned that the function $y(x)$ reaches a maximum at the same value of x as the function $y(x)$. From (IV-16):

$$\ln \varphi = \ln A - \frac{2z_1^2}{N_1} - \frac{2z_2^2}{N_2}$$

This value has an extremum when the derivative with respect to z_1 is equal to zero. For the first derivative:

$$-\frac{4z_1}{N_1}+\frac{4(z-z_1)}{N_2}=0$$

The second derivative

$$-4\left(\frac{1}{N_1}+\frac{1}{N_2}\right)$$

is negative and, consequently, the extremum is a maximum. Thus, the most probable configuration is the one for which the following relationship is fulfilled:

$$\frac{z_1}{N_1}=\frac{z-z_1}{N_2}=\frac{z_2}{N_2}, \quad \text{i.e. } F_1=F_2$$

Here we have obtained a very curious relationship. This indicates that the most probable state of the system is established in such a manner that the separation factor appears to equalize the different numbers of particles in the flow. Thus, we have obtained a result confirming experimentally the invariance of the degree of fraction separation in relation to the composition of the initial mixture. This conclusion shows the statistical principle of this empirical result which was obtained a long time ago but has not as yet been explained.

If at the maximum of the examined product z_1 and z_2 are equal to respectively z_{m_1} and z_{m_2}, the resultant relationship may be written in the form:

$$\frac{z_{m_1}}{N_1}=\frac{z_{m_2}}{N_2}=\frac{z}{N}$$

Consequently,

$$(\varphi_1\varphi_2)_{\max}=\varphi_1(N_1;z_{m_1})\varphi_2(N_2;z-z_{m_2})=\varphi_1(N_1;0)\varphi_2(N_2;0)e^{\frac{-2z^2}{N}}$$

It is assumed that

$$z_1=z_{m_1}+\varepsilon; \qquad z_2=z_{m_2}-\varepsilon$$

Here ε is a measure of deviation of z_1 and z_2 from their maximum values z_{m_1} and z_{m_2}. Consequently, it is clear that

$$z_1^2=z_1^2+2z_{m_1}\varepsilon+\varepsilon^2$$

$$z_2^2=z_{m_1}^2+2z_{m_2}\varepsilon+\varepsilon^2$$

Taking this into account

$$\varphi_1(N_1;z_1)\varphi_2(N_2z_2)=(\varphi_1\varphi_2)_{\max}e^{\left(-\frac{4z_1\varepsilon}{N_1}\right)\left(\frac{2\varepsilon^2}{N_1}\right)+\left(\frac{4z_2\varepsilon}{N_2}\right)\left(\frac{2\varepsilon^2}{N_2}\right)}$$

Since

$$\frac{z_{m_1}}{N_1}=\frac{z_{m_2}}{N_2}$$

the number of states in the configuration, characterised by the deviation of ε from maximum, is

$$\varphi_1(N_{1:}m_1+\varepsilon)\varphi_2(N_2;m_2-\varepsilon)=(\varphi_1\varphi_2)_{\max}e^{\left(-\frac{2\varepsilon^2}{N_1}\right)\left(\frac{2\varepsilon^2}{N_2}\right)} \qquad \text{(IV-17)}$$

In order to understand the effect of this dependence, it is assumed that $N_1 = N_2 = 10^{10}$ and $\varepsilon = 10^6$. Consequently $\frac{\varepsilon}{N}=10^{-4}$. For this small deviation from the equilibrium value we have $\frac{2\varepsilon^2}{N_1}=\frac{2\ 10^{12}}{10^{10}}=200$.

The product $\varphi_1\varphi_2$ represents the fraction equal to $e^{-400}\approx 10^{-179}$ of its maximum value. It is therefore clear that the decrease is very strong and, consequently, $\varphi_1\varphi_2$ should be a function of z_{m_1} with a very sharp peak.

This means that almost all frequently observed values z_1 and z_2 are very close to z_{m_1} and z_{m_2}. The result for a number of permissible states of two systems, situated in the same flow, may be generalised taking the lifting factor into account.

Using the same considerations as previously, the following equation is obtained for self-similarity of the joint system:

$$\varphi(N;I)=\sum_{I_1}\varphi_1(N_1I_1)\varphi_2(N_2;I-I_1),$$

where summation is carried out with respect to the values of I_1 which are smaller than or equal to I. Here $\varphi_1(N_1;\ I_1)$ is the number of permissible states of system I at I_1. The configuration of the joint system is determined by the values of I_1 and I_2. The number of permissible states is represented by the product $\varphi_1(N_1;I_1)\ \varphi_2(N_2;I_2)$, and the sum with respect to all configurations gives $\varphi(N;I)$. It is required to find the highest term in this sum. For the extremum, the corresponding differential must be equal to 0:

$$d\varphi = \left(\frac{\partial \varphi_1}{\partial I_1}\right)_{N_1} \varphi_2 dI_1 + \left(\frac{\partial \varphi_2}{\partial I_2}\right)_{N_2} \varphi_1 dI_2 = 0$$

We take into account that $dI_1 + dI_2 = 0$. Dividing this equation by $\varphi_1\varphi_2$ and taking into account that $dI_1 = -dI_2$, we obtain,

$$\frac{1}{\varphi_1}\left(\frac{\partial \varphi_1}{\partial I_1}\right)_{N_1} = \frac{1}{\varphi_2}\left(\frac{\partial \varphi_2}{\partial I_2}\right)_{N_2}$$

Taking into account that $\frac{d}{dx}\ln y = \frac{1}{y}\frac{dy}{dx}$, the previous equation may be written in the form

$$\left(\frac{\partial \ln \varphi_1}{\partial I_1}\right)_{N_1} = \left(\frac{\partial \ln \varphi_2}{\partial I_2}\right) \tag{IV-18}$$

This gives a very important relationship for statistical examination of the discussed problem to which we shall return more than once. At present, the following should be noted:

Firstly, the derivative of the logarithm of the number of self-similar states of each system with respect to the lifting factor means the most probable configuration of the system, and this is the most important property of the dependence (IV-18);

Secondly, the two systems are in equilibrium with each other when the joint system is in the most probable configuration, i.e. when the number of permissible states is maximum;

Thirdly, we shall pay attention to the quantity situated in the numerator of equation (IV-18)

$$H = \ln \varphi \tag{IV-19}$$

According to Boltzmann's classic definition, this value is in fact entropy. According to this definition, entropy is a measure of disorder or indeterminacy of the system, i.e. as the value of H increases, the value of φ also increases. This definition corresponds to the dependence (IV-19) in the sense that as the number of permissible self-similar states in the system increases, the entropy becomes higher.

However, this entropy has not been derived for the ideal gas in relation to its temperature. It was derived for the characteristic of the two-phase flow in the conditions of solid phase separation.

Thus, we have introduced a new concept into the theory of two-phase flows which has important physical meaning and at the same time forms a bridge to the statistical mechanics of the ideal gas.

This theory is based on the assumptions according to which the gas molecules represent ideal spheres of the same diameter, enclosed in a closed volume and having the velocity unambiguously determined by temperature of the medium. On the basis of examination of this type of system taking into account collisions of the molecules of the gas amongst each other and with the walls of the chamber, it was possible to develop an elegant statistical theory which is confirmed by the entire set of the data available on gas systems. Recently, the interest in this theory has increased in two aspects. On the one hand, a large number of fundamental studies, expanding and investigating more extensively the theory proposed by L. Boltzmann, have been published. On the other hand, the concept of this theory are now used successfully in other areas of investigation of non-gas systems, for example, such as solids, nuclear matter, magnetism, quantum optics, polymerisation, etc.

We also use several concepts of statistical mechanics connecting them, naturally, with the specific conditions of the examined problem. At the same time, the specific features of the two-phase flow requires principle rethinking of these concepts. The main difficulties when examining these types of flows from the position of this statistical approach to the gas systems have three main aspects.

Firstly, in statistical mechanics the gas systems are examined in a limited volume, and all possible directions of movement of the particles are regarded as equally probable. In the two phase flow, the closed volume cannot be discussed and, in addition to this, the resultant movement of the flow has a preferential direction.

Secondly, the principal parameter of the statistical approach is the potential energy of continuum of the particles, determined by temperature. It is clear that in the two-phase flow it is completely irrational to discuss the problem of determination of the potential energy of moving particles, as regards temperature, because its variation usually has no effect on the main parameters of this type of flow.

Thirdly, and most importantly for the given problem, we shall ignore the generally accepted parameters of the system (temperature, energy, heat capacity, work, etc.) and introduce new parameters determining the degree of fraction separation, the conditions of movement of the medium, the probability of direction of movement of the particles, etc. In this formulation, the problems have never been examined from the position of statistical mechanics.

At the same time, it should be mentioned that in examination of this problem, the main concepts and methods of the statistical approach will be used. Although the examined process is studied for the first time from the position of statistical mechanics, the main assumptions of thermodynamics and the theory of the ideal gas will be used because it was shown that there is a sufficiently reliable meaning and physical analogy between these processes. This examination yields certain relationships whose structure resembles the relationships of thermodynamics. Although they themselves and the parameters included in them have a completely different physical meaning, they will be referred to using notations conventional in statistical mechanics, for example, the Gibbs factor, the Boltzmann factor, entropy, the statistical sum for the two-phase flow, etc. Taking these preliminary comments into account, the main concepts of the method will now be explained.

It has been established that for the examined process, entropy is a function of the number of particles of the system and the lifting factor, i.e.:

$$H = f\,(N;I)$$

The relationship of entropy with other processed parameters will be examined slightly later and, this time attention will be given to the main properties of this new entropy.

5. MAIN PROPERTIES OF ENTROPY CHARACTERISING THE TWO-PHASE SYSTEM

These properties will be examined in a specific sequence.

1. Entropy is equal to zero when the state of the system is fully determined and unambiguous. If the separation factor has the value, for example, $z = \frac{N}{2}$, it means that $\varphi = 1$. In this case, $H = \ln\varphi = 0$.

2. It will be attempted to determine the physical meaning of equation (IV-18). In principle, it is the quantity equal to derivative of entropy with respect to the lifting factor, being the same for both systems, i.e.

$$\frac{\partial H}{\partial I} = \frac{1}{\chi}$$

On the other hand, I and z differ by a constant, and according to (IV-16) it may be written that:

$$\frac{\partial H}{\partial z} = -\frac{4z}{N} = \frac{1}{\chi}$$

By analogy with gas dynamics, parameter χ plays the role of a factor making the process random. Its dimension should correspond to that of the lifting factor, i.e. it should be equal to $\left[\frac{\text{m}^2}{\text{sec}^2}\right]$. Here, it is quite justified to assume that the parameter is proportional to the square of the velocity of the upward flow, i.e. $\chi \approx w^2$.

3. When the randomizing factor of two systems, which are in contact, is exactly the same, contact permits spontaneous exchange of particles between them, although the systems are in equilibrium. The number of states of the first system is φ_1 and either of them can be realised simultaneously with any of the permissible states of the second system φ_2, i.e. the total indeterminacy of the two isolated systems will be higher or equal to the indeterminacy of the combined system.

This means that

$$H_\Sigma \leq H_1 + H_2$$

Thus, at mixing in the flow the entropy increases.

4. At the same time, for a steady process, the entropy has the additivity property. It was known that $H = \ln(\varphi_1\varphi_2)$ and consequently, $H = \ln\varphi_1 + \ln\varphi_2 = H_1 + H_2$. The aspect in which the additivity properties are justified will now be analysed. It has been determined that the number of states in a configuration, characterised by some deviation ε, is (IV-17). To facilitate considerations, it will be assumed that $N_1 = N_2 = \frac{N}{2}$. Consequently, it may be written that

$$\varphi(N;z) = \sum_{\varepsilon} \varphi_1(N_1; z_{w_1} + \varepsilon)\varphi_2\left[N_2(z_{w_2} - \varepsilon)\right] = (\varphi_1\varphi_2)_{\max} \int_{-\infty}^{+\infty} d\varepsilon e^{-\left(\frac{8\varepsilon^2}{N}\right)}$$

where the sum with respect to the deviations ε is replaced by the integral. Denoting $\frac{8\varepsilon^2}{N} = x^2$, then

$$\int_{-\infty}^{+\infty} d\varepsilon e^{-\frac{8\varepsilon^2}{N}} = \sqrt{\frac{N}{8}} \int_{-\infty}^{+\infty} dx e^{-x^2} = \sqrt{\frac{N}{8}}\sqrt{\pi} = \sqrt{\frac{\pi N}{8}}$$

and, consequently

$$\ln\varphi(N;z) = \ln(\varphi_1\varphi_2)_{\max} + \frac{1}{2}\ln\frac{\pi N}{8}$$

which differs from the value $\ln(\varphi_1\varphi_2)_{\max}$ by the value of the order of $\ln N$. It is known that the order of $\ln(\varphi_1\varphi_2)_{max}$ is equal to N, since the order of magnitude $(\varphi_1\varphi_2)_{max}$ is equal to 2^N. This means that at $N \gg 1$ $\ln N$ can be ignored in comparison with N.

This shows that the entropy of a combined system may be assumed to be equal to the sum of entropies of the systems included in it on the condition that the latter have the most probable configuration.

5. The expression for entropy will now be determined. In all cases, for any narrow size class, the number of self-similar states of the system is expressed as follows:

$$\varphi = \frac{N!}{\frac{N}{2}!\frac{N}{2}!} e^{-\frac{2z^2}{N}}$$

Consequently

$$H = \ln\varphi = \ln N! - 2\ln\frac{N}{2}! - \frac{2z^2}{N}$$

Taking into account the Stirling equation $\ln n! \approx N(\ln n - 1)$, this expression may be reduced to the form

$$H = N(\ln N - 1) - N\left(\ln\frac{N}{2} - 1\right) - \frac{2z^2}{N} \qquad \text{(IV-20)}$$

According to (IV-19) and (IV-20), one obtains:

$$H = N\ln 2 - \frac{2z^2}{N} \qquad \text{(IV-21)}$$

Since $\frac{4z}{N} = \frac{1}{\chi}$, finally:

$$H = N\ln 2 - \frac{z}{2\chi}$$

Thus, it has been shown how the entropy of the system is associated with the main parameters of the flow: the number of particles and the separation factor.

6. TRANSVERSE TRANSFER IN AN UPWARD TWO-PHASE FLOW

Attention will now be given to the steady flow in a system whose

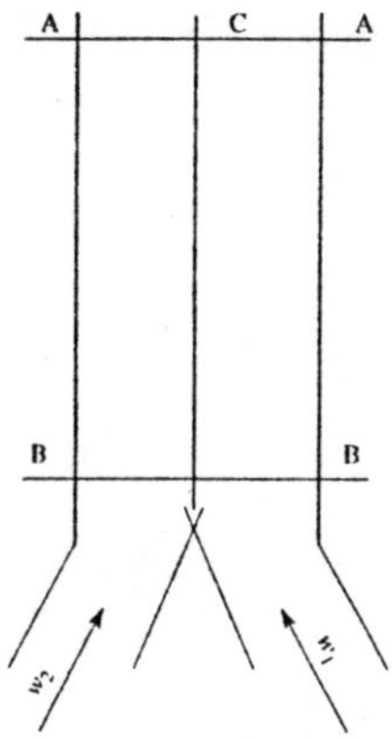

Fig. IV-2. Channel with a longitudinal partition C.

configuration is shown in Fig.IV-2. The special feature of this arrangement is the presence of a partition longitudinal in relation to the flow. This partition divides the system consisting, it is assumed, of particles of the same size class, into two isolated flows.

It is assumed that the first of these flows is characterised by parameters N_1; I_1 and the second one N_2; I_2. It should be mentioned that the dynamic circumstances of the flow for both parts are not necessarily identical, i.e. $\chi_1 \neq \chi_2;(w_1 \neq w_2)$. It may easily be seen that the following relationships hold for this stationary statistical system:

$$N = N_1 + N_2 = const \tag{IV-22}$$

$$z = z_1 + z_2 = const$$

If the partition is now removed, this gives a combined system in which mutual exchange of the particles is possible. It has been shown that the most probable configuration of the combined system is the one for which the number of permissible states is maximum if the randomizing factor of both systems is identical.

This maximality may be determined by analysis of the product of the number of permissible states of individual systems with respect to independent variables, characterising both systems. From the relationship

$$\varphi = \varphi_1\varphi_2 = \varphi_1(N_1;I_1)\varphi_2(N - N_1;I - I_1)$$

The extremum condition may be written in the form

$$d(\varphi_1\varphi_2) = \left(\frac{\partial\ \varphi_1}{\partial\ N_1}dN_1 + \frac{\partial\ \varphi_1}{\partial\ J_1}dI_1\right)\varphi_2 + \left(\frac{\partial\ \varphi_2}{\partial\ N_2}dN_2 + \frac{\partial\ \varphi_2}{\partial\ J_2}dI_2\right)\varphi_1 = 0 \tag{IV-23}$$

Taking (IV-22) into account it may be written that:

$$dN_2 = -dN_1$$

$$dI_2 = -dI_1$$

Consequently,

$$\frac{\partial\ \varphi_2}{\partial\ N_1} = -\frac{\partial\ \varphi_2}{\partial\ N_2};\frac{\partial\ \varphi_2}{\partial\ I_1} = -\frac{\partial\ \varphi_2}{\partial\ I_2}$$

Dividing both parts of (IV-23) by the product $\varphi_1\varphi_2$ and taking into account the resultant relationships, one obtains:

$$\left(\frac{1}{\varphi_1}\frac{\partial\ \varphi_1}{\partial\ N_1} - \frac{1}{\varphi_2}\frac{\partial\ \varphi_2}{\partial\ N_2}\right)dN_1 + \left(\frac{1}{\varphi_1}\frac{\partial\ \varphi_1}{\partial\ I_1} - \frac{1}{\varphi_2}\frac{\partial\ \varphi_2}{\partial\ I_2}\right)dI_1 = 0$$

This expression reflects the condition of mutual equalisation or equilibrium of both systems. This dependence may be simplified to the form:

$$\left(\frac{\partial\ \ln\varphi_1}{\partial\ N_1} - \frac{\partial\ \ln\varphi_2}{\partial\ N_2}\right)dN_1 + \left(\frac{\partial\ \ln\varphi_1}{\partial\ I_1} - \frac{\partial\ \ln\varphi}{\partial\ I_2}\right)dI_1 = 0 \qquad \text{(IV-24)}$$

Evidently, the equalisation condition of two systems will be fulfilled when the expressions in the brackets have the values equal to zero because the second expression in the brackets in the equilibrium condition is equal to zero, as established previously. Thus, equation (IV-24) gives

$$\frac{\partial\ H_1}{\partial\ N_1} = \frac{\partial\ H_2}{\partial\ N_2} \quad \text{and} \quad \frac{\partial\ H_1}{\partial\ I_1} = \frac{\partial\ H_2}{\partial\ I_2}$$

The second condition is known, it is solved as $\chi_1 = \chi_2$, i.e. the values of the randomising factors in both parts of the system are equalised.

The first condition is new. The notation will be introduced:

$$\frac{\partial\ H}{\partial\ N} = \frac{\tau}{\chi}$$

where τ is a parameter having the meaning of the mobility factor. H and N are dimensionless quantities and, therefore, the right hand part should also be dimensionless. Thus, another condition of the steady process is added here. In combining two systems at the same flow velocity, the additional new condition of the steady flow is form:

$$\frac{\tau_1}{\chi_1} = \frac{\tau_2}{\chi_2} \tag{IV-25}$$

i.e. the two systems, which can exchange particles, come to equilibrium when the ratios of their mobility factors to the randomizing factor become equal. It has been assumed that the randomizing factor is proportional to the square of the mean flow velocity, i.e. $\tau = w^2$. The mobility factor characterises the particles and design of equipment, but its dimension should be equal to the square of velocity. This unique characteristic of the particle is presented by the quantity which fully determines all aspects of the behaviour of the particle in relation to a specific flow. It is clear that this parameter is proportional or equal to the local velocity of the flow at a specific point of the cross section

$$\tau \cong w_i^2 \tag{IV-26}$$

7. DETERMINATION OF THE MAIN STATISTICAL RELATIONSHIPS FOR THE SEPARATION PROCESS

A system which in the static state has a constant number of particlces N_0 will be examined. At a specific flow velocity of the medium the lifting factor of this system is characterised by quantity I_0. This system will be conditionally divided into two parts. The larger part will be referred to as apparatus, the small one as a zone. The zone is the part of the volume of a vertical channel of small height and overlapping the entire cross section of the channel. The height of the zone is assumed to be small but sufficient for holding a large number of particles, but insufficient with respect to height for any significant change of composition, concentration and other process parameters. To simplify examination, the separated zone will be placed on the upper edge of the system, although in principle it may be chosen in any part of apparatus and this has no effect on the correctness of conclusions.

Another restriction for the height of the zone will be made. It is selected so small that all particles which move upwards in this zone, leave its limits, i.e. are extracted from apparatus.

The statistical properties of this zone will be examined, taking into account the position of this zone in contact with apparatus. Contact means that the flow velocities in them are equal or at least have a rigid link determined only by the ratio of the appropriate efficient sections. In addition to this, it is necessary to assume that the randomising factor and the mobility factor are equal. Apparatus and the zone exchange particles.

If the number of particles in the zone is $N(N \ll N_0)$, in apparatus it is $(N_0 - N)$. If the zone has the lifting factor E, the value of this parameter for apparatus will be $(I_0 - E)$.

Using the results in the previous paragraph, it is possible to determine the probability of the zone being detected, in the given examination, in the i-th condition with lifting factor E_i and containing N particles.

Probability $P(E_i;N)$ is proportional to the number of permissible states of apparatus and not of the zone because when fixing the state of the zone, the number of permissible states of the entire system is equal to the number of permissible states of apparatus:

$$P(N;E_i) \sim \varphi\left[(N_0 - N);(I_0 - E_i)\right]$$

In this equation, the proportionality coefficient is not known. The approach used usually by investigators for overcoming this difficulty will be used: the ratio of the probabilities of the zone being in two states will be determined:

$$\frac{P(N_1;E_1)}{P(N_2;E_2)} = \frac{\varphi(N_0 - N_1;I_0 - E_1)}{\varphi(N_0 - N_2;I_0 - E_2)}$$

(IV-27)

By definition of entropy for the entire apparatus:

$$\varphi(N_0;I_0) = e^{H(N_0;I_0)}$$

Taking this into account, equation (IV-27) may be written as the difference of the entropies

$$\frac{P_1[N_1;E_1]}{P_2[N_2;E_2]} = e^{\Delta H} = e^{[H(N_0-N_1)(I_0-E_1)]-[H(N_0-N_2)(I_0-E_2)]} \qquad \text{(IV-28)}$$

This equation may be expanded into a Taylor's series. It should be remembered that

$$f(y+c) \approx f(y) + c\frac{df}{dy} + \ldots$$

Thus, for the apparatus:

$$H(N_0 - N;I_0 - E) = H(N_0;I_0) - \left(N\frac{\partial H}{\partial N_0}\right)_{I_0} - E\left(\frac{\partial H}{\partial I_0}\right)_{N_0} + \ldots..$$

For the difference of entropies with the accuracy to the first order:

$$\Delta H \approx [(N_0 - N_1) - (N_0 - N_2)]\left(\frac{\partial H}{\partial N_0}\right)_{I_0} +$$

$$+[(I_0 - E_1) - (I_0 - E_2)]\left(\frac{\partial H}{\partial I_0}\right)_{N_0} = \qquad \text{(IV-29)}$$

$$= -(N_1 - N_2)\left(\frac{\partial H}{\partial N}\right)_{I_0} - (E_1 - E_2)\left(\frac{\partial H}{\partial N}\right)_{N_0}$$

Using the definition of the introduced new factors

$$\frac{1}{\chi} = \left(\frac{\partial H}{\partial I}\right)_N ; \frac{\tau}{\chi} = \left(\frac{\partial H}{\partial N}\right)_J$$

the equation (IV-29) may be written in the form

$$\Delta H = \frac{(N_1 - N_2)\tau}{\chi} - \frac{(E_1 - E_2)}{\chi} \qquad \text{(IV-30)}$$

It is interesting to note that ΔH relates to apparatus, and N_1; N_2; E_1; E_2 to the zone.

Thus, the changes taking place in the zone predetermine the variation of entropy of the entire apparatus. This is also evident on the intuitive level, because all and only objects which leave the zone predetermine the value of fractional separation sought in this evaluation.

Taking this into account, dependence (IV-30) gives an exceptionally important relationship from the viewpoint of the statistical approach to the problem:

$$\frac{P_1[N_1; E_1]}{P_2[N_2; E_2]} = \frac{e^{\frac{N_1\tau - E_1}{\chi}}}{e^{\frac{N_2\tau - E_2}{\chi}}} \qquad \text{(IV-31)}$$

The structure of this relationship is identical with the relationship obtained by Gibbs when examining the thermodynamics of elementary particles of an ideal gas. Although this equation includes completely different parameters, determining the examined process, they only will be referred to as the Gibbs factor for the two-phase flow.

Another relationship, referred to as the Boltzmann factor, is known in thermodynamics. It is obtained from the Gibbs factor when fixing the number of particles ($N_1 = N_2 = N$). In this case, the expression, identical with the Boltzmann factor, has the form

$$\frac{P_1(I_1)}{P_2(I_2)} = \frac{e^{-\frac{E_1}{\chi}}}{e^{-\frac{E_2}{\chi}}} = e^{-\frac{E_1 - E_2}{\chi}} \qquad \text{(IV-32)}$$

The results make it possible to take further steps in examining the analogy between the investigated process and thermodynamics. Several other parameters of exceptional importance because of their meaning, will be investigated.

If the dependence characterising the Gibbs factor, is summed up with respect to all states of the zone and with respect to all particles, one obtains an expression referred to as the large statistical sum:

$$M(\tau;\chi) = \sum_N \sum_i e^{\frac{N\tau - E_i}{\chi}} \qquad \text{(IV-33)}$$

This sum is a normalising multiplier transforming the relative probabilities to absolute ones, i.e. it plays the role of the previously unknown proportionality coefficient. In chemical kinetics, the large statistical sum is often expressed using the so-called parameter of 'absolute activity'. In the present case, it is written in the form:

$$\lambda = e^{\frac{\tau}{\chi}}$$

and by analogy it will be referred to as the parameter of 'absolute mobility' of the system, and the large sum in this case is

$$M = \sum_N \sum_i \lambda^N e^{-\frac{E_i}{\chi}} \qquad \text{(IV-34)}$$

In the relationships, one can determine that the probability of the zone being in i-th state is determined as follows:

$$P(N_i;I_i) = \frac{e^{\frac{[N_i\tau - I_i]}{\chi}}}{M}$$

Using this approach, it is possible to determine a number of important parameters, for example, the mean number of particles in the zone. The number of particles in the zone may change owing to the fact that the zone exchanges particles with the apparatus and some of the particles leave the system at the top. To obtain the mean value, each Gibbs factor in the large sum should be multiplied by N

$$\langle N\rangle=\frac{\sum_{N}\sum_{i}Ne\frac{N\tau-E_i}{\chi}}{M} \qquad \text{(IV-35)}$$

The mean value for the lifting factor of the zone by analogy is:

$$\langle E_i\rangle=\frac{\sum_{N}\sum_{i}E_i;e\frac{N\tau-E_i}{\chi}}{M} \qquad \text{(IV-36)}$$

If the number of particles in the zone is constant, the normalising sum should be represented by the value analogous to the Boltzmann factor. This gives the dependence

$$Z=\sum_{i}e^{-\frac{E_i}{\chi}}$$

This dependence is referred to simply as a statistical sum. It also plays the role of the proportionality coefficient between probability and the Boltzmann factor, i.e.

$$P(E_i)=\frac{e^{-\frac{E_i}{\chi}}}{Z} \qquad \text{(IV-37)}$$

At a fixed number of particles, the mean value of the lifting factor in the zone is:

$$\langle E_i\rangle=\frac{\sum_{i}E_ie^{-\frac{E_i}{\chi}}}{Z}=\frac{\chi^2}{Z}\frac{\partial Z}{\partial \chi}=\chi^2\frac{\partial \ln Z}{\partial \chi} \qquad \text{(IV-38)}$$

Averaging here is carried out with respect to an ensemble of states of the zone which is in contact with apparatus, but has a constant number of particles in the steady process.

To understand mass exchange with the cell, attention will be given to a boundary case in which the zone contains constantly only one particle of some fixed size class, and then N identical independent particles of the same class will be examined. The statistical sum will be determined for a single particle. It is obvious that one particle has a total of two possible states, when its velocity is oriented upwards or downwards. For these two possible states:

$$Z = 1 + e^{-\frac{I}{\chi}} \qquad \text{(IV-39)}$$

The mean value of the lifting factor for a single particle is:

$$\langle E \rangle = \frac{0 \cdot 1 + Ee^{-\frac{E}{\chi}}}{Z} = \frac{Ee^{-E/\chi}}{1 + e^{-E/\chi}} \qquad \text{(IV-40)}$$

The mean value for a system of N particles will be N times larger and is:

$$\langle E \rangle = \frac{NEe^{-\frac{E}{\chi}}}{1 + e^{\frac{E}{\chi}}} = \frac{NE}{e^{\frac{E}{\chi}} + 1}$$

These relationships make it necessary to introduce another element of the examined model, i.e. its cells. However, this will be carried out later, and at the moment it will be attempted to determine entropy by this procedure.

Taking the logarithm of equation (IV-37) gives:

$$\ln P(E_i) = -\frac{E_i}{\chi} - \ln Z$$

Consequently,

$$E_i = \chi(\ln P + \ln Z) \qquad \text{(IV-41)}$$

This holds for time-stabilised systems.

The mathematical expectation of the lifting factor is:

$$I = \sum_i E_i P_i$$

where P_i is the probability of the system being in the i-th state. P_i is determined by the Boltzmann factor. It had been determined that

$$dH = -\frac{I}{\chi}$$

and consequently

$$\chi dH = -\sum_i E_i dP_i = -\chi \sum_i (\ln P_i) P_i - \tau \ln Z \sum_i dP_i$$

However, probabilities are normalised with respect to unity, i.e. $\sum_i P_i = 1$; therefore $\sum_i dP_i = 0$. Consequently:

$$\chi dH = -\chi \sum_i (\ln P_i) dP_i$$

Here it can be shown that

$$d\sum_i (P_i \ln P_i) = \sum_i (\ln P_i) dP_i = \sum_i \ln(P_i) dP_i$$

Taking this into account it may be written that

$$\chi dH = \sum E_i dP_i = \chi d(-\sum_i P_i \ln P_i)$$

The following equation is obtained for entropy variation

$$dH = d\left[-\sum_i P_i \ln P_i\right] \tag{IV-42}$$

and entropy is

$$H = -\sum_i P_i \ln P_i \tag{IV-43}$$

For the particle oriented upwards $P_i = 1$, otherwise $P_i = 0$. Therefore, $H = -1$; $\ln 1 = 0$. It is therefore clear that in transition from (IV-42) to (IV-43) no additional constant appears. It should be mentioned that (IV-43) is the definition of entropy according to Boltzmann. If the zone has equiprobable permissible states, then for each state:

$$P = \frac{1}{\varphi}$$

Consequently,

$$-P_i \ln P_i = -\frac{1}{\varphi} \ln \frac{1}{\varphi} = -\frac{1}{\varphi}(\ln 1 - \ln \varphi) = \frac{1}{\varphi} \ln \varphi$$

and

$$H = \sum_i \frac{1}{\varphi} \ln \varphi = \ln \varphi$$

which is in complete agreement with the initial definition of entropy.

8. SEPARATION WITH LOW CONCENTRATION

When examining the behaviour of a single particle in the zone of apparatus it was concluded that it is necessary to examine the cellular model. It is assumed that the entire volume of apparatus is divided into rectangular cells in such a manner that volume of each cell incorporates no more than one coarse particle. If the examined size class contains N particles, it may be assumed that N cells in apparatus are occupied and others are free. Thus, it is assumed that the number of cells is considerably larger than the number of particles.

A system formed by a single cell will be examined. In this case, the apparatus is represented by all remaining cells, occupied by $(N-1)$ particles.

From the definition of the large sum for a single cell:

$$M = 1 + \lambda e^{-\frac{E}{x}} \tag{IV-44}$$

The first term corresponds to the case in which the cell is not occupied and the lifting factor here is equal to zero. The second term relates to an occupied cell, where $M = 1$ and $I = E$.

The mean occupation of the cell is:

$$n(E) = \frac{1}{\lambda^{-1} e^{\frac{E}{x}} + 1} \tag{IV-45}$$

This form gives the mean number of particles of the examined class per cell. Its value will always change from zero to unity. It should be mentioned that for coarse particles, the parameter of mean occupation of the cell coincides with the probability because the cell may be either occupied or free.

Previously, the parameter of absolute mobility was determined as follows:

$$\lambda = e^{\frac{\tau}{\chi}}$$

Substituting this equation into (IV-45) gives

$$\langle n(E) \rangle = \frac{1}{e^{\frac{E-\tau}{\chi}} + 1} \tag{IV-46}$$

This dependence for the mean occupation of the cell will be denoted by

$$f(E) = \langle n(E) \rangle = \frac{1}{e^{\frac{E-\tau}{\chi}} + 1} \tag{IV-47}$$

This function shows the mean number of particles for a single cell with parameter (E) oriented upwards. The value $f(E)$ is always between zero and unity.

As regards fine particles, it is necessary to assume that several particles may be situated simultaneously in a single cell. If a comparable ratio of the dimensions of coarse and fine particles is introduced, then a large number of fine particles may be included in a single cell. For definiteness, it is assumed, since the size boundary is not limited at the bottom, that it may include any large number of particles. A single cell in the form of the zone of apparatus will be investigated. The number of fine particles in the cell is n.

It should again be stressed that from the viewpoint of behaviour of coarse particles, the cell can be either occupied or free, and for fine particles n can vary in a wide range.

The large sum for the fine particles is written in the form:

$$M = \sum_{n=0}^{\infty} \lambda^n e^{-nE/\chi} = \sum_{n=0}^{\infty} (\lambda e^{-E/\chi})^n \qquad \text{(IV-48)}$$

Denoting $\lambda e^{-E/x} = X$, then (IV-48) will hold on the condition that $X < 1$

$$M = \sum_{n=0}^{\infty} X^n = \frac{1}{1-X} = \frac{1}{1-\lambda e^{-E/\chi}}$$

The mean number of particles in the cell is:

$$\langle n(E) \rangle = \frac{\sum_{n=0}^{\infty} nX^n}{\sum_{0}^{\infty} X^n}$$

After several transformations, one obtains

$$\langle n(E) \rangle = \frac{X}{1-X} = \frac{1}{X^{-1}-1} = \frac{1}{\lambda^{-1} e^{E/\chi} - 1}$$

Consequently

$$n(E) = \frac{1}{e^{\frac{E-\tau}{\chi}} - 1} \qquad \text{(IV-49)}$$

For fine particles, the occupation of the cell does not coincide with the probability of a specific cell being filled. Dependences (IV-46) and (IV-49) differ by only ±1, but have completely different physical meanings.

Here, there is another statistical definition of the difference between the fine and coarse particles. The particles, whose distribution laws are described by equation (IV-46), may be regarded as coarse, and other particles, distributed in the flow in accordance with equation (IV-49) in contrast to the former, must be regarded as fine particles.

Thus, the mean occupation of the cell with a specific potential is described by a single dependence

$$\langle n(E)\rangle = \frac{1}{e^{(E-\tau)/\chi} \pm 1} \quad \text{(IV-50)}$$

The plus sign relates to the distribution of coarse particles, the minus sign defines fine particles.

Since $n(E)$ is always lower than unity, it may be assumed that, in specific conditions, the first term of the denominator may have the values considerably higher than unity. In this case, the dependence (IV-50) for an individual cell will be represented by the equation

$$n(E) \approx e^{-(E-\tau)/\chi} = \lambda\, e^{-E/\chi} \quad \text{(IV-51)}$$

In statistical mechanics, the equation is referred to as the classic distribution function. The physical meaning of this case of distribution is that the mean occupation of any cell in the examined system is considerably smaller than unity. This condition fully characterises the conditions of fractioning with the change in concentration in the area of the working zone, i.e. up to $\mu = 2$ kg/m^3. In this case, it may be assumed with sufficient reliability that both the coarse and fine particles have in the limit the common distribution function (IV-51), although their physical realisations, as shown, differ in principle.

Since the object of the present examination is the turbulent regime of the flows, it may be assumed that the region of separation in this case will be restricted by relatively coarse particles (examination of the distribution of fine particles will be investigated later).

For the mean occupation of the cell, the dependence (IV-47) was obtained. Taking into account that for a single particle according to the definition

$$E = gd$$

It may be shown that

$$E = w_{50}^2$$

For the specific point of the cross section of the channel in which the cell is situated.

The mean occupation of the cell in the zone characterises the degree of fraction separation

$$f(E)=\frac{1}{e^{\frac{E-\tau}{\chi}}+1}=\frac{1}{e^{\frac{w_{50}^2-w_i^2}{w^2}}+1} \tag{IV-52}$$

The hovering velocity is proportional to the particle size, i.e. $w_{50}^2 \approx gd$. Taking this into account, the dependence of the type

$$f(x)=\varphi(d)$$

will be analysed. Figure IV-3 shows an approximate pattern of the structure of the flow. The values of w_1 in different points of the cross sections differ. If the point at which $w_i = w_{50}$ is examined, then at this point

$$f(x)=\frac{1}{1+1}=\frac{1}{2}$$

which corresponds to the physical meaning of the separation process showing the optimum separation conditions. At points where $w_i > w_{50}$, i.e. for coarser particles, $\frac{w_i^2}{w_{0.5}^2}>1$, i.e. in the expression e^x the value $x < 0$ and the dependence $f(x)>\frac{1}{2}$. For points where $w_i < w_{50}$, i.e. for smaller particles in expression e_x the value $x < 0$ and the dependence $f(x)<\frac{1}{2}$. This dependence accurately corresponds to the form of the distribution curve of the type $F_f(x = f(d)$. At $w_{50} - w$, i.e. in the optimum conditions, equation (IV-52) gives $f(E)=\frac{1}{2}$. This may be interpreted as the optimum distribution with respect to a specific size class.

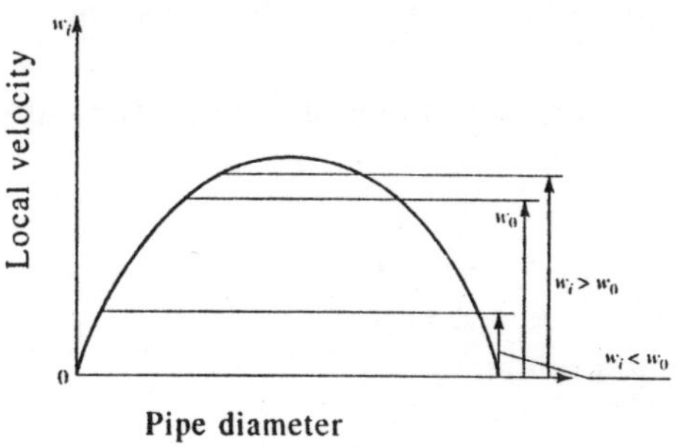

Fig.IV-3. Analysis of the $f(x)$ dependence with the flow structure taken into account.

On the other hand, the relationship $\frac{w_{50}^2}{w^2}$ is proportional to the dependence $Fr = \frac{gd}{w^2}$.

It was found by an empirical procedure that the Froode criterion is the controlling parameter for the examined class of the processes. Here it was possible to show for the first time the role of this factor from the position of a purely theoretical approach. All this provides a facility for further development of the theory of the process. The dependence (IV-52) reflects extensively the physics of separation because the resultant curve corresponds to the form of the separation curve of the type $F(x) = f(d)$. We shall return to dependence IV-52. It can be written in the following form

$$f(E) = \frac{1}{e^{\frac{w_{50}^2}{w^2} - \frac{w_i^2}{w^2}} + 1}$$

The second part in the denominator determines the structure of the two-phase flow, i.e. the geometrical parameters of the design of apparatus. For example, for a two-phase flow in a circular pipe it can be assumed that this structure is parabolic, i.e.

$$w_i = 2w\left[1 - \left(\frac{r}{R}\right)^2\right]$$

where R is the current radius, $0 \le r \le R$, R is the radius of the vertical pipe and consequently

$$\frac{w_i}{w} = 2\left[1 - \left(\frac{r}{R}\right)^2\right]$$

This shows that in a steady process only the designed parameters determine this relationship.

The first term in the denominator determines the set of the flow parameters. In fact

$$\frac{w_{50}^2}{w^2} = \frac{4gd(\rho - \rho_0)}{3\lambda\rho_0 w^2} = c\frac{gd}{w^2} = cFr$$

where c is the set of constant parameters.

References

1. Gibbs J.W. Elementary principles in statistical mechanics, developed with special reference to the rational foundation of thermodynamics, Yale Univ. Press 1902.
2. Tolmon R.C. Principles of statistical mechanics. Oxford Univ. Press (1938)
3. Landau L.D. and Lifshits E.M., Statistical physics, Nauka, Moscow (1964).
4. Kittel C. Thermal Physics, John Wiley and Sons, Inc., New York (1977).
5. Brillouin L. Science and information theory, Academic Press Inc, New York (1956)
6. Chambodal P.P., Evolution et applications du concept d'entropie, Dunov, Paris (1963).
7. Boltzmann L., Lectures in gas theory (Russian translation), Gostekhizdat, Moscow (1956).
8. Smoldyrev A.B., Pipeline transport, Nedra, Moscow (1970).
9. Barsky M.D., Fractionation of powders, Nedra, Moscow (1980).

Chapter V

KINEMATIC FUNDAMENTALS OF THE PROCESS

1. MECHANICAL INTERACTION OF PARTICLES

In the previous chapter, interesting relationships were obtained for the examined process. However, they are relatively abstract and require detailed verification. Therefore, it will be attempted to approach this process from a slightly different side. Attention will be given to the mass movement of particles in their physical realisation taking into account the presence of mechanical contact interaction of solid particles in a two-phase flow.

Taking into account the mechanical interaction of particles in a flow greatly complicates the considerations regarding the mechanism of gravitational separation.

This interaction results in a constant redistribution of the velocities of different size classes as a result of inhibition of fine particles and acceleration of larger particles in the direction of movement of the medium, leads to changes in the trajectory of movement of the individual particles, increases the radial component of their velocity generated by different migration effects. Identical results may be obtained not only by direct contact interaction of particles but also by affecting them through a moving medium, especially if they are closely spaced.

Under the same regime parameters, the frequency of such an interaction depends greatly on the physical properties of the material, primarily on the particle size. It is evident that the number of interactions amongst particles of different sizes increases with increasing concentration of the material in the flow, but it is not clear what is the dynamics of this increase. There are three aspects of the model examined here.

First, only the pair-wise interaction of particles is analysed, because triple or larger numbers of simultaneously colliding particles have a considerably smaller probability.

Secondly, attention is given to the interaction of particles of different size because the effect of collisions of two identical particles having approximately the same velocity, is negligible.

Thirdly, the problem will be simplified by assuming that the interaction of the particles does not lead to the formation of aggregates.

All these three boundary conditions greatly simplify the actual process, but at present there is insufficient experimental data for more accurate explanation. It will be assumed that the results obtained in this type of examination reflect the relevant phenomenon in the first approximation, which is sometimes sufficient for drawing important conclusions.

The collisions of two bodies in the mechanics is regarded as an impact phenomenon. According to classic considerations, impacts are accompanied by the development of high forces acting over a short period of time during which the finite change of the velocity takes place without any significant displacement of the colliding bodies.

In the flow, the trajectories of movement of the particles deviate from the straight trajectory and the velocity of the particles changes as a result of the effect of different reasons. It is therefore necessary to carry out averaging with respect to details of interaction in such a manner as to retain the unique information which is of interest in the given examination on the probability that two particles with velocities of v_i and v_j accelerate at the start of interaction after interaction with velocities of v_i' and v_j' respectively.

In a dispersed flow, there are two types of impact interactions of particles. The first type includes all impacts amongst the particles which are referred to as internal impacts of the investigated system. The second type includes impact interactions of particles with walls of the apparatus which may be regarded as outer walls for the given system.

It is well known that the variation of the sum of the momentums of the system is equal to the sum of impact momentums of external forces. Internal impacts in a system do not change the total momentum, but only distribute the latter between the individual particles.

For derivation, section Δl will be defined hypothetically in the vertical ascending cylindrical flow restricted by solid walls.

In the stationary conditions of the classification process, the concentration of the material in the vertical counter flow differs in different sections of the flow and gradually decreases with exit from the area of supply into both sides.

Therefore, this section should be relatively large in order to include a large number of particles of both fractions and should be sufficiently small in comparison with the scale variation of the velocities and the concentrations of the dispersed material.

It may be assumed that in the steady process this section contains, at every moment of time, a constant number of solid particles. Attention will be given to a unidimensional system obtained as a result of projection of the velocity of particles on the axis of the flow.

It is assumed that the amount of the solid phase transferred to section Δl per unit time is:

$$M = M_i + M_j$$

where M_i is the mass of fine particles, kg/s; M_j is the mass of coarse particles, kg/s.

The following notations will be introduced: r_i; m_i; v_i is the radius, mass, and projection of the mean axial component of the velocity for a fine particle; r_j; m_j; v_j are the radius, mass and identical component of the velocity for a coarse particle.

It should be mentioned that for the conditions of gravitational classification in this reference system, the mean axial velocity of the particle may differ not only in magnitude but also in direction.

At any moment of time in every unit of length of the examined section there are coarse particles whose weight is:

$$\Delta g_j = \frac{gM_j}{v_j} = \frac{G_j}{v_j} \tag{V-1}$$

where g is gravitational acceleration, m/s^2; G_j is the consumed part of the j-th component in the composition of the mixture, kg/s.

The weight of these particles in the entire examined section Δl is:

$$\Delta G_j = \Delta g_j \Delta l = \frac{G_j \Delta l}{v_j} \tag{V-2}$$

similarly, for fine particles it can be written that:

$$\Delta G_i = \frac{G_i \Delta l}{v_i} \tag{V-3}$$

Irrespective of whether the coarse particles move in the direction of fine particles or against them, they appear to be constantly 'pierced' by the fine particles.

It is well known that not all fine particles fall into the fine product, because some of them are included in the yield of the coarse product.

Therefore, it should be assumed that the coarse particles, situated in the investigated volume, are 'pierced' by not all fine particles but only by some of them:

$$\Delta G_i^{'} = z\Delta G_i$$

where z is a coefficient proportional to the degree of fractional extraction of the fine particles.

Two particles can collide only when they meet on a corresponding area referred to as the cross section of collisions.

During the unit time, a coarse particle may collide with those fine particles whose centers at the given moment of time are situated inside a cylinder whose base is represented by the cross section of collisions, and the height is the difference of the path traveled by these particles per unit time, i.e.

$$h = v_i - v_j$$

To determine the probability of collisions of the particles $P(x)$ where the coarse particles can be regarded as stationary, and the fine particles as moving with relative velocities.

The value of $P(x)$ may be determined as the ratio of all collision areas in a single section of apparatus to the size of this section, i.e.

$$P(x) = \frac{4\sum S}{\pi D_e^2} \tag{V-4}$$

where

$$\sum S = n_j^{'}(r_i + r_j)^2 \pi \tag{V-5}$$

Here D_e is the equivalent diameter of the cross section of the flow; $n_j^{'}$ is the mean number of the coarse particles in some section of the flow.

The mean number of the particles in the cross section of the flow may be determined from the equation

$$n_j^{'} = \frac{\Delta G_j 2r_j}{m_j g \Delta l} \tag{V-6}$$

These particles can collide during the unit time only with those fine particles which are situated at distance h from this layer. Their number can be determined as follows:

$$n_i^{\cdot} = \frac{\Delta G_i (v_i - v_j)}{m_i g \Delta l} z \qquad \text{(V-7)}$$

Not all fine particles take place in collisions, only some of them, with this fraction determined by the probability of collisions, i.e.

$$\Delta n_i^{\cdot} = P(x) n_i^{\cdot} = \frac{4(r_i + r_j)^2 n_j n_i (v_i - v_j)^2 2 r_j z}{D_e^2} \qquad \text{(V-8)}$$

The dependence (V-8) determines the mean number of the fine particles which interact in the investigated flow with the coarse particles situated in the fixed cross section of the flow.

The total number of collisions in the entire examined section is:

$$\Delta N = \frac{\Delta n_i^{\cdot} \Delta l}{2 r_j} = \frac{4z(r_i + r_j)^2 (v_i - v_j)^2 n_i n_j \Delta l}{D_e^2} \qquad \text{(V-9)}$$

This number is proportional to the cross section of the collision, the number of coarse and fine particles, situated in the examined section of the flow, its length, and also to the difference in the velocities of the particles.

In this dependence, n_i and n_j correspond to the number of particles of the two fractions in the unit height of the investigated flow:

$$n_i = \frac{\Delta G_i}{\Delta l m_i g}; \qquad n_j = \frac{\Delta G_j}{\Delta l m_j g} \qquad \text{(V-10)}$$

Taking (V-3) and (V-10) into account:

$$n_i = \frac{G_i}{v_i m_i g}; \qquad n_j = \frac{G_j}{v_j m_j g}; \qquad \text{(V-11)}$$

The transition from the number of particles to their concentration may be carried out on the basis of the following considerations.

In the investigated process, the flow rate of the medium is

$$Q = F \rho_0 w \qquad \text{(V-12)}$$

where Q is the weight flow rate of the medium, kg/s; F is the cross sectional area of apparatus, m^2; w is the velocity of the flow, m/s; ρ_0 is the density of the medium, kg/m^3.

Similarly, the consumption productivity of each of the examined size classes may be represented by:

$$G_i = F v_i \rho_i; \; G_j = F v_j \rho_j \qquad \text{(V-13)}$$

where ρ_i; ρ_j has the physical meaning of the mass of the corresponding solid particles in the unit volume occupied by these particles.

Relating equation (V-13) to (V-12) gives:

$$\rho_i = \rho_0\mu_i;\ \rho_j = \rho_0\mu_j$$

where μ_i; μ_j is the weight concentration of the solid particles per unit weight of air, kg/kg;

Taking this into account

$$n_i = \frac{F\rho_0}{m_i}\mu_i \qquad n_j = \frac{F\rho_o}{m_j}\mu_j \tag{V-14}$$

Taking these relationships into account, equation (V-9) can be written in the following form

$$\Delta N = \frac{\pi(r_i + r_j)^2 Vz\rho_0^2}{m_i m_j}(v_i - v_j)^2 \mu_i \mu_j$$

where V is the volume of the examined zone.

Consequently, the total number of collisions of the particles in some volume is directly proportional to their concentration in the flow, the difference in the velocities of the particles of different size classes, and also the size of this volume.

Correspondingly, the total number of collisions in the unit volume:

$$N = \frac{\pi(r_i + r_j)^2 \rho_0^2 z}{m_i m_j}(v_i - v_j)^2 \mu_i \mu_j \tag{V-15}$$

For a non-steady process of movement of the particles in which their velocity differs from the mean velocity to either side, the number of interactions has the values in a specific range, and the mathematical expectation of distribution this range is (V-15). Equation (V-8) shows the number of fine particles interacting with the coarse ones per unit time in a single cross section of the flow. It is thus possible to determine the number of impacts from the side of the fine particles which is applied on the average to a single coarse particle. This value is:

$$N_j = \frac{\Delta n_i'}{n_j'} = \frac{\pi(r_i + r_j)^2 z\rho_0^2}{m_i}(v_i - v_j)^2 \mu_i \tag{V-16}$$

Knowing this value, it is possible to calculate the mean distance passed by the particle between two interactions. During the time Δt the particle travels some path $v_i\,\Delta t$. The distance travelled by the particle between two collision can be determined as the ratio of the path travelled by that particle, to the number of collisions of the particle in this path:

$$\lambda_j = \frac{v_i \Delta t}{N_j \Delta t} = \frac{m_i (w - w_{oi})^2}{\pi (r_i + r_j)^2 z \rho_0^2 (v_i - v_j)^2} \cdot \frac{1}{\mu_i} \qquad \text{(V-17)}$$

In this equation, the variable parameters are w and μ. Equation $(v_i - v_j)^2$ can be calculated with the known degree of accuracy as follows:

$$\left[(w - w_{oi}) - (w - w_{0j})\right]^2 \approx (w_{oj} - w_{0i})^2 = const$$

Consequently, it can be written that:

$$\lambda_j = c \frac{(w - w_{0j})^2}{\mu_i} \qquad \text{(V-18)}$$

where c are all constant parameters.

This shows that, with a certain degree of approximation, the mean free path length of the particles of some size class in the flow is inversely proportional to the concentration of particles of another class in this flow.

2. FORCES FROM THE INTERACTION AMONGST PARTICLES OF DIFFERENT SIZE CLASSES

In collision, every fine particle reduces its velocity in the axial direction on the average by the value Δv_i, which increases the velocity of each coarse particle by Δv_j.

To determine the corresponding variation of the velocities of particles of different classes, it is necessary to examine the mechanism of redistribution of the velocities for two separate particle. Since only the axial variation of the velocity of the solid particle is of interest in this case, it is sufficient to confine examination to a direct impact. The impact between two solids is referred to as direct if at the moment of impact they do not rotate and the velocities of their centres c_1 and c_2 are directed along the line c_1c_2 in the direction normal to the colliding surfaces at the contact point.

It is assumed that two spheres with the mass m_i and m_j collide at the moment of time t_0. During a very short period of time $t_1'-t_0$, during which the impact takes place, the line of the centres may be regarded as stationary. The algebraic value of the velocities of the particles prior to the impact will be denoted by v_i and v_j, after impact v'_i and v'_j. The general features of the phenomena taking place at the contacting particles during impact will be analysed.

Starting from the moment t_0 when the particles come into contact, they are deformed around the contact point. In this case, their centres continue to converge to until the moment t'_1 when the distance between them becomes the smallest.

At the time $t'_1 - t_0$ of the first phase of interaction between the particles, a reaction tending to separate the particles occurs.

The work of the reactions during this time will be negative and the kinetic energy of the system decreases.

At the moment t'_1 the velocity of both particles are equalised, their centres no longer converge, and the value of deformation becomes maximum. Starting from this moment, the mutual reactions of the particles will continue to operate until both particles acquire the initial shape.

At some moment of time t_1 they will contact only at a single point.

During the second phase $t_1 - t'_1$, the kinetic energy of the system will increase because the work of reactions is positive and this results in the movement of particles away from each other with the velocities differing from the initial velocities.

During impact, very high forces develop as a result of their short duration. Therefore, when examining two colliding particles as an isolated system, the conventional forces, such as the gravitational force may be ignored.

According to the theorem of the velocity of the centre of gravity of the system, it may be assumed that the velocity of the common centre of gravity of the two particles does not change because no external impact pulses have been applied to this system, i.e.

$$v_0 = \frac{m_i v_i + m_j v_j}{m_i + m_j} = \frac{m_i v'_i + m_j v'_j}{m_i + m_j}$$

hence,

$$m_i v_i + m_j v_j = m_i v'_i + m_j v'_j \tag{V-19}$$

In order to determine the velocity of the particles after an impact, it is necessary to explain the properties of the colliding bodies.

From the viewpoint of impact interaction, all the bodies may be divided into three groups: absolutely inelastic, absolutely elastic and those having intermediate properties.

The effect of these properties on special features of the impact interaction of the bodies will be investigated. The absolutely inelastic bodies after an impact remain in contact, i.e. the velocity acquired by these bodies as a result of interaction is the same:

$$v_i' = v_j'$$

or

$$v_0 = \frac{m_i v_i + m_j v_j}{m_i + m_j} = v_i' = v_j'$$

In the given case, the impact phenomenon is reduced to the first phase, and the moment of time t_1' coincides with t_1. This is accompanied by a loss of kinetic energy. In collision of absolutely elastic bodies there is no loss of energy, i.e.

$$\frac{m_i v_i^2}{2} + \frac{m_j v_j^2}{2} = \frac{m_i v_i'^2}{2} + \frac{m_j v_j'^2}{2}$$

This relationship together with equation (V-19) makes it possible to determine unambiguously the velocity of the particles after the impact:

$$v_i' = v_j + \frac{m_i - m_j}{m_i + m_j}(v_i - v_j)$$

$$v_j' = v_i - \frac{m_i - m_j}{m_i + m_j}(v_i - v_j) \qquad \text{(V-20)}$$

The variation of the velocities for both particles is

$$\Delta v_i = v_i' - v_i = \frac{2m_j(v_j - v_i)}{m_i + m_j} \qquad \text{(V-21)}$$

Thus, under the condition of the absolutely inelastic bodies the relative velocity of these bodies becomes equal to zero, and in the case of absolute elastic bodies, this velocity only changes its sign because according to equation (V-20)

$$v_i' - v_j' = v_i - v_j$$

For non-absolutely elastic bodies it may be accepted that:

$$v_i' - v_j' = k(v_i - v_j)$$

To find the finite velocities, the equation of the momentum must be added to the equation:

$$m_i v_i + m_j v_j = m_i v_i' + m_j v_j'$$

The transformation of this system and the corresponding solutions make it possible to determine

$$\Delta v_i = \frac{m_j (k+1)(v_i - v_j)}{m_i + m_j} \qquad \text{(V-22)}$$

$$\Delta v_j = \frac{m_i (k+1)(v - v_j)}{m_i + m_j} \qquad \text{(V-23)}$$

The general impact pulse, acting on a cluster of fine particles per unit time, related to the unit volume of the apparatus, can be determined from the equation

$$F_m = \Delta N \cdot m_i \Delta v_i$$

where $\Delta N_i m_i$ is the mass of the fine particles, colliding with the coarse ones.

Taking (V-15) into account, it may be written that:

$$F_m = \frac{\pi (r_i + r_j) \rho_0^2 z \alpha (v_i - v_j)^2 \mu_i \mu_j (k+1)}{m_i + m_j} \qquad \text{(V-24)}$$

where $\alpha < 1$.

This coefficient takes into account the difference of the actual effect in collisions of the particles from the direct impact.

Using a simular substitution, it is possible to determine the value of the force acting on the cluster of coarse particles, whose value will be equal to that found for the fine particles and reversed in respect to direction.

In the examined dependence, z and α are random parameters whose mean-probability value can be determined only by experiments. It should be mentioned that in every specific case they should be given some mean constant value.

To simplify the dependence, the set of the constant parameters in the equation (V-24) will be denoted by

$$C_1 = \frac{\pi (r_i + r_j)^2 \rho_0^2 z \alpha (1+k)}{(m_i + m_j)}$$

Taking this into account

$$F_m = C_1 \mu_i \mu_j (v_i - v_j)^2 \qquad \text{(V-25)}$$

This dependence will be analysed. In steady movement with a known degree of accuracy it may be assumed that:

$$v_1 = w - w_{0i}; \qquad v_j = w - w_{0j};$$

and consequently,

$$v_i - v_j = w_{0i} - w_{0j} \approx const \tag{V-26}$$

This means that the value of the variation of the velocity of two colliding particles in a steady regime with the known degree of accuracy is independent of the velocity of the flow of the medium and is determined only by the hovering velocities of these particles, i.e. by their dimensions.

According to equation (V-26), the general force of interaction of the fine and coarse particles in the final analysis is also independent of the velocity of the medium.

Therefore, for the steady regime of movement of the particles, the value of the force of interaction is a function of only the concentration of particles of different classes, i.e.

$$F_m = C_2 \mu_1 \mu_2$$

3. FORCES DUE TO THE INTERACTION OF PARTICLES WITH THE CHANNEL WALLS

Among the studies carried out in recent years to investigate the relationships governing the hydrodynamics of two-phase systems, only a small number of studies have been concerned with the problems associated with the interaction of a discrete phase with the channel walls. At the same time, the nature of movement of solid particles is determined to a large extent by the collisions of these particles with the walls restricting the flow. To confirm this fact, it is sufficient to refer to the well-known effect of wear of pipelines during pneumatic transport and walls of apparatus in gravitational classification. Experiments show that the movement of particles in a dispersed flow is not parallel to its axis. The presence in the flow of different disturbing random factors causes the particles to acquire the radial velocity component.

Consequently, the velocity of particles of any size may be assumed to consist of two components. The relationship between the mean values of the radial and axial components of the velocity is determined by the specific separation conditions. The radial component is a reason for disordered impact interaction of the particles with the channel walls.

Every impact with a wall results in the loss of the kinetic energy of movement of the particle. The value of this loss depends on the elastic properties of the dispersed material and the solid wall and also on the state of their surface at the contact point. After an impact, the particle loses part of the component of the axial velocity of its movement. This loss is then compensated by the carrying energy of the flow leading to the acceleration of the particles to the initial values of the axial component of the velocity. In this case, the velocity of the particles may also increase as a result of collisions of the particles with faster particles. This collision results in the momentum exchange as a result of which the slowly moving particles are accelerated and the fast ones slow down. The slowed-down particles again use energy for acceleration from the flow. If the length of apparatus is sufficiently large, after some period of time the same particle may again collide with a wall because the reasons generating the radial components of the velocities of the particles continue to act.

These considerations show that the interaction of the particle of the two-phase flow with the walls of apparatus is of the jump-like, pulsating nature.

With a large increase of the concentration, the radial displacement of the particles decreases because the trajectories of the particles will become quite similar to the straight trajectory parallel to the axis of the channel.

This does not take place as a result of the elimination of the radial component in the velocity of the particles but, as a result of 'extinction' of this component as a result of their mass interaction, starting from a specific concentration of the material. In the final analysis, this results in the redistribution of particles of different classes in the radial direction.

For a polydispersed material, the number of impacts of coarse particles on the wall decreases in comparison with the movement of the monofraction from the same coarse particles. This may be explained by a decrease in the degree of freedom of these particles, pressed to the wall by the fine particles, moving in the centre of the flow with higher velocities. In addition to this, the mechanical interaction of the coarse particles with rapidly moving fine particles leads to this effect. This interaction results in an increase of the axial component in the velocity of the coarse particles, thus increasing the velocity and the path between two consecutive collisions of the coarse particle on the wall, i.e. decreases their frequency.

In the case of a stable grain size composition of the solid phase, the frequency of the impacts of the particle on the channel walls increases with an increase in the velocity of the air flow and the consumption concentration of the solid phase. In this case, the effect of concentration differs: at $\mu = 1 \div 2.2$ kg/m^3 the rate of increase of the number of impacts of the particles is larger than at $\mu = 2.2 \div 5$ kg/m^3. The most marked increase in the number of impacts on the walls takes place when the concentration increases from 0 to 1.5 kg/m^3 which is an almost total range for gravitational classification. In this range of the variation of concentration, the examined relationship is distinctively linear. The increase in the flow velocity results in an increase of the radial component on the velocity of the particles thus increasing the frequency of the impacts on the wall.

Thus, another force, formed as a result of the interaction of the particles with the apparatus walls, acts against movement of each narrow class. The magnitude of this force will now be determined.

To determine this force, an element of a hollow apparatus Δl will be defined (Fig. V-1). The cross section of apparatus will be denoted by F, its hydraulic diameter by D_e.

It is assumed that in the steady process, the total amount of the material, passing through this section in both directions per unit time at some flow velocity w is ΔG.

This is the amount of the material which can be recorded in the volume restricted by the levels A and B:

$$\Delta G = \sum_i \Delta G_i$$

where ΔG_i is the weight of the i-th fraction passing per unit time through a given section, kg/s; i is the number of different size classes in a mixture.

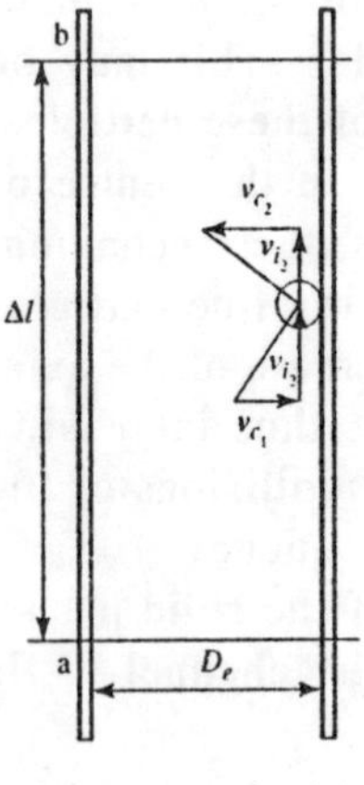

Fig. V-1 Transformation of the components of the velocity of a particle at its interaction with the wall.

The gravitational force of the investigated narrow size class, which is within the section Δl, may be determined by the equality:

$$\Delta G = \sum_i \Delta G_i; \qquad \Delta G_i = \sum_i n_i \Delta m_i g$$

where Δm_i is the mass of the particle of the i-th size in section Δl; n_i is the number of particles of the i-th size in section Δl.

For this section:

$$\Delta G_i = F v_i \rho_i q \tag{V-27}$$

taking into account equation (V-27)

$$\rho_i = \frac{\Delta G_i}{g F v_i} = \rho_0 \mu_i \tag{V-28}$$

The weight of the solid particles of each i-th size class in the examined section of the flow with height Δl is $\rho_i F \Delta l q$.

It is not possible to determine the force experienced by the wall in collision with each individual particle.

To understand the mechanism of this phenomenon, it is sufficient to determine the mean force arising from collisions with walls of many particles of the same size, if their mean velocities are known and if it is assumed that the collisions are completely elastic. In this case, the force, acting on the wall, may be determined on the basis of the second Newton law. It is equal to and opposite in the sign to the variation of the momentum of the particles colliding with the wall per unit time.

Regardless of the orientation of the velocity of the particle in space, it can always be divided into three components. One of these components is normal to the wall of the apparatus, the other one is parallel to the flow aixs. Figure V-1 shows the mean axial component of the velocity of the examined size class prior to collision v_{i_1} and the radial component v_{r_1}. If the particle with the mass m_i has the radial component of the v_{r_1}, the corresponding component of the momentum is $m_i v_{r_1}$.

After an impact, the particle acquires the radial component v_{r_2} whose value is predetermined by the elasticity of the properties of the particle and the wall and by the condition of the surface at the contact point. The reduced coefficient of recovery (elasticity) of the particle and the wall will be denoted by K_1 and, consequently, we may write that:

$$v_{r_2} = -K_1 v_{r_1} \tag{V-29}$$

The variation of the momentum at a collision is:

$$\Delta \rho = m_i v_{r_i} - m_i v_{r_2} = m_i (1 + K) v_{r1}$$

The weight of particles of the examined narrow size class, which reach the wall of the apparatus per unit time, is proportional to the content of the material in every volume unit of the apparatus, the value of the radial component and the value of the wall, i.e.

$$\Delta G_i = \varphi \rho_i g v_{r_i} B \Delta l \qquad \text{(V-30)}$$

where φ is the coefficient of proportionality; B is the perimeter of apparatus.

The total number of the particles of the given size class, reaching the walls of the apparatus per unit time, will be denoted by n_0.

The total variation of the momentum of all these particles per unit time is:

$$\Delta P = \sum_n \Delta \rho = \sum_n m_i v_{r_1} (1 + K_1) = n_0 m_i v_{r_i} (1 + K_1)$$

During unit time, the wall can be reached only by the particles whose distance from the wall is not greater than v_{r_1}, i.e. those that are enclosed in the volume of the cylinder with a base $B\Delta l$ and the generating line equal to v_{r_1}. The mean number of the particles in unit space is:

$$n = \frac{\rho_i}{m_i} = \frac{\Delta G_i}{F \Delta \ln m_i} \qquad \text{(V-31)}$$

In the examined volume the number of these particles:

$$N_0 = n v_{r_1} B \Delta l = \frac{\Delta G_i \, v_{r_i} B}{F m_i g} \qquad \text{(V-32)}$$

It may be assumed that as a result of the random nature of the examined process, half the particles in the given space travel in the direction of the wall, half away from it.

Because of the absence of direct experimental data confirming this assumption, for the general case it can be written that:

$$n_0 = \psi N_0$$

where $\psi > 1$ (according to the static meaning of the process ψ = 0.5). The mean force, experienced by the surface as a result of an impact per unit time, is:

$$P_i = \psi \frac{\Delta G_i v_{r_i} B}{F m_i g} m_i v_{r_i} (1 + K_1) \qquad \text{(V-33)}$$

The force per unit surface from this effect is:

$$f_i = \frac{P_i}{B\Delta l} = \psi \frac{\Delta G_i v_{r_i}^2 (1+K_1)}{gF\Delta l} \quad \text{(V-34)}$$

if it assumed that $K = 1$, $\psi = 0.5$, then

$$f_i = \frac{\Delta M_i}{v} \cdot v_{ri}^2$$

It is evident that

$$f_\Sigma = \sum_i f_i = \frac{\sum_i \Delta M_i v_{ri}^2}{V}$$

This means that the specific force, experienced by the wall from the side of the particles of the dispersed flow, is proportional to the product of the mass of these particles in unit volume per square of the radial component of the velocity for each size class.

This force, acting in the normal direction on the wall of the apparatus, generates a friction force whose specific value is:

$$\tau_i = \mu f_\varepsilon \quad \text{(V-35)}$$

where μ is the friction coefficient.

Analysis of the literature sources shows that the value of the radial component of the velocity of the particles is proportional to other parameters – the velocity of the particles (inertial force) and the concentration of the material in the flow, i.e.

$$v_{r_1} = \varphi(v_i; \mu_i)$$

For each specific case (μ_i = const) within the limits of the concentration of the solid phase, characteristic of gravitational classification, it can be written that:

$$v_{r1}^2 = \psi v_i^2$$

where

$$\tau_i = \psi\varphi\mu \frac{\Delta G v_i^2 (1+K_1)}{gF\Delta l} \quad \text{(V-36)}$$

If in this relationship all constant coefficients are denoted as

$$\lambda_1 = \frac{1}{2}\psi\varphi\mu(1+K_1), \quad \text{(V-37)}$$

then

$$\tau = \lambda_1 \frac{\Delta G_i v_i^2}{2g\Delta l}$$

This equation was derived by experiments and is well known in calculations of pneumatic transport, indicating that the conclusion was correct.

The general force of resistance to the movement of the particles of the *i*-th size in the examined section of the flow is:

$$T_i = \lambda_1 \frac{\Delta G_i}{2g} \cdot \frac{v_i^2 B\Delta l}{F\Delta l} = \lambda_1 \frac{\Delta G_i v_i^2}{2gD_e} \qquad \text{(V-38)}$$

the friction resistance due to the effect of all size classes, is

$$T = \sum_i T = \frac{\lambda_1}{2gD_e} \sum \Delta G_i v_i^2 \qquad \text{(V-39)}$$

The friction coefficients in the form of the dependence

$$\lambda_z = \lambda_1 \frac{v_i}{w}$$

were determined by experiments for the conditions of pneumatic transport by Gasterstadt et al.

This analysis shows that the value of this coefficient cannot be determined from the general pressure drop in a transport pipeline, as is sometimes the case. This determination may greatly increase the value of the required coefficient because it includes not only the friction loss but also other losses.

The calculations, presented in this chapter show that the results of gravitational separation are determined by the nature of behaviour in the classification conditions of all size classes forming the polydispersed materials.

4. EQUATION OF THE DYNAMIC MODEL

In theoretical investigations, attention is usually given to two types of forces acting on particles of a narrow size class in the conditions of gravitational classification: the gravitational force and the aerodynamic drag from the side of the flow.

Detailed examination of the mechanisms of process indicates that when constructing a dynamic model of the process, it is important to take into account two further forces formed as a result of the interaction of particles of different size classes with each other and with the walls restricting the flow.

The overall dynamic system, consisting of four previously mentioned components, reflects more efficiently the general pattern of separation, but does not reflect its overall complicated nature.

It should be mentioned that all components of this system are distinctively random.

The general equation of movement of the particles of a narrow size class, assuming that each particle moves autonomously, can, in the conditions of gravitation classification, be written in the following form in accordance with the previously derived relationships:

$$-\Delta G_i + \Delta G_i \frac{(w-v_i)^2}{w_0^2} - \lambda_1 \frac{\Delta G_i v_i^2}{2gD_e} \pm K\Delta G_i = \frac{\Delta G_i}{g} \cdot \frac{dv_i}{dt} \quad \text{(V-40)}$$

since the value of the dynamic drag, to which this class is subjected in the turbulent flow of the medium, is determined by the following equation on the basis of the principle of independence of the effect of the forces

$$R = m_i g \frac{(w-v_i)^2}{w_0^2} n_i = \Delta G_i \frac{(w-v_i)^2}{w_0^2} \quad \text{(V-41)}$$

where v_i is the mathematical expectation of the velocity of particles of size i in the flow of the medium, m/s; w_0 is the hovering velocity of the particles of the i-th size.

After transformations, equation (V-40) gives

$$\left[1 - \lambda_1 \frac{w_0^2}{2gD_e}\right] - 2\left(\frac{v_i}{w}\right) + \left[1 + (1 \pm K)\left(\frac{w_0}{w}\right)^2\right] = \frac{w_0^2}{gw^2} \cdot \frac{dv_i}{dt} \quad \text{(V-42)}$$

Equation (V-42) gives the similarity criterion for the examined process. For this purpose, as usually accepted, it is sufficient to relate the appropriate coefficients of the differential equation. The same parameters can also be obtained in determination of the roots of the examined system.

$$\left[1 - \lambda \frac{w_0^2}{2gD_e}\right] = const$$
$$\left[1 - (1 \pm K)\left(\frac{w_0}{w}\right)^2\right] = const \quad \text{(V-43)}$$

Consequently,

$$\frac{gD_e}{w_0^2} = \frac{\lambda_1}{2} + const$$

i.e. for the developed turbulent flow

$$Fr_0 = const$$

The second condition can be expanded assuming for the same flow that:

$$w_0 = K_2\sqrt{g_0 d}$$

Consequently, for the same conditions

$$\frac{gd}{w^2} = \frac{1}{K_2^2(1 \pm K)} = const$$

i.e. $Fr = const$

Here we have again obtained the relationship indicating that the Froude criterion is a generalising parameter for the separation processes.

References

1. Barsky M.D., Revnivtsev V.I. and Sokolkin Yu.V., Gravitational classificationof granular materials, Nedra, Moscow (1974)
2. Muschelknfutz M.E., Teoretische und experimentale Untersuchungen uber die Druckverluste, VDJ-Forschung-geselschaft, 25 (1959).
3. Smoldyrev A.E., Pipeline transport, Nedra, Moscow (1974)
4. Barsky M.D., Fractionation of powders, Nedra, Moscow (1980).

Chapter VI

EMPIRICAL FUNDAMENTALS OF THE PROCESS

1. SPECIAL FEATURES OF SEPARATION IN MOVING FLOWS

All the above considerations relate to equilibrium classification, i.e. to flows in hollow circular or rectangular pipes. Until recently, one of the controlling factors of organisation of high efficiency separation was assumed to be the best possible homogeneity of the classification conditions. This means that the sufficient efficiency of separation in a hollow body apparatus may be obtained only if it is relatively long (high). However, the practice of recent years shows that in some types of perturbation of the flow, it is possible to obtain a strong effect in apparatus of small height.

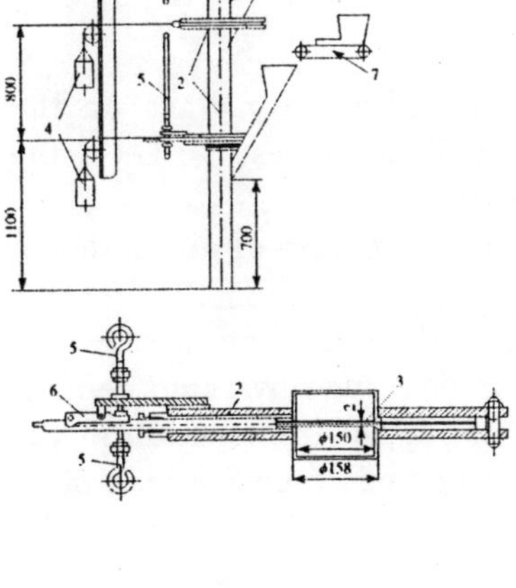

VI-I Diagram of equipment for examining the distribution of different fractions of the material along the height in separation conditions.

To explain this problem, special investigations were carried out. For this purpose, specialised equipment, Fig. VI-1, was developed.

The main element of equipment is a vertical pipe with a diameter of 150 mm, height 7 m. The pipe consists of eight separate sections (1) with a height of 800 mm, connected together with flanges (2) with built-in rapidly acting gates (3). The movement of each gate and rapid overlapping of the pipe channel are carried out by means of a spring and counterweight (4). When rotating the controlling lever, the common connecting rod (5) rotates the latches of the gate (6) and they simultaneously and rapidly overlap the cross-section.

Equipment operates under a rarefaction generated by the valve (8). The air flow rate is measured with a double diaphragm and controlled with the slide valve (9). The uniform feed of the material is ensured by the feeder (7) whose productivity is 11kg /min.

Size class, mm	2.5	2.5–1.6	1.6–1.0	1.0–0.85	0.85–0.63	0.63–0.40	0.40–0.315	0.315–0
Partial residues, %	0.26	35.49	45.65	8.77	4.36	1.18	0.19	4.1

The material for the experiments was crushed quartz whose fraction composition is given below:

Experiments were carried out using the following procedure. The valve was activated and a specific velocity of the air flow was set. This was followed by the start of operation of the feeder. After reaching the steady regime, the cross section of the channel was simultaneously overlapped by all gates. Subsequently, starting from the lower gate, the distribution of the material along the height of the pipe was determined:

$$g_i = \frac{G_i}{\sum_i G_i} 100\%$$

here g_i is the fraction of the weight of some size class on the i-th gate in relation to the total weight on all gates, %; G_i is the weight of the material at the i-th gate, kg.

The resultant dependence for different flow rates of air is shown in Fig. VI-2.

The graphs show that for all cases without exception the content of the material in the rising flow is governed by some general relationship. In movement away from the area of introduction of the material into the flow, its amount increases and reaches a maximum

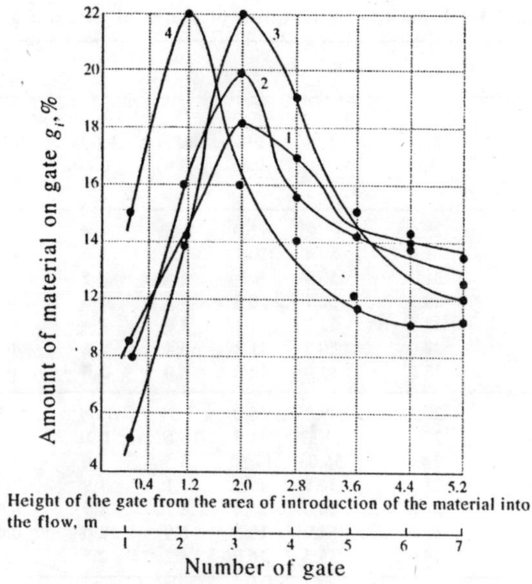

Fig.VI-2. Dependence of the distribution of the material along the height of a pipe at different velocities of the airflow: 1) 10.8 m/s; 2) 8.85 m/s; 3) 8.05 m/s; 4) 6.2 m/s.

value at a specific level. Subsequently, the content of the material starts to decrease down to the values close to stable ones.

When examining the general relationship, it is necessary to exclude the result obtained on the seventh gate because this gate could have been affected by the close by rotation of the flow.

At a high velocity of air (from 10.8 to 8.05 m/s) the maximum content of the material in the flow was obtained at the height corresponding to the position of the third gate. A decrease of the velocity decreases the height at which the maximum of the examined dependence is reached. For example, at a velocity of the air flow of 6.2 m/s this maximum is displaced to the second gate. Consequently, there is some relationship between the velocity of air and the height of establishment of the maximum content of the material in the flow. However, the nature of this dependence could not be determined in our experiments because the height changed in a discrete manner over a wide range, and this was not the subject of the investigation.

Of special interest is the distribution of the material in the fractions in each gate, expressed in per cent, for different rates of the air flow (Table VI-1).

Comparison of the data in Table VI-1 with the fraction characteristic of the initial material shows the mechanism of separation in the rising

Table VI-1 Fraction distribution of material in each gate

Flow velocity, m/s	Number of gates	Size classes, mm 2,5-1,6	1,6--1,0	1,0--0,85	0,85--0.63	0,63--0,40	0,40--0,315	Bottom residues, mm
	1	25.9	47.45	11.0	5.62	1.9	0.5	7.39
	2	38.6	50.4	10.2	5.2	1.6	0.4	0.75
	3	31.3	50.9	9.3	4.4	1.5	0.4	1.31
10.8	4	29.7	52.1	10.8	4.5	1.6	0.3	0.82
	5	28.6	52.7	11.1	4.9	1.4	0.4	0.88
	6	27.9	50.7	11.8	5.4	1.8	0.3	1.20
	7	26.0	52.0	12.0	6.0	2.0	0.5	1.14
	1	20.3	49.8	15.0	7.54	2.71	0.54	3.98
	2	27.9	51.57	11.5	5.6	1.91	0.5	0.85
	3	23.9	55.03	12.55	5.5	1.6	0.4	1.0
8.85	4	21.7	54.8	13.1	7.1	1.9	0.5	0.97
	5	21.0	56.0	13.0	6.7	1.9	0.41	1.15
	6	19.6	54.9	15.3	6.9	2.0	0.44	1.1
	7	19.0	55.3	16.0	8.2	2.5	0.43	1.05
	1	14.85	51.5	16.95	9.6	3.5	1.2	2.62
	2	30.0	47.2	12.8	6.8	2.24	0.4	0.76
	3	24.9	51.5	14.2	6.4	2.2	0.41	0.8
8.05	4	19.1	56.5	13.3	7.32	2.24	0.59	1.08
	5	13.8	57.4	15.72	8.65	2.74	0.7	1.41
	6	13.0	57.2	17.15	8.15	2.9	0.61	1.05
	7	11.15	55.0	17.1	10.5	4.23	0.82	1.41
	1	1.42	45.5	20.4	13.9	6.6	1.7	10.0
	2	0.8	43.96	24.8	17.0	6.97	1.73	5.4
	3	0.8	32.2	27.25	21.45	10.0	2.6	5.68
6.2	4	0.8	32.8	27.2	21.4	9.85	2.33	5.32
	5	0.9	29.95	28.15	22.8	9.73	2.9	5.94
	6	0.0	26.4	26.5	22.6	11.4	2.7	9.73
	7	1.0	30.2	28.5	23.8	10.9	2.7	3.77

air flow.

In all cases, the fraction characteristics of the material, starting from the second gate, change only slightly, although each of them contains a different amount of material.

A large fraction difference in all experiments was typical only of the material on the first gate, and a smaller amount on the second gate.

The process is initially intensive and reaches a specific effect at a small height of apparatus (6–8 gauges) and subsequently its intensity rapidly decreases.

The result indicates that the process of gravitational separation starts from the area of introduction of the material into the flow and has a distinctive exponential form. On the basis of the experiment it is possible to draw very important conclusions on the mechanism of the process.

Firstly, the process of separation is almost completed at a limited height of the hollow body apparatus. A simple increase of the height of apparatus has only a slight effect on the classification results. To increase the efficiency of the process it is necessary to take special measures.

Secondly, the region of the most intensive variation of the fraction composition coincides with the region of transition (non-steady) regime.

All these results indicate that the generally accepted concept of the effect of height on the results of separation has not been confirmed.

An important generally accepted fact of high efficiency organisation of the process is the all-out laminarisation of the separation conditions; to reach this, a large number of measures were taken, such as transverse grids, guiding apparatus, division of the flow into elementary jets, etc. To verify the accuracy of this assumption, it would be necessary to formulate experiments in which it would be possible to compare the results of separation in laminar flows and the flows with artificial turbulence at the corresponding velocities of the medium. These investigations were carried out with the simplest hydraulic classifier.

A stable laminar flow can be obtained in a vertical pipe of a small diameter when supplying water from the bottom through a special device-vortex. The experiments were carried out in equipment shown

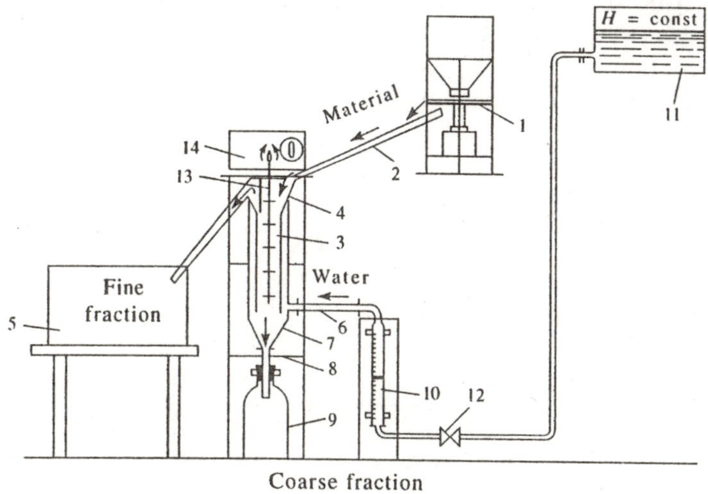

VI-3 Diagram of experimental equipment for hydraulic classification: 1) tray feeder; 2) trough; 3) classifier; 4) loading funnel; 5) settling tank; 6) feeding pipe; 7) conical collector; 8) pipe; 9) container for the coarse fraction; 10) flow-rate meter; 11) constant pressure vat; 12) regulating valve; 13) rod; 14) vibrator.

in Fig. VI-3. The classifier is a pipe, open from two sides, diameter 40 mm, length 350 mm. The vibration device consists of a universal shaking machine and a metallic bar freely suspended to an eccentric shaft carrying out vibration movements. The angle of rotation of the rolling shaft is 20°. Wire rods with a diameter of 3 mm were brazed to the bar with a diameter of 6 mm along the entire length, in 20 mm intervals. The shaking machine was placed in the upper part of equipment in such a manner that the rod was situated inside the apparatus. The frequency of vibrations of the vibrating device was set through potentiometer for regulating the rate in the range 50–500 rpm.

The first series of experiments was carried out with a non-working vibrator in the laminar regime of movement of the medium and processed using the proposed procedure. When the vibrator was switched and water supplied, artificial turbulisation of the movement of the flow took place. Two series of experiments were carried out with artificial turbulisation of the flow in which the frequency of the vibration of the vibrating device was $\omega_1 = 6.2$ 1/s and $\omega_2 = 5.33$ 1/s. The amplitude of vibrations in both series was the same, 12 mm. With the variation of the velocity of the flow in a wide range it was possible to determine the optimum attainable efficiency of classification for different values of the boundary sizes. The results of these experiments as shown in Fig. VI-4. Comparison of these curves shows that the transition from the laminar regimes of separation to the vibration regimes also increases the effect of separation in relation to the entire range of the boundary sizes. This means that the transition to the non-steady regimes of movement of the medium results in an in-

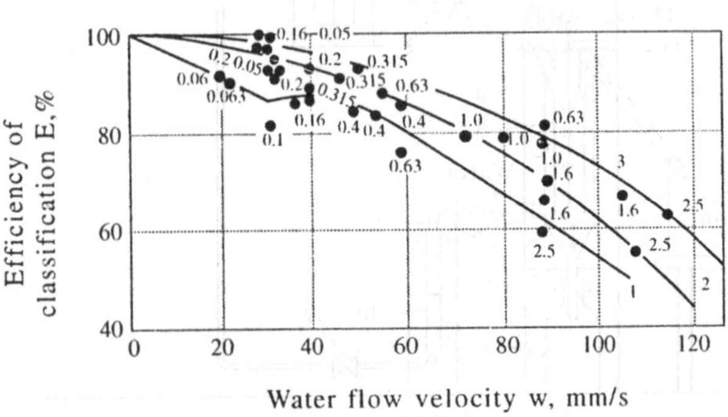

Fig.VI-4. Dependence of the optimum efficiency of the hydraulic classifier on the velocity of the rising flow of water and the boundary separation size: 1) $\omega = 0$; 2) $\omega = 372$ 1/s; 3) $\omega = 500$ 1/s (the numbers on the curves indicate the boundary separation size, in mm).

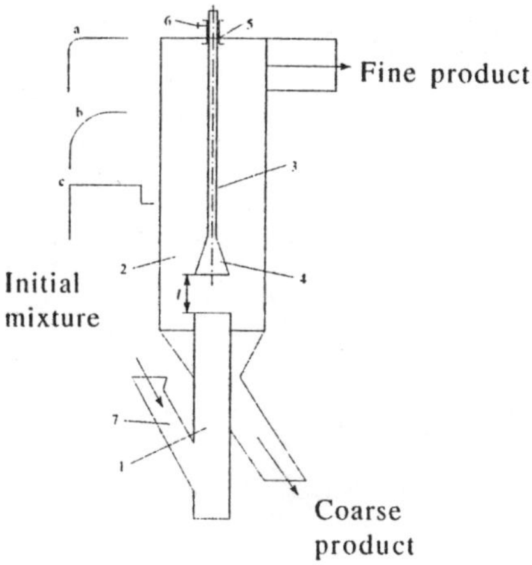

Fig.VI-5. Diagram of equipment for examining the effect of deceleration and rotation of the flow on the efficiency of air classification.

crease in the separation effect.

A perturbation of the medium may be achieved by different methods: deceleration, changes of the direction of the flow, application of vibrations, etc.

The effect of deceleration of the flow for the straight-flow classification was carried out in equipment shown in Fig. VI-5. This equipment consists of vertical pipe (1), entering the rectangular chamber (2) from the bottom. On the axis of equipment on a special rod there is the reflecting cone (4) whose distance from the outlet of the pipe (1) can be smoothly changed. This is achieved by moving the rod (3) in a sealing device situated on the upper lid of apparatus. The position of the rod is fixed by the bolt (6). The diameter of the base of the cone corresponds to the diameter of the orifice in the pipe.

During operation, the chamber is under a rarefaction generated by a special fan. The material is supplied through a nozzle into the pipe (1), trapped by the air flow and carried upwards. When entering into the shaft, the particles are divided on the basis of their size as a result of the changes in the cross section of the flow. Experiments were carried out with crushed quartz. The efficiency of classification in these experiments was determined only in relation to the shaft of the straight flow apparatus. Six series of experiments were carried out with different distances of the reflecting cone from the edge of the pipe. These distances in the experiments were assumed to be equal to 27, 55, 110, 165 and 220 mm, respectively. To compare the results, one series of experiments was carried out with a hollow shaft from which the reflecting cone was removed. In each experimental series, the air flow velocity was changed over a wide range, so it

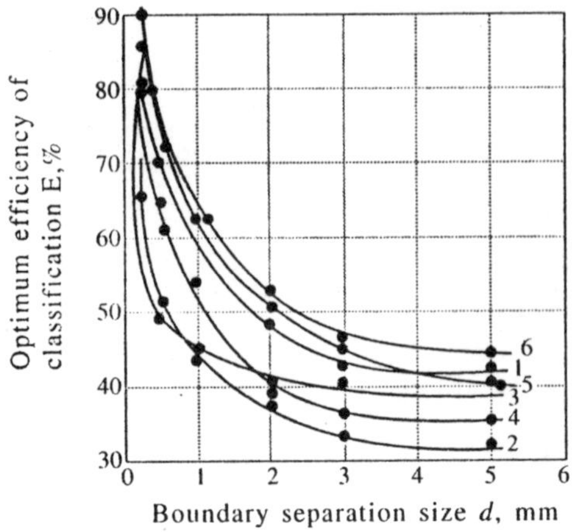

Fig. VI-6. The effect of the position of the cone (distance l) on the efficiency of classification. 1) without insert; 2) l = 27 mm; 3) l = 55 mm; 4) l = 110 mm; 5) l = 165 mm; 6) l = 220 mm.

was possible to determine by experiment the optimum conditions for each boundary size.

The optimum obtainable efficiency of classification for all boundary separation sizes in relation to the position of the reflecting insert shown in Fig.VI-6.

The small distance between the cone and the edge of the pipe, corresponding to 27, 55 and 110 mm, results in a decrease of the efficiency in comparison with the hollow shaft. Already in these experiments, the increase of the distance resulted in an increase of the efficiency of classification. An increase of this distance to 165 and 220 mm resulted in a large increase of the efficiency for all boundary separation sizes in comparison with the hollow channel, and at l = 220 mm, the efficiency slightly decreased. Thus, the results in these two experiments confirm the assumption according to which the perturbation may improve the quality of the process.

However, not every perturbation of the flow results in this effect, only those organised in a special manner.

This is also indicated by the results of experiments with the determination of the conditions of rotation of the flow on the efficiency of classification. All these investigations are of great importance for the efficient organisation of the process. Analysis of the results obtained in previous experiments shows that the efficiently organised rotation of the flow should create suitable conditions for the separation of the material. To explain this problem, a series of experiments in

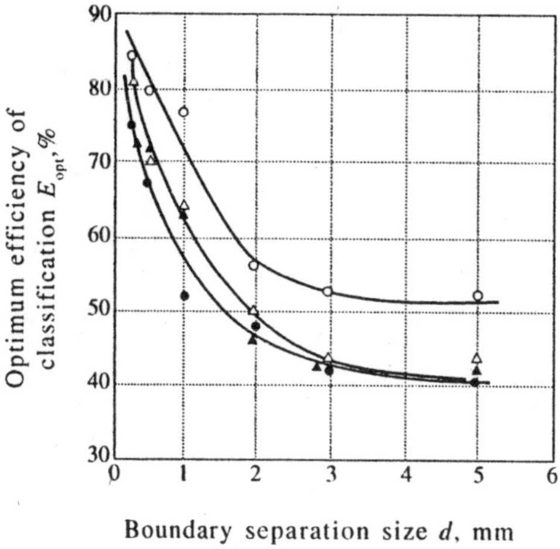

Fig. VI-7 Dependence of the optimum efficiency of classification on the method of rotation of the flow: o) large rotation radius (b); Δ) rotation $\alpha = 90°$ (c); •) small radius of rotation (a); ▲) rotation with a trap (c).

which the conditions of flow rotation were varied, was carried out.

For these experiments, the upper lid of the apparatus and the reflecting cone were removed and replaced with a lid with different inserts forming a rotation with a trap, at an angle of 90° (Fig. VI-5) with small and large curvature radii (Fig. VI-5a, b). The values of the optimum efficiency of classification for different boundary separation sizes, obtained in experiments, are shown in Fig. VI-7. The experiments show that the largest effect is ensured by the smooth rotation of the flow with a large radius. In all other cases, the efficiency of the process decreases, evidently as a result of the mixing phenomenon, caused by the rapid rotation of the flow resulting in an increase of the degree of contamination of the coarse product with fine particles.

Thus, rapid rotation is efficient in the process of separation in which it is required to obtain a homogeneous product in the yield of the fine product. This measure should be taken in, for example, separators of shaft mills in electric power stations where grinding and separation are carried out to produce a narrow coal fraction.

Smooth rotation of the flow, included, for example, in the design of Zigzag air classifiers and hydraulic classifiers with a wave-shaped form of the chamber, greatly increases the separation effect.

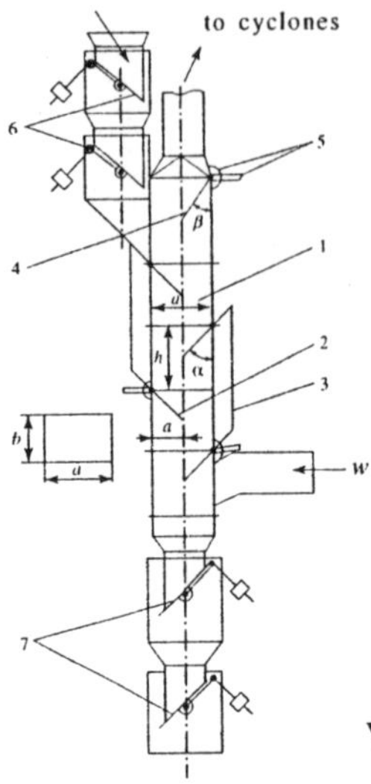

VI-8 Air classifier with inclined shelves.

2. CASCADE PRINCIPLE OF ORGANISATION OF SEPARATION

Since both deceleration and rotation of the flow have a beneficial effect on the classification results, it was obviously interesting to combine the effect of these two factors in a single apparatus.

In development of this apparatus it was taken into account that it is necessary to remove periodically the material from the wall of the channel of the working chamber into the centre of the flow.

The apparatus in which the displaced material is capable of being removed constantly from the walls without using any complicated mechanical devices, is impossible to construct. Therefore, it was decided to restrict examination to a device in which the effect of removal of the material from the periphery of the flow is periodic and multiple. This also solved the problem of the periodic perturbation of the two phase flow.

The simplest device of this type is a vertical hollow pipe in which inclined devices are distributed in the staggered order. The apparatus is a vertical rectangular section chamber containing inclined shelves. For the first group of the experiments, the spacing of the shelves was equal to the side of the cross section (Fig. VI-8).

The apparatus was a vertical shaft (1) with inclined shelves (2). The initial material travels into the shaft through the receiving bunker with gate (6) and is blown by an air flow from the bottom. The coarse product is unloaded through the gate (7). The classification proc-

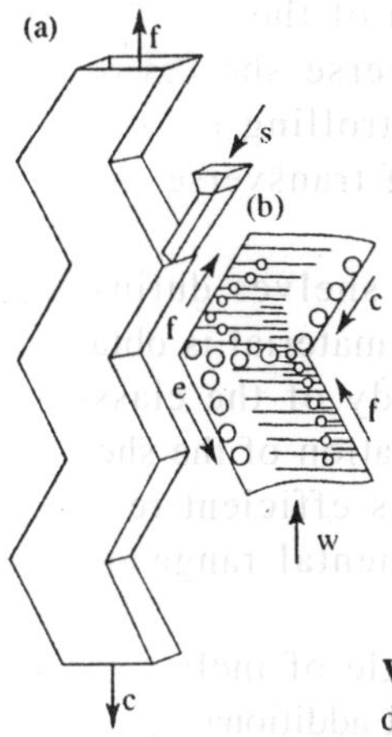

VI-9 Cascade classification of the the zigzag type. a) general diagram; b) displacement of different classes.

ess is regulated by changing the flow rate of air with a throttling valve (4) connected to a handle (5), and also by the position of the shelves which are connected in pairs with connecting rods (3) for changing their angle of inclination.

In Germany, a similar design of classifier was developed (Fig. VI-9) which is also referred to widely as a zigzag classifier.

Because it is important to carry out comparative tests of a cascade separator and a hollow shaft, and also determine the effect of the angle of inclination on the nature of the process, the shelves are fixed on rotating axes. Consequently, the angle of inclination of the shelves in relation to the vertical α can be changed from 0 to 90°. The apparatus is made of sections. By selecting the appropriate number of the sections it is possible to change its height and also the position of the area of introduction of the material into the apparatus.

To determine the effect of the position of the shelves on the results of the classification, experiments were carried out where the angle of inclination was varied over a relatively wide range.

To determine the general relationship of the process, these investigations were repeated with different materials greatly differing in density (gypsum rubble 2350 kg/m^3, magnetic iron ore 4490 kg/m^3, colophony 1070 kg/m^3).

Four series experiments were carried out using gypsum rubble; in these experiments, the angle of inclination of transverse shelves was 0; 22.5; 45; 67.5°

For each boundary size, the variation of efficiency was examined as a function of the velocity of the air flow through the classifier.

The air flow rate for each experiment was determined in relation to the efficient section of the shaft of the classifier.

The efficient section is the horizontal section of the shaft of the classifier from the non-fixed ends of the transverse shelves to the opposite wall. This section was selected as controlling because its size unambiguously characterises the position of transverse shelves of fixed length.

At α = 67.5°, the material hanged from the shelves during the experiments. In this case, uniform descent of the material is obtained as a result of light rhythmic tapping on the body of the classifier throughout the experiment. At this angle of inclination of the shelves, the tests were carried out, regardless of the less efficient removal of the material, in order to expand the experimental range with a given diameter.

Since the service of the classifier with an angle of inclination of the shelves of α = 67.5° is almost impossible without additional measures, and in this case the efficiency of separation was reduced, the results of these experiments were not processed subsequently, and in experiments with other materials, this position of the shelves was not used.

The results of this group of experiments are presented in Table VI-2.

Table VI-2 shows that in the transfer of the shelves from position α = 0° to α = 22.5°, the optimum efficiency decreases for almost all values of the boundary size. When transferring the shelves to the position α = 45°, the optimum efficiency of separation is the highest.

All the experimental data, obtained in classification of different materials, confirm the assumption according to which the perturbation of the flow, organised in an appropriate manner, improves the ef-

Table VI-2 Optimum value of efficiency of classification (%) for different materials in relation to the angle of inclination of inclined shelves

Angle of inclination of transfer shelves, deg	Boundary separation size, mm							Material
	7	5	3	2	1	0.5	0.25	
0	54.26	63	76.5	86.36	94	93.9	97.1	Gypsum rubble
22.5	39	49	64	79	89	91.3	94.2	
45	54.9	66.2	76.3	87	90.6	94	98	
67.5	48	61	70	80	90.01	91.3	96.1	
0	46	56.4	72.5	82.9	93.36	94.9	98.75	Magnetic iron ore
22.5	29	40	59	70	86	93	9798	
45	50.7	57.9	72.5	88	93.3	97		

Fig.VI-10. Nature of the flow in an air cascade classifier with inclined transverse shelves.

ficiency of separation for all size classes. In particular, it should be mentioned that not every position of the shelves in the flow makes it possible to obtain this effect ($\alpha = 22.5°$). Experiments were carried out to determine the position of transverse devices resulting in intensification and efficient organisation of the process ($\alpha = 45°$). In this case, it is always possible to obtain the best separation for different materials. In this case, for this position of the shelves, the process of separation is completely different in comparison with the hollow charge. Special films were taken when examining the special features of the mechanism of this transfer.

Figure VI-10 shows a photograph indicating the nature of distribution of the two-phase flow during its upward movement through a cascade classifier. The photograph shows that in this apparatus the flow of the medium is not an integral unit but breaks up into individual components, distinctive vortices, characterised by mutual directed mass transfer ensuring a high efficiency of separation.

Thus, the air separator with a rising flow was used for the development of a multi-step classifier with circulation zones in which there is directional exchange of the particles.

In a conventional classifier with a rising flow, separation is a purely equilibrium process, i.e. for particles of the boundary size it is necessary to select velocity of the medium which balance their force of gravity. The particles whose size is below the separation boundary are carried upwards and the larger particles fall. In fact, this separation is complicated by the superimposition of a larger number of stochastic factors on the examined process. Nevertheless, the nature of these phenomena remains unchanged in principle.

In a cascade classifier with transverse shelves, the material moves in a different manner. Inside each step there is a stable vortex with a horizontal axis. Almost all solid materials and a small part of the flow of the medium take part in this vortex movement. A large part

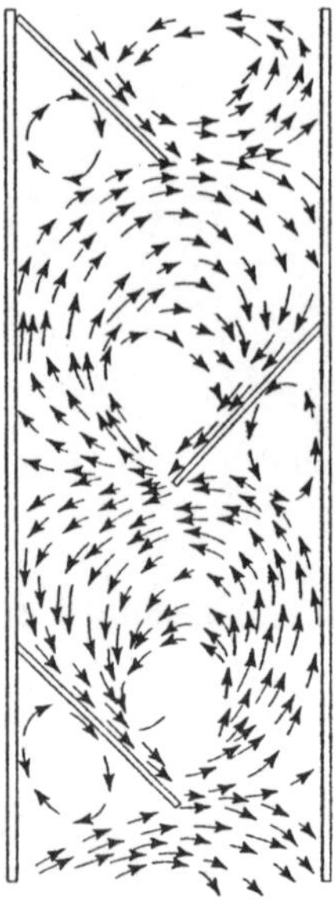

FigVI-11. Movement of flows in a cascade classifier.

of air takes part in the zigzag rising movement. The single act of classification takes place as follows (Fig. VI-11). Falling from the shelf, the hard material is deflected in the direction from the opposite wall and intersects the flow in the transverse direction. This is accompanied by the redistribution of particles in such a manner that some of them, enriched with the fine product, travel upwards, and the others descend.

This process takes place along the entire distance from the shelf to the wall and ends at the wall by the separation of the material into two flows. One of them travels upwards and again intersects the flow of the medium, leaving from below the upper shelve, and the other one is closed on the underlying shelf and intersects the flow falling from the shelf.

The distribution of the particles, achieved in the transverse flow, is greatly intensified by the effect of distribution of the material into rising and falling vortices in reflection of the flow from the wall. Consequently, the cascade classifier is not an apparatus operating on the basis of the equilibrium principle, but it is a multi-step separator in which the general flow is divided into individual zones in which the fine and coarse particles move as the counterflow, and each step is characterised by the directional exchange of the particles.

Single acts of separation do not lead to any distinctive division of the material, because the nature of movement of the particles depends on many random factors in the reflection, the intersection

Rectangular section Circular section

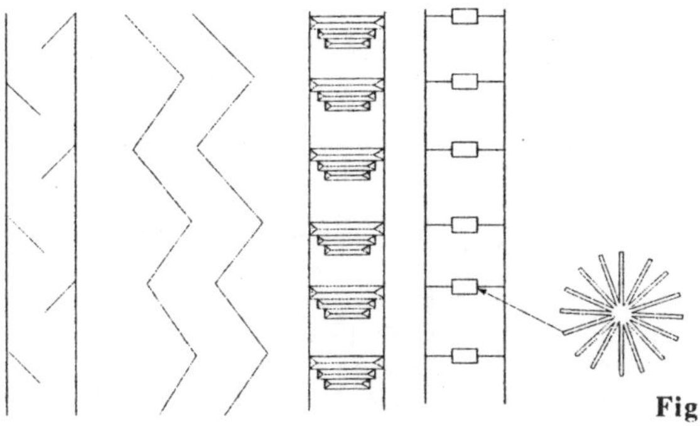

Fig.VI-12. Cascade classifiers.

of the flow, vortex movement, etc.

The process of separation in the cascade classifier is characterised by the fact that each particle of the solid material may move several times upwards or downwards, passing from one zone into another. This shows that the possible deviation in the movement of the particle from the regular direction may be corrected by an increase in the number of stages. Therefore, the efficiency of the process should increase with an increase in the number of separation stages. The results of these investigations indicate that pneumatic separation even in the case of the larger boundary size in upper apparatus of limited height is capable of ensuring efficiently high efficiency of classification. These experimental facts are not included in the framework of the generally accepted theoretical considerations, providing for the necessity for strict homogenising of the separation conditions in order to ensure a high quality process. Improvement of the course of the process as a result of the organised perturbation of the flow is not a partial case characteristic only of cascade classifier, but is a general relationship for the entire class of the gravitational separation processes. Naturally, this apparatus is not the only one. The cascade scheme of organisation of separation may be organised also with other inserts (Fig. VI-12).

As shown by the investigations, the cascade classifiers are characterised by high separating capacity in comparison with other types of apparatus.

The Japanese investigator J. Ueda treats the development of cascade classifiers as the most significant achievement of the technology of fractionation in recent years.

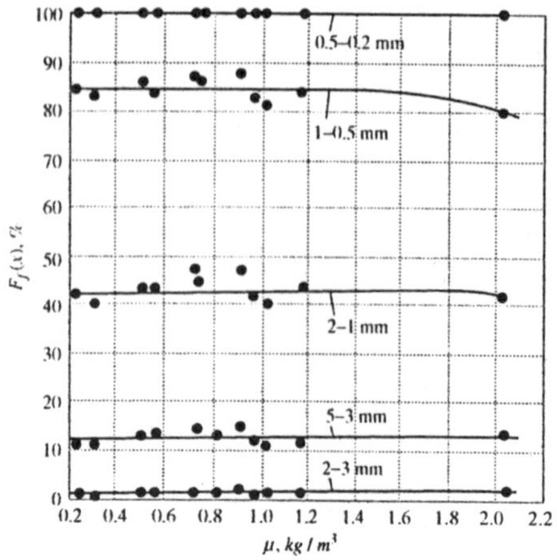

Fig. VI-13. Dependence of the fractional extraction of different narrow size classes on the concentration of the material in the cascade classifier (z = 7; i^* = 4; w = 6.2 m/s; material – ground quartzite, ρ = 2650 kg/m³).

3. EFFECT OF THE CONCENTRATION OF THE SOLID PHASE

The concentration of the solid phase in the flow is a parameter controlling the main relationships of the process, because the productivity of the classifying devices is unambiguously associated with the dimensions of apparatus and the concentration of the bulk material in the flow.

Extensive experimental investigations were carried out to explain the effect of concentration on the results of separation using air classifiers of different design and different materials. In all cases, the results were qualitatively identical. They are illustrated by the dependence shown, for example, in Fig. VI-13.

The dependences characterising the relationship of the degree of fraction separation of each narrow size class and the concentration of the material in the flow for different air classifiers are identical. A characteristic feature of the experimental determined relationships is the presence of a section parallel to the concentration axis. Within the limits of this section, characterised by different boundaries in relation to the design of apparatus, the value of the achieved efficiency and the fractional extraction are almost independent of the concentration of the material in the flow.

When moving outside the limits of the section towards lower concentrations, the efficiency of separation slightly increases. With an increase of concentration outside the limits of this section the efficiency of classification decreases. The concentration corresponding to the first section of the detected dependence is not high and is no interest for practice. In addition to this, the classification within

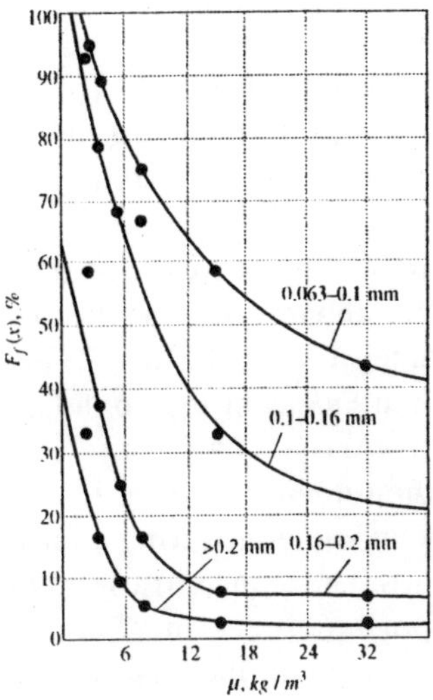

Fig.VI-14. Dependence $F_f(x) = f(\mu)$ for casting sand in a cascade classifier (z = 7; i^* = 4; w = 3.83 m/s).

the limits of the concentration smaller than 0.1 kg/m³, does not ensure stability of the process, and its smallest change results in sharp jumps in the quality of the separation products. Evidently, the second section of the examined dependences is of special interest. Within the limits of this range, separation is stable and its results are independent of the variation of the concentration of the solid phase. Therefore, it may be concluded that within the limits of this range, the process of gravitational classification is self-similar in relation to the concentration, or that this parameter (concentration) has been degenerated.

In particular, it should be mentioned that the self-similarity of the concentration completely coincides with the section of variation of this parameter in which the separation curves are invariant in relation to composition.

Usually, the experimental investigations of gravitational classification are restricted by the concentration limit of 2 kg/m³. It was interesting to determine the effect of this parameter outside the limits of this boundary. For this purpose, special investigations were carried out with a tray cascade apparatus with 7 cleaning stages (z = 7) with different organisation of the introduction of the material into the classifier (i = 4.2). The consumption concentration in these experiments was varied from 2.5 to 36 kg/m³. The dependence of the fractional extraction into the fine product on the content of the solid phase in the flow for z = 7; i = 4 is shown in Fig. VI-14. This graph shows that with increasing concentration, the degree of fraction separation decreases. In this case, it should be mentioned that this decrease is monotonic, but not linear. For each class, there is a specific concentration, and if this concentration is exceeded, this has only a slight effect on the change of this parameter. It should be mentioned that all these experiments were carried out at a constant rate of the air flow.

Consequently, it may be concluded that in the absence of strict requirements of the quality of powders, the separation may be organised at higher concentrations so that the dimensions of the classifying device may be greatly reduced. To maintain the values of the boundary size, it is necessary to increase the flow velocity.

Explanation of the mechanism of the phenomena, leading to the relationships of this type, should be found in the controlling factor of the process which depends on the variation of concentration. In this case, it was shown that the most important is the mechanical interaction of the particle in the flow. Evidently, an increase in the frequency of interaction of the particles of different classes has a negative effect on the separation results, because this results in the penetration of the fine product into the yield of the coarse material and of the coarse product into the yield of the fine one.

Thus, it should be accepted that the working range of the concentration for the effective separation of the powders are the values of this parameter, corresponding to the self-similar region of its variation. Within the limits of this range, the effect of concentration is insignificant.

As regards the increase of the productivity of the classifiers, results of tests of apparatus with attachments are of great interest. Attempts have been made to construct apparatus in such a manner that the internal elements of the apparatus are distributed almost continuously over the entire volume of the apparatus or some part of the apparatus.

The investigations were carried out on an apparatus with a circular cross section (diameter 100mm), consisting of 9 conventional sections (H = 900mm). The materials were introduced into the third section (from the top). The material for classification was quartzite. Two types of attachment were investigated: a garland chain attachment (made of staples) and an attachment of inflated rubber balls with a diameter of 15–30mm. In the first case, the degree of filling of the apparatus (sections) was $\varphi \approx 5\%$, in the case of the spherical attachment it was $\varphi \approx 25\%$.

Five series of experiments were conducted.

Series I – a chain attachment, consisting of n = 1000 staples, suspected in the first two sections (i = 1, 2).

Series II – a chain attachment, including n = 4000 staples, suspended

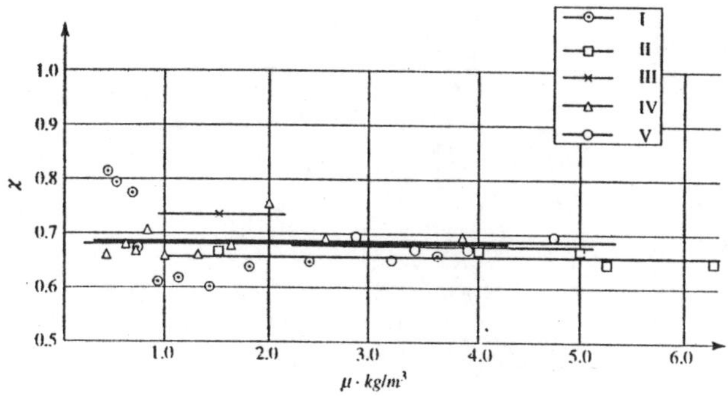

Fig. VI-15 Dependence of the Eder–Mayer criterion on the consumption concentration of the material.

throughout the entire volume of the apparatus and excluding the section for introduction of the material (i = 1, 2; 4 ÷ 9).

Series III – a free spherical attachment (n = 80–100 spheres) between two wire meshes in the two upper sections (i = 1, 2).

Series IV – semi-free spherical attachment in the two upper sections, secured on a flexible filament to the lower mesh. The upper mesh was not used (i = 1, 2).

Series V – a semi-free spherical attachment (i = 1, 2).

Experiments were carried out at different air velocities in apparatus in a wide range of the consumption concentration of the material. The experimental results were used to determine the fraction extraction into the fine products $Ff(x)$ of all narrow size classes of the particles and separation curves were plotted. These curves were evaluated using the Eder–Mayer criterion $\chi = {}^{75}\!/_{25}$. The effect of the consumption concentration of the material on the quality of separation is represented by the graphical dependence (Fig. VI-15) which shows that in the examined range to μ = 6 kg/m³ the consumption concentration of the material has no effect on the process.

4. PHENOMENON OF EQUIVALENCE IN THE PARTIAL SEPARATION OF THE SOLID PHASE BY TURBULENT FLOWS

The criteria determined in the previous chapters were used for describing the most general relationships of the process in the experiments. They proved to be valid for apparatus of almost any configuration and height in the separation of greatly varying natural powder materials.

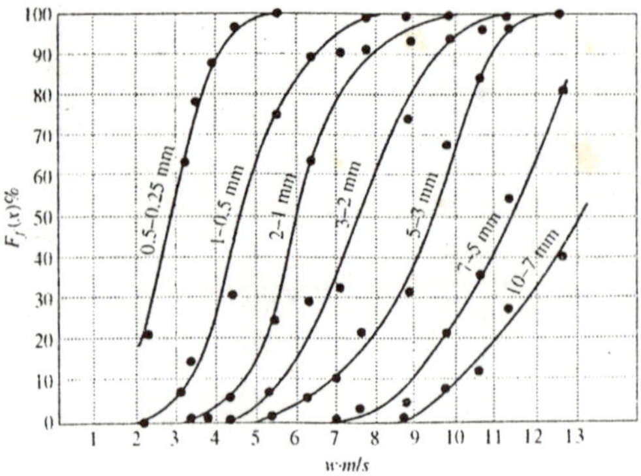

Fig.VI-16. Dependence of the fraction extraction of different size classes on the velocity of the flow (z = 4; i^* = 1).

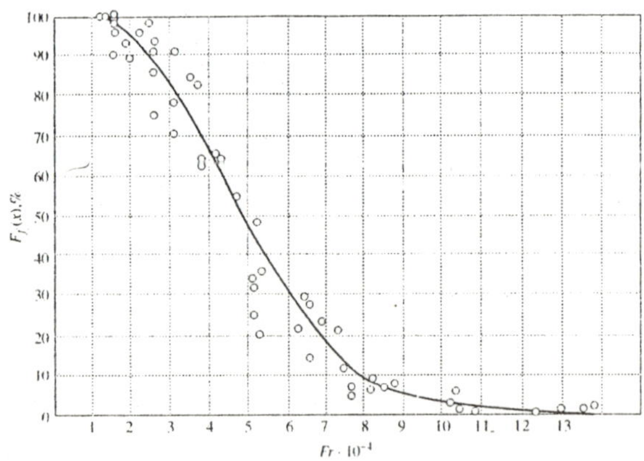

Fig.VI-17. Affinisation of separation curves using Froude criterion (z = 4; i = 1).

These relationships are based on the phenomenon of equivalence in the partial separation of different size classes in relation to the entire set of the regime parameters, discovered by the authors of the present book.

In initial publications, this phenomenon was referred to as the affinity of diffraction separation curves. Its principle may be described most efficiently on a specific example.

The fraction separation curves, obtained for different size classes, for example, in a cascade shelf apparatus at z = 4; i = 1 (Fig. VI-16) merge into a single line, if on the axis we plot the appropriate values of the modified Froude criterion, as shown in Fig. VI-17. The controlling importance of the Froude criterion Fr in this type of process was shown in the two previous chapters.

It should be mentioned that identical results were obtained in experimental examination of more than 150 types of air gravitational classifiers with real industrial powders representing polyfraction mixtures with a wide range of the size with the separation boundary from 50 μm to 10 mm. This range of the boundaries is characterised by the developed turbulence regimes.

It should be noted that this phenomenon was verified and confirmed only for self-similar values of the consumption concentration of solid matter.

The principle of equivalence, detected in the experiments, is that each apparatus divides the powder material entering this apparatus in a single manner on the basis of the same curve, characteristic of this apparatus. The nature of this distribution is not affected by the regime parameters, nor by the grain size characteristics, nor by the concentration of the solid phase within the limits of the working range of the values because at any velocity of the flow, for any boundary distribution size, the powders of any composition are separated in accordance with a single curve.

The determination of statistical substantiation of the general nature of this dependence for the entire class of the processes has made it possible to obtain a large amount of information on the physical fundamentals of the two-phase rising flow in the separation conditions.

In addition to the single nature of separation, mentioned previously, the conditions of equal extraction of different size classes become obvious. This is achieved if the controlling parameter of the process is constant,

$$Fr = \text{const}$$

This fact creates promising conditions for predicting the results of separation and controlling the course of the process.

The resultant universal dependence, being the single and invariant dependence in relation to all previously mentioned physical factors of the process, reflects by the nature of its position in the separation system

$$F_f(x) = f(Fr)$$

only the design of the separation system in which the process was realised. Each apparatus is characterised by its own, single separation curve.

Consequently, it may be concluded that the curve of this type contains the largest amount of information on the separating capacity of the classifying system. Therefore, the curve can be used for unambiguous and objective evaluation of this capacity.

Here, it is possible to use a completely different procedure, compared with that used from the moment of publishing of Hancock's studies (1915) to formulate and solve the problem of optimisation and comparison of the separation systems as a result of introducing a completely new parameter which evaluates unambiguously the design of apparatus from the viewpoint of efficiency of organisation of the separation process in the apparatus.

This also results in the reversed conclusion according to which the comparison of the separating capacity of different classification devices may be verified not only on the basis of the universal dependences but also conventional fraction separation curves. For this purpose, it is not essential to examine the entire family of the separation curves for different narrow size classes. This gives complete information on the separating capacity of compared systems. In this case, it is important to make one principle assumption. The conventional (non-universal) separation curves should be compared in the optimum separation conditions in relation to the given size class.

This phenomenon is used for making a conclusion on the required limiting information on the process and separation equipment as a whole. In principle, this complete and comprehensive information is present in each individual experiment carried out in the apparatus. On the basis of the analysis of the products of separation of any experiment in comparison with the initial composition, it is possible to restore (or obtain) the entire universal curve.

In order to express quantitatlvely the characteristic of this type, it is necessary to find the approximating equation for this affine dependence.

Since the curve was not previously known, this problem is formulated for the first time. As regards the approximation of conventional separation curves, a large number of attempts of such a type have been made from the moment of publishing studies by Dutch engineer Tromp (1935). The most characteristic of these attempts will now be examined. Tromp himself and a large group of his followers have been sticking up to now with the analogy of the separation curves with a curve of the overall law of normal distribution. This analogy is based on the S-shaped form of these curves. The controlling characteristic of the separation curves is represented by one of the parameters of the normal distribution law. For example, Terra evaluates the completion of the separation process by the so-called mean-probability deviation.

Similar characteristics were also proposed by Mayer, Drissen, Grumbrecht, Eder and others. However, it is clear that the curve of fractional separation and the normal distribution have nothing in common in both the physical plan and even in the external appearance, since the separation curves are never symmetrical. For the optimum separation conditions of all narrow size classes in a single system, the controlling parameter is the value of the Froude parameter $Fr_{0.5}$ = const, unique for each apparatus. Consequently, it is possible to select the flow velocity of the medium for any narrow size class, since

$$w_{0.5} = \sqrt{\frac{gd}{Fr}}$$

The resultant universal separation curves and their approximation make it possible to carry out an objective evaluation of the separating capacity of a specific apparatus. A number of preliminary comments will now be made.

In ideal separation (the process is completed) the examined curve transforms into a line normal to the axis of the sizes. In separation of the initial material into parts in any ratio without changes in the fraction composition (zero completion of the process), curve $F_f(x)$ degenerates into a straight line, parallel to the size axis.

Thus, as the universal curve becomes steeper, the separating capacity of the appropriate apparatus increases. From the purely geometric position, the curvature of curves of this type may be characterised by the angle of inclination of the tangent to some point, for example, corresponding to $Fr_{0.5}$, for which ${}_ff(x)$ = 50%. A parameter, differing by a constant from $Fr_{0.5}$ will be introduced, where k is the tangent of the angle of inclination of the tangential line:

$$\psi = kFr_{0.5}$$

This parameter unambiguously evaluates the curvature of the universal curve and, consequently, the separation capacity of the classifier.

Since this parameter was introduced for the first time, it is referred to as the 'criterion of completeness of separation', which reflects most accurately its meaning.

5. RELATIONSHIP BETWEEN THE HOVERING VELOCITY OF PARTICLES OF THE BOUNDARY SIZE AND THE OPTIMUM VELOCITY OF THE FLOW AT CLASSIFICATION

In technical issues, the hovering velocity relates to the controlling parameter of the optimum organisation of separation. Previously, in Chapter III, we showed a principle difference between the hovering

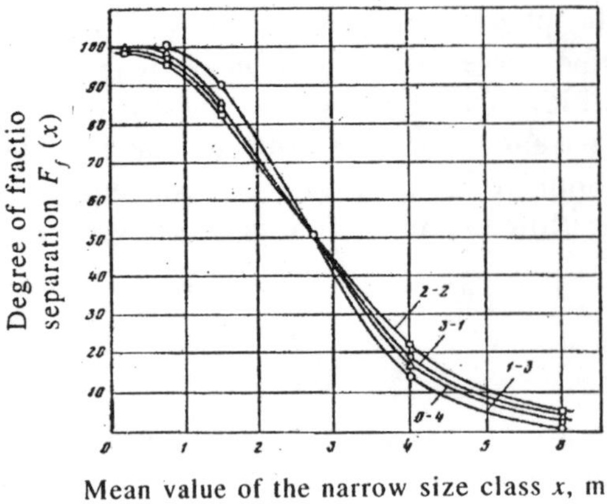

Fig. VI-18. Dependence of the degree of the fractional extraction of different narrow size classes on the ratio of the direct flow and counter-flow classification in the optimum conditions.

Table VI-3 Determining parameters at different levels of material supply into the apparatus

Place of material supply	$Fr_{0.5} \cdot 10^2$
i=1	0.0449
i=2	0.043
i=3	0.027
i=4	0.034

velocity of the particle and its finite velocity of deposition in a stationary medium. Doubts regarding the validity of this claim arose when examining the effect of the area of introduction of the initial material into a cascade classifier.

The experiments were carried out with the supply of crushed quartzite gradually into the first stage ($z = 4$, $i = 1$), the second stage ($z = 4$, $i = 2$), the third stage ($z = 4$, $i = 3$) and the fourth stage ($z = 4$, $i = 4$) with the air flow rate varying over a wide range.

The experimental data was used to plot the universal separation curve, and the optimum conditions were determined for each case. The results of this separation are presented in Table VI-3.

This shows that for different systems there are different values of the optimum rates of separation for the same size class since

$$w_{0,5} = \sqrt{\frac{gd}{Fr_{0,5}}}$$

Table VI-4

Material feeding stage	1	2	3	4	5	6	7	8
Optimum extractability of a narrow class in each stage. K	0.33	0.42	0.455	0.47	0.5	0.51	0.55	0.585

The efficiency of classification also differs here. It is clearly indicated by the graph in Fig. VI-18. This graph summarises separation curves obtained in all examined cases, optimum for a single separation size. This shows that a classifier with initial feed into the central part (i = 3) is the most efficient apparatus.

This shows that in addition to the aerodynamic properties of material, for optimum organisation of separation it is essential to take into account the special design features of apparatus. We carried out special experiments for more detailed examination of this problem.

In a cascade shelf apparatus consisting of ten stages, the same material was classified. The separation conditions remained the same and only the area of supply of the materials to the apparatus was changed.

To obtain optimum separation for a specific narrow size class ($F(x)$ = 50%), it is important to ensure different rates of flow in relation to the area of supply.

Naturally, this resulted in different values of extraction of a fixed narrow class recalculated to a single stage. The results of these experiments are in Table VI-4.

In each specific case, the flow velocity of the medium was different and increased monotonically with increase in the depth of the supply area and approached the equilibrium rate only at the central introduction of material into the apparatus (i = 5). The optimum flow rate varied several times, deviating far from the values of the hovering velocity. Thus, the optimum velocity $w_{0.5}$ is not the hovering velocity of particles of the boundary size in apparatus and it is the velocity ensuring the uniform distribution of the examined class in both outputs in relation to the shape and length of the channel and also the area of supply of material into the apparatus. This means that the hovering velocity or deposition velocity, being at present the main object of examination, does not determine the optimum separation conditions. The optimum rate of separation is determined not only by the properties of solid particles and the flow, but also by the design of apparatus. This must

Table VI-5 Initial grain size composition of the powder in partial residues

No	Powder name	Partial residues on sieves, mm						
		2.5	1.5	1.0	0.75	0.43	0.2	0
1	Polyvinyl chloride	10.1	20.9	28.5	16.3	15.3	7.97	0.93.
2	Potassium salt	13.7	34.1	33.9	5.0	5.3	5.0	3.0
3	Crushed gypsum	4.1	29.5	23.6	10.3	12.9	11.2	8.4
4	Quartzite	7.2	27.8	21.2	10.3	15.78	12.8	4.9
5	Clinker	0.2	19.4	25.5	11.7	14.5	11.9	16.8
6	Magnetic iron ore	7.2	26.0	22.3	11.3	13.8	13.3	5.6
7	Alloy No. 1	0.6	10.1	26.8	18.6	25.4	15.2	3.3
8	Granulated grey iron	5.3	34.4	39.7	10.7	6.8	2.4	0.7
		0.25	0.2	0.15	0.12	0.088		0
9	Alloy No. 2 12.6	41.8	7.7	7.5	29.7		0.7	

always be taken into account, and in a general case, this velocity is not equal to the hovering velocity of particles of the boundary size.

6. NATURE OF THE EFFECT OF THE DENSITY OF SEPARATED MATERIALS ON THE MAIN PROCESS PARAMETERS

The transition to the problems of gravitational enrichment, i.e. separation of the materials on the basis of density, becomes possible on the condition of explaining the effect of this parameter on the fractionation process. Special experiments were carried out for this purpose.

The experimental objects were tray cascade classifiers with $z = 7$; $i = 2$ and $z = 7$; $i = 4$.

To expand the range of variation of the examined parameter, the following materials with a density (kg/m^3) were selected:

Granulated polyvinyl chloride 1070
Potassium salt (granulated) 1980
Crushed gypsum 2270
Ground quartzite 2675
Coarse-ground cement clinker 3170
Magnetic iron ore 4350
Granulated cast iron 7810
Granulated alloy (No.1) 6210
Granulated alloy (No.2) 8650

All these materials, with the exception of potassium salt and cement clinker, are characterised by spherical particles. It may be noted that the density range selected for the experiments basically overlaps the

Table VI-6 Parameter $Fr_{0.5}$ in the classification of materials with different densities

Cascade apparatus type	Material	Main parameters $Fr_{0.2} \times 10^2$
$z = 7; i = 2$	Polyvinyl chloride	0.046
	Potassium salt	0.024
	Crushed gypsum	0.0265
	Quartzite	0.028
	Clinker	0.0245
	Magnetic iron ore	0.0175
	Alloy No. 1	0.0105
	Granulated grey iron	0.0075
	Alloy No. 2	0.0095
$z = 7; i = 4$	Polyvinyl chloride	0.045
	Potassium salt	0.0275
	Crushed gypsum	0.027
	Quartzite	0.0265
	Clinker	0.025
	Magnetic iron ore	0.015
	Alloy No. 1	0.010
	Granulated grey iron	0.008
	Alloy No. 2	0.009

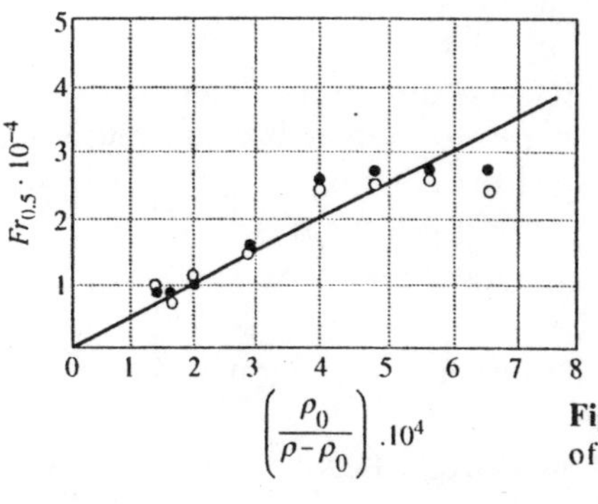

Fig VI-19 Dependence of $Fr_{0.5}$ on the density of separated powders.

range characteristic of enrichment for both ore and non-ore materials.

The grain size composition of these materials is shown in Table VI-5. As indicated by the Table, all the materials, with the exception of alloy No.2, have the same size ranges. The size range for this alloy is shown in the next to last line of the Table.

Each material was used for experiments in a wide range of variation of the velocities in both classifiers.

The results of these investigations were used to plot the dependences of the type

Tab. VI-7 Parameter $Fr_{0.5}$ in separation of mixtures with different densities

Apparatus characteristic	Component name	Defining parameters $Fr_{0.5} \times 10^2$
$z = 7$ $i = 2$	Magnetic iron ore Quartzite	0.0175 0.029
$z = 7$ $i = 4$	Magnetic iron ore Quartzite	0.0155 0.025

$$\lg F(x) = f(Fr)$$

Values of $Fr_{0.5}$, determined from these graphs, are shown in Table VI-6. The Table was used for plotting the dependence $Fr_{0.5} = f(\frac{\rho_0}{\rho - \rho_0})$.

It is shown in Fig. VI-19. Consequently, it may be written that:

$$Fr_{0.5} = 0.51 \frac{\rho_0}{\rho - \rho_0} \quad \text{(VI-1)}$$

For both apparatuses ($z = 7$, $i = 2$ and $i = 4$), the values of the parameters were almost identical. This may be explained by similar positions of introduction of the material.

In addition to these investigations, the same systems were used for separating mixtures of materials with different densities. Mixtures of magnetic iron ore and quartzite were prepared for this purpose. Using a magnet, it was easy to separate the components of the mixture in classification products resulting in reliable analysis.

The results of these determinations are given in Table VI-7.

The following conclusions can be made on the basis of this Table:

Firstly, each material forming the mixture is separated separately, it has its own controlling parameters;

Secondly, when separating the mixtures, the parameter obtained for each of the components is close to the values characteristic of the separation of each component separately in the same system.

This circumstance indicates the independence of separation of each component of the mixture in relation to each other.

All this provides sufficient information for transition to generalised separation parameters. This concerns primarily the universal nature of the separation curves in fractionation on the basis of the size of the particles. This universal nature is achieved, as is well known, by means of the Froude criterion:

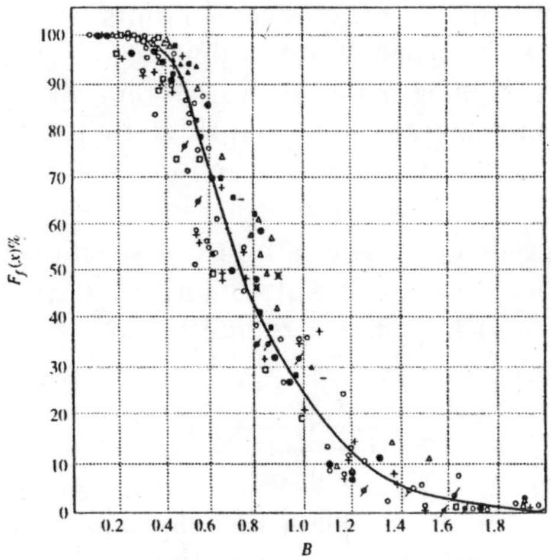

Fig.VI-20. Universal dependence $F_f(x) = f(B)$ at the separation of material of different densities and their mixtures.

$$Fr = \frac{gd}{w^2}$$

It may be mentioned that this parameter was defined in examining different theoretical aspects of the problem, but it is not the pure form of the Froude criterion because it includes the size related to the particle, and the rate related to the flow of the medium. However, externally, it is similar to the given parameter and, consequently, it was given the general term. The point of course is not in the name, but in the fact that this parameter reliably 'operates', as shown in the gravitational separation of the materials with respect to sizes. It was shown possible to unify the separation curve not only taking into account the different size of the particles but also their different density. This parameter was determined:

$$\frac{gd(\rho - \rho_0)}{w^2 \rho_0} = B \qquad \text{(VI-2)}$$

It is no longer similar to the Froude criterion and has the form characteristic for the determination of the hovering velocity of particles.

This parameter was used to process the results of experiments with nine materials of different density (Table VI-5) fractionated both separately and in a mixture. The results of this processing are also identical and correspond to the dependence shown in Fig. VI-20. These dependences show an immutable conclusion on the universal nature of the separation curves in relation to this parameter. The slightly larger scatter of the experimental points may be related to both the difference in the shape of the particles and a small difference of the density for the same material in different size classes.

The values of parameters $B_{0.5}$ (identical to parameter $Fr_{0.5}$), determined on the graph, are identical and equal to

$$B_{0.5} = 0.55$$

The value $B_{0.5}$ corresponds to the conditions of separation into halves of the particles of different size and density in a specific apparatus, i.e. the particles of boundary size and boundary density. It can be written that

$$B_{0.5} = \frac{gd\,(\rho - \rho_0)}{w_{0.5}^2 \rho_0} \tag{VI-3}$$

It should be attempted to explain the physical meaning of this parameter. The flow rate of the medium, ensuring the hovering velocity of a specific particle, is

$$w_{0.5} = \sqrt{\frac{4gd(\rho - \rho_0)}{3\lambda\,\rho_0}} \tag{VI-4}$$

from this, we single out a complex in the right part of equation (VI-3):

$$\frac{gd(\rho - \rho_0)}{w_{0.5}^2 \rho_0} = \frac{3}{4}\lambda \tag{VI-5}$$

Thus, the generalised meaning of parameter B is that it corresponds to the same degree of fractional separation. We will now try to determine physical quantities whose constant values predetermine the same degree of fractional separation.

As is often the case, an answer to this complicated problem may be provided on the basis of relatively elementary considerations.

The general equation of equilibrium of some specific particle in the flow can be written in the form:

$$\frac{\pi d^3}{6}(\rho - \rho_0) = \frac{\pi\, d^2}{4}\lambda \frac{(w - v)^2}{2g}\rho_0 \tag{VI-6}$$

consequently,

$$gd\,\frac{\rho - \rho_0}{\rho_0} = \frac{3}{4}\lambda\ (w - v)^2$$

Both parts of the equation will be divided by the square of the flow rate:

$$\frac{gd}{w^2}\frac{\rho-\rho_0}{\rho_0}=\frac{3}{4}\lambda\left(\frac{w-v}{w}\right)^2$$

In accordance with the latter equation, it may be written that:

$$B=B_0\left(\frac{w-v}{w}\right)^2=B\left(\frac{w_{0.5}}{w}\right)^2 \quad \text{(VI-7)}$$

This expression is a connecting equation between the actual value of B and the value of $B_{0.5}$ fixed for the given conditions.

Consequently, the relationship between the velocity of the flow and the separation result is quite obvious:

a) at $w = w_{0.5}$ $F_m(x) = 50\%$
b) at $w > w_{0.5}$ $F_m(x) < 50\%$
c) at $w < w_{0.5}$ $F_m(x) > 50\%$

The constancy of the degree of fractional separation of different classes of size and density is determined by the constancy of parameter B which is a generalising parameter of gravitational separation.

It should be stressed that all the results are valid for air separation methods when the flow is characterised by the developed turbulence. The transition to low-velocity flows, for example, separation in an aqüeous medium, evidently requires the appropriate correction of this result.

7. FRACTIONATION OF VERY FINE POWDERS

The specific nature of the separation problem of very fine powders is that these powders start to behave in the flow slightly differently than larger ones.

Small particles start experiencing the effect of fine-scale flow vortices.

Separation in this size range is organised at a low velocity of the media flows for which the interaction of the finest particles with the flow might not be characterised by turbulent regimes. Evidently, this imposes large differences on the main relationships of the process in comparison with those detected previously for larger particles. In this respect, they become more similar to separation in viscous liquids where the interaction (flow around particles) with the medium does not take place basically in turbulent conditions.

The fractionation of this type is usually realised in centrifugal fields. However, their shortcomings make it possible to produce a high quality powder product as a result of the fuzzy separation boundary in the

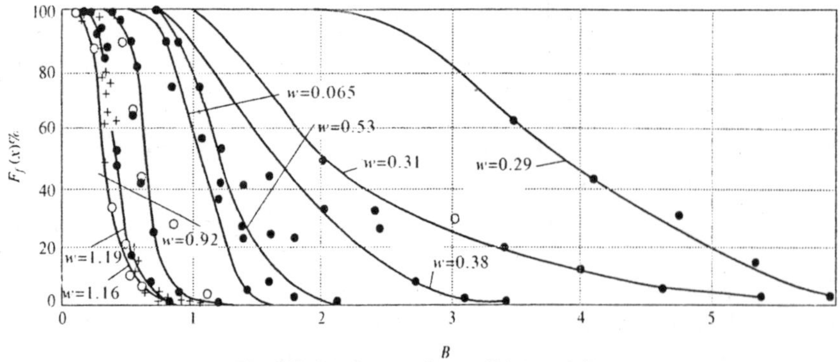

Fig.VI-21. Dependence $F_f(x) = f(B)$.

size range. We studied the possibilities of gravitational separation in the size range in the vicinity of 10 microns.

These investigations were carried out on an aluminium powder used for the preparation of paints. Its specific weight was 2600 kg/m^3.

These experiments were carried out in a tray cascade classifier consisting of nine stages with the central input ($z = 9$, $i = 5$).

Taking into account the fact that in this size range all relationships will be completely different in comparison with those for larger particles, initially we carried out experiments to detect the effect of the concentration of material in the flow. As explained, higher concentrations can be considered in this case. Three series of experiments were carried out at a flow rate of $W = 0.53$ m/s with consumption concentration of the solid phase of 2.75; 6; 14.3 kg/m^3.

They showed that in the given range of variation of the concentration, the effect of concentration is only slight, due to primarily the low flow rates.

Main experiments were carried out within the limits of these values of the concentration at an air flow rate in the range from 1.46 to 0.29 m/s. In this experimental series some of the experiments were repeated up to three times. The results of these experiments, processed in the identical co-ordinates, are shown in Fig.VI-21. This graph indicates that the resultant curves after normal processing do not become affine. Here, it should be mentioned that at higher velocities of 1.46 and 1.19 m/s they almost completely merge, but at lower velocities these curves move away from each other and the rate of this movement increases with a decrease in the velocity of the flow. This is a new, previously not encountered element of the process. Does it mean that for different particles in the given size range the particles have their own relationships, or is there anything in their behaviour?

Tab VI-8 Main parameters of the process of aluminium powder classification in a cascade classifier at z = 9, i = 5

Process parameters				
No.	Flow rate m/s	Boundary grain size $X_{0.5}$mm	$B_{0.5}$	Reynolds number $Re_{0.5}$
1	1.46	0.078	0.35	8.14
2	1.19	0.056	0.41	4.76
3	0.92	0.050	0.65	3.28
4	0.65	0.043	1.1	1.99
5	0.53	0.032	1.23	1.21
6	0.38	0.020	1.6	0.54
7	0.31	0.015	2.0	0.33
8	0.29	0.013	3.9	0.26

The following dependence was plotted on the basis of these experimental data:

$$F_f(x) = f(x)$$

and was used to determine the value of $X_{0.5}$ for each flow rate. The values of $B_{0.5}$ were determined for each velocity. The results of these determinations are summarised in Table VI-8.

On the basis of the data in columns 2 and 3 in the Table calculations were carried out to determine the values of the Reynolds number for the boundary size and the values were placed in the final column. Of greatest interest here is the dependence of $B_{0.5}$ on the Reynolds number Re. It is necessary to answer the question how

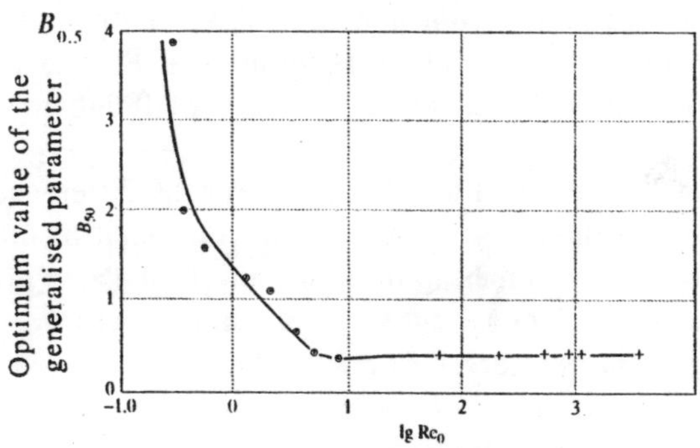

Fig.VI-22 Dependence of the optimum value of the generalised parameter ($B_{0.5}$) of the Reynold's criterion.

these experimentally determined relationships are linked with the relationships detected previously in the separation of coarser powders. For this purpose, the same tray apparatus (z = 9; i = 5) was used for additional experiments with a quartzite powder with a specific density of ρ = 2670 kg/m^3, a particle size from 0.1 to 3 mm, with the air flow rates of 4.7; 5.57; 6.67; 7.3; 7.89 m/s. The results of these experiments were processed using an appropriate method and are plotted in the graph in Fig. VI-22 (experimental points are indicated by crosses).

As shown by the distribution of these points, they form an affine dependence which coincides with the experimental data for fractionation of the aluminium powder at air flow velocities of 1.46 and 1.19 m/s. On the basis of the Table VI-8 and Fig. VI-21 we determine the dependence of parameter $B_{0.5}$ on the Reynolds number Re (Fig. VI-22).

Comparison of the dependences for $B_{0.5}$ and $w_{0.5}$ shows that:

$$B_{0.5} = \frac{3}{4}\lambda$$

i.e. has the meaning of the coefficient of resistance of the boundary size particles. This provides a key to understanding the determined relationships. The dependence of the type $B_{0.5} = f(\text{Re})$ is very similar to that of $\lambda = f(\text{Re})$ type for a single particle.

This dependence at high values of Re has constant values of $B_{0.5}$ which ensures the unambiguous affinization of the separation curves in relation to the Froude criterion or the generalised parameter B. This range corresponds to the turbulent interaction of the particles and the flow. At the transition to the laminar processes, this relationship is disrupted and there is no affinization in relation to these parameters. The transition from one regime to another in the given apparatus takes place at the Reynolds number of Re ≈ 5 which corresponds to a boundary size (at ρ = 2600) of 0.046 mm (46 microns).

Here, one can provide the third definition for the difference between the fine and coarse particles directly from the position of gravitational separation (the two other definitions were formulated previously).

Coarse particles are those whose separation curves are affinated in relation to the parameters Fr or B.

The fine particles are those whose separation curves are not affinated in relation to the given parameters.

It should be mentioned that the value $\text{Re}_{kp} \approx 5$ in the dependence of the type

$$B_{0.5} = f(\text{Re})$$

corresponds to the transition from one separation regime to another for the entire range of the investigated classifiers.

Taking into account the generalised nature of parameter B, the value dkp can be determined for any material. Here it is necessary to examine the need to introduce into practice of examination of the two-phase flow a new parameter of the size of the particles, namely, the hydrodynamic size which takes into account the density of the solid phase and the moving medium. This parameter may be determined from the equation

$$d_q = d\frac{\rho - \rho_0}{\rho_0} \tag{VI-8}$$

It has a linear size and can be used to simplify greatly the previously examined relationships, if this parameter is used everywhere, for example, to write generally recognised criteria in the form

$$Fr = \frac{gd_q}{w^2}; \qquad \text{Re} = \frac{d_q w}{\upsilon} \tag{VI-9}$$

In this interpretation, the dimensionless similarity criteria are generalised more extensively and this must greatly simplify the understanding of the accumulated experimental material. The relationships used as a basis for such a conclusion, were obtained in examining the separation processes in an air flow. However, verification of these

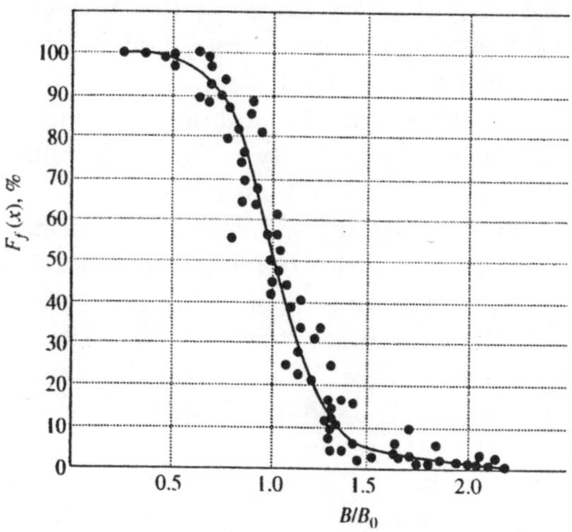

Fig.VI-23 Affinized dependence $F_f(x) = f\left(\frac{B}{B_{0.5}}\right)$.

relationships also for other media, for example, for water, completely confirm their validity.

The results show that for all dependences, shown in Fig. VI-21, there is a general relationship manifested at their affinization. Reducing all curves to a single curve became possible when the following method was applied. The value $B_{0.5}$ was determined for each curve and, subsequently, for each curve the value of the ordinate was multiplied by the value reciprocal to $B_{0.5}$.

The dependences obtained in this manner, are summarised in a single graph and shown in Fig. VI-23. As indicated by the graph, the new dependence is affine (universal).

Thus, in the entire range of the sizes, the dependence of type

$$F_f(x) = f\left(B/B_{0.5} \right),$$

or, which is the same

$$F(x) = (Fr_{0,5}/Fr_{0,5}) \qquad \text{(VI-10)}$$

is affine.

This dependence is more general than

$$F_m(x) = f(F) \qquad \text{(VI-11)}$$

The latter is a partial case of the dependence of the previous relationship, when for the entire range of variation of the sizes

$$Fr_{0.5} = \text{const}$$

Thus, in the entire separation range of the powders we established the general relationship of the process regardless of the separated medium. For air processes at particle sizes greater than 50–60μm, the dependence is greatly simplified and acquires the form identical to (VI-11). The dependence of the type (VI-10) is valued in almost the entire range of hydraulic classification.

Thus, the data, obtained here by a purely empirical method, make it possible to expand greatly the range of considerations on the mechanisms of the process when separating powders with a size from 10 mm to 5–10 μm in air or water media. For finer classes, it is necessary to carry out identical investigations to find general relationships. This is a very complicated task not as much due to the difficulties in determining the particle size as to the need to leave the generally accepted methods of separation for other unusual methods. In this direction, intensive investigations are carried out at present in many countries, and we shall return to them at the end of this book.

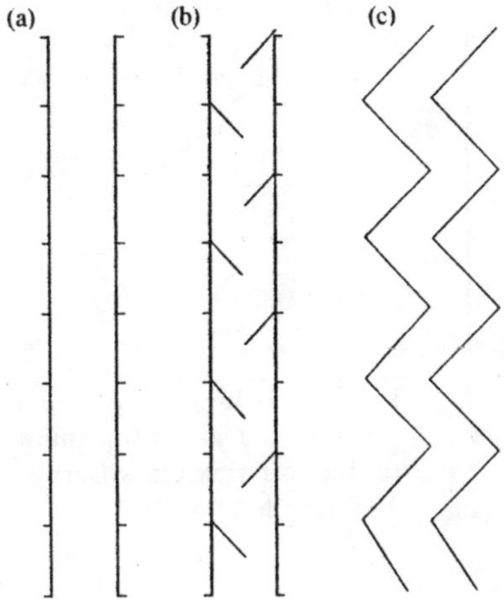

Fig.VI-24 Diagram of air classifiers: a) equilibrium; b) cascade with trays; c) 'zigzag' cascade.

8. RELATIONSHIP OF THE SEPARATION CAPACITY OF APPARATUS WITH ITS HEIGHT

The height of cascade apparatus may be changed by two methods: both by an increase in the number of single-type stages, and as a result of the change of the distance between the stages without changing their number.

It was interesting to compare in the experiments the cascade classifiers, Fig. VI-24, of different height constructed on the basis of the two methods.

To ensure that these experiments were sufficiently reliable, general and objective, comparison of the results was carried out in the same conditions. For comparison it is efficient to use the apparatus realising a new principle of organisation of the process, and supplement in the experiment by an appropriate equilibrium classifier.

The dependence of the type

$$F_f(x) = f(w)$$

was constructed for each group of experiments. It is well known that the optimality condition with respect to the separation curve is:

$$F_f(x) = 50\%$$

If on each of the examined dependences for the examined size class we determine the flow rate, corresponding to these conditions, and determine from the graphs the values of the degree of fractional separation for other classes, the following dependence is obtained for the optimum regime

$$F_f(x) = f(x)$$

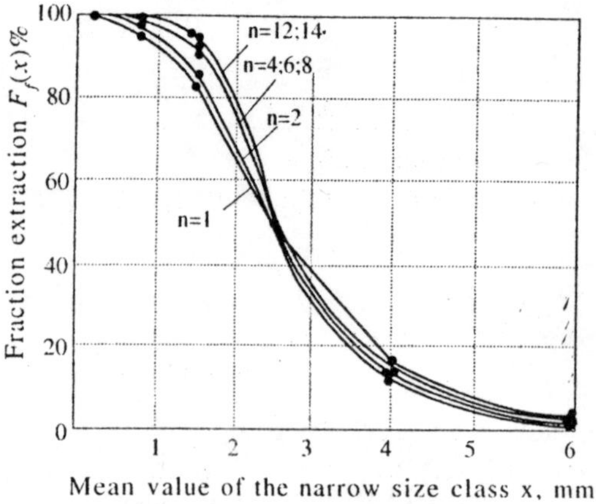

Fig. VI-25 Dependence $F_f(x) = f(x)$ at the optimum regime for tray classifiers of different height.

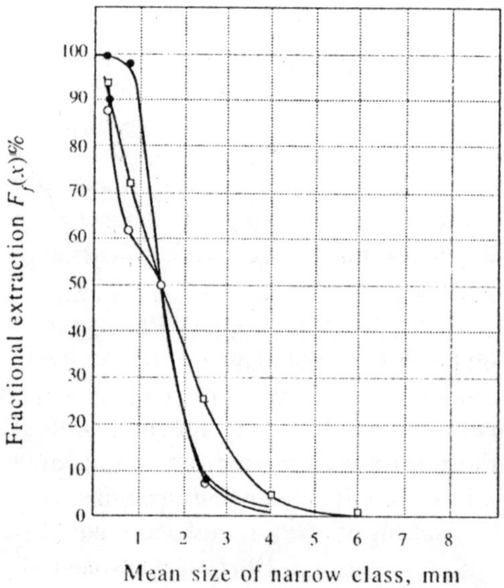

Fig.VI-26. Dependence $F_f(x) = f(x)$ for different apparatus of the same height (n=8) in the optimum conditions.

For each type of apparatus such dependences were reduced to a single graph. As an example, Fig. VI-25 shows such a dependence for a tray classifier. The graph shows a general relationship of the effect of the height of the apparatus on the quality of separation. In all cases, with an increase in the height of apparatus (number of stages) the curvature of the curves increases monotonously, i.e. the separation effect increases. It is interesting to compare the separating capacity of the compared systems (Fig. VI-26). The higher separating capacity is shown by the tray cascade. This is followed by

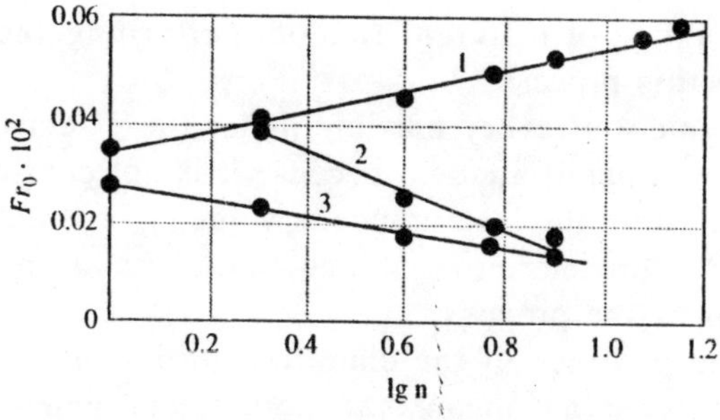

Fig. VI-27. Dependence of parameter $Fr_{0.5}$ on the number of separation stages for different classifiers: 1) trays; 2) 'zigzag'; 3) equilibrium.

the zigzag-type classifier. The efficiency of a hollow apparatus with a right-angled cross section is slightly lower.

It should be noted that some of the curves on this graph coincide. Evidently, this is due to the fact that the distance between them is in their range not exceeding the experiment accuracy. Fig. VI-27 shows the values of $Fr_{0.5}$ in relation to the height of apparatus. It may be shown that the variation of this parameter in relation to the height of different cascade classifiers is described by the relationship

$$Fr_{0.5} = Fr_{0.5(1)} \pm m \ln z$$

where $Fr_{0.5(1)}$ is the value of parameter $Fr_{0.5}$ for an apparatus in a single stage; m is a coefficient which depends on the design of apparatus.

In all cases, with exception of the tray apparatus, an increase in the size of apparatus decreases the values $Fr_{0.5}$, i.e. increases the optimum air flow rate.

The cascade classifier is classified by a different relationship. With an increase in the number of stages, the value of parameter $Fr_{0.5}$ slightly increases, i.e. the optimum air flow rate decreases. This fact confirms that the process of cascade separation is based on the mechanism which differs principally from equilibrium classification and from classification in zigzag-type equipment.

9. LAYER SEPARATION – THE BASE OF THE MECHANISM OF SEPARATION OF PARTICLES IN THE FLOW

It will now be attempted to formulate some general theoretical concept of the process. Separation may be regarded as a mass process in which myriads of particles of different sizes take part simultaneously. A wide range of different random perturbing factors is superimposed on this process.

The displacement of every particle under the effect of the flow and perturbation is purely random because it is not possible to show for the particle either the instantaneous velocity or the direction of the velocity. This predetermines the complete chaotic nature of the general pattern of the process.

However, the presence of the chaotic disorder in the movement of the particles, does not mean that there are no general relationships in the behaviour of a dispersed continuum. On the contrary, the internal rigid function relationships are manifested only through general randomisation, as established, for example, in the development of the kinetic theory of gases.

L. Boltzmann showed that the distribution of the particles of an ideal gas in relation to the height or the level of potential energy is governed by a hypsometric law in accordance with the equation:

$$n = n_0 e^{-\frac{mg(h-h_0)}{KT}} = n_0 e^{-\frac{\Delta U}{KT}} \qquad \text{(VI-12)}$$

where n; n_0 is the concentration of the particles of the ideal gas on the levels h and h_0; m is the mass of the particles of the ideal gas; K is the Boltzmann constant; T is absolute temperature; ΔU is the increase of the potential energy of the particle at its transition from level h_0 to h.

The value of n_0 in the Boltzmann equation expresses the concentration of particles typical of the equilibrium state. We examine the exponent in dependence (VI-12). Here, the numerator expresses the potential energy of the particles situated at a set distance from the equilibrium position, and the denominator gives the value characterising the kinetic energy of the system whose value predetermines the pattern of the given specific distribution.

Temperature T is the variable parameter in the nominator of the exponent in equation (VI-12). This parameter determines the degree of randomisation in the movement of the particles of the ideal gas, leading in the final analysis to a specific statistical distribution of these particles along the height or, which is the same, with respect to velocity (Maxwell's law). As shown previously, its analogue in the examined process is the square of the velocity of the carrying flow. The randomisation effect is predetermined by the turbulisation of the flow, collisions of the particles between themselves and with the walls, and also by the nonuniformity of the fields of velocity and concentration. With an increase of the velocity of the flow, this effect from these factors increases, and this is reflected in the nature of behaviour of the particles.

On the basis of these considerations it is possible to make a conclusion on the presence of an external analogy between the two compared phenomena. We examine the internal relationship between them.

The kinetic theory of the ideal case is based on the presentation according to which the molecules are solid, spherical particles of the same size. The velocity of the molecules and the mean kinetic energy of the system are determined by the external energy effect – the temperature of the medium, and the nature of distribution – by the mechanical interactions of the particles of the gas (collision) with each other.

According to the Boltzmann law, at a constant temperature, the particle concentration increases with a decrease of the potential energy of their position. It is well known that the minimum of the potential energy corresponds to the stable position of the mechanical system. This shows that the particles of the ideal gas are characterised by the maximum concentration in the most stable positions. The most stable position is the one in which all particles would be situated in the absence of the factors disrupting the distribution. Equation VI-12 shows that at $T = 0$ all the particles would be distributed on the level of h_0. At high values of the randomising parameter ($T \to \infty$) the concentration of the gas particles at height is equalised.

Boltzmann showed that this law remains valid not only for a homogeneous field of the gravitational force but also for the distribution of the particles of the ideal gas in any nonuniform force field. This distribution corresponds to many physical phenomenona based on dispersed matter (Van-Hoff law, Pearson law, etc.).

The attributes identical with the previously examined distribution are also characteristic of the process examined here.

In gravitational classification, the solid particles move as a result of the external energy carried by the moving flow. The separation process is constantly affected by the fields of gravitational forces. The results of separation always differ from the ideal situation. This is caused by the presence of different perturbing factors.

All these considerations confirm some analogy of the examined phenomena, but there is also a large difference between these processes, i.e. the instantaneous state of the particles in the flow of the medium is not stable because all particles have a directional velocity at any moment of time. The presence of displacement of the medium in gravitational classification complicates the pattern of the simple hypsometric distribution of particles along the height.

The stable position of a mechanical system corresponds to a state which the system could acquire if the entire set of the stochastic factors would have been excluded. The limiting velocity corresponds to this condition for a two-phase flow. This ideal velocity for every solid particle differs with a different degree of approximation from the value of the velocity determined from the determinate equation:

$$v = w - w_{0.5}$$

The difference will increase with an increase in the strength of the effect of different random factors.

The movement with such a velocity is stable because the resultant of all forces, acting in this case on the particles of a narrow range, is equal to zero. If in the conditions of gravitational classification all particles of different size classes will be capable of acquiring such velocities, the results of such a process could be ideal. However, the actual separation always has a final result which differs from the ideal one.

Evidently, the behaviour of a set of identical particles in the flow is determined by the tendency of each particle to acquire a steady velocity, characteristic of the flow conditions. However, the difference between the real process and the determined process results in some probability distribution of the velocities of these particles in relation to the steady velocity. This is confirmed by the results of experimental measurement for monofractions.

In this distribution, the relationship of the number of particles with the value of the difference in the velocity of their movement in the flow and the steady velocity for the given class has the form identical with the Boltzmann law. This means that as the velocity of the particles increases above the steady velocity, the number of particles having this velocity decreases, and vice versa. This distribution, in contrast to the Boltzmann law, has a positive density of probability to either side of the steady velocity.

For a polyfraction mixture, the form of the velocity distribution of the particles is evidently the same for each narrow size class. But, each class has their own steady velocity. The velocity of all particles of the same class is distributed in relation to this velocity, as a result of the presence of perturbation factors. The distributions of different classes are superimposed on each other.

Regardless of this situation, this analysis shows absolutely clearly the concept of the dominant tendency of the behaviour of the polyfraction mixture in the moving flow. As a result of the tendency of the particles to the steady regime of movement, the tendency of this process is the probability of layer separation of the particles with respect to their steady velocities or size.

If the steady velocity in the flow of the medium for different classes has different direction, the flow ensures that the separation process takes place. If all steady velocities have the same direction, these flows ensure the transport regimes.

It should be mentioned that as a result of the nonuniformity of the random factors, the density of distribution of the velocities of the particles in relation to their mathematical expectation is not stable with respect to time. The variation of this density may affect the value of the mathematical expectation of the examined distribution.

If a set of particles with different aerodynamic characteristics is placed in a rising flow, then, evidently, at the initial moment, the number of particles showing a tendency to the opposite directional movement, is maximum in the volume occupied by them. In this case, the probability of the mechanical interaction between them is also maximum. After a certain period of time, the volume, occupied by the material in the flow, starts to increase as a result of a directional movement to both sides. In this movement, the particles tried to obtain their steady velocity. This is prevented by the effect of random factors leading to the probability distribution of the velocity of the particles in relation to this velocity. With a decrease in the concentration of the material in the initial volume, the effect of several factors, such as, for example, collision of particles, decreases. This increases the intensity of the layer separation effect which will be more effective with an increase in the holding time of the material in the flow. This is confirmed by the practice of gravitational classification, showing that with an increase of the holding time of the material in the classification zone or with an increase in the height of the apparatus, the separation effect increases.

The experimental results indicate that the process of gravitational separation has a distinctive exponential nature. This confirms our assumptions on the nature of layer separation in gravitational classification. Consequently, very important conclusions can be made regarding the mechanism of the process.

The separation process is almost completed at a limited height of the hollow apparatus. A simple increase of the apparatus height has only a small effect on the classification results. To increase the efficiency of the process, it is necessary to take special measures capable of increasing the effect of layer separation.

This confirms that the requirements to obtain generally accepted theoretical considerations on the necessity of carrying out separation only in steady regimes are not justified.

The exponential nature of the layer separation process confirms the experimental conclusion according to which the process is most intensive in the initial moments and then decreases in efficiency.

The holding time of the material in the classification zone depends on the height of apparatus and the flow rate of the medium. Considerations regarding the exponential nature of the process also confirm the conclusion according to which in the apparatus of moderate height it is possible to produce a sufficiently effective process even for an increased boundary size in separation. All this is in good agreement with the experimental data.

References

1. Boltzman L., Lectures in the theory of gases, Gostekhizdat, Moscow (1956)
2. Barsky M.D., Revnivtsev V.I. and Sokolkin Yu.V., Gravitational classification of granular materials, Nedra, Moscow (1974)
3. Barsky M.D., Fractionation of Powders, Nedra, Moscow (1980)

Chapter VII

MATHEMATICAL MODELS OF REGULAR CASCADES

1. PROPORTIONAL MODEL

The system of separation of the materials by the cascade principle is illustrated most convincingly by a proportional model, reflecting the mass exchange taking place between stages.

The value, characterising the degree of separation of a narrow size class in an individual cascade, can be presented in a simplified form:

$$K = \frac{r_i^*}{r_i}$$

where r_i is the initial content of the particles of a narrow size class on the i-th stage of purification; r^*_i is the number of particles of the same size class, transferred from the i-th stage to stage $(i\text{-}1)$; K or k is the coefficient of distribution.

At the same structure of the stages for each size class the distribution coefficient K will be constant. This coefficient does not depend on the number of the stage.

The proportional model of the distribution of particles of some fixed size class along the height of the apparatus at the supply to stage i^* is shown in Fig. VII-1,a.

The nature of classification is controlled by the area of supply of the material into the flow. It is assumed that the initial mixture contains some number of the particles of the j-th size. The initial content of these particles will be regarded as unity. The degree of fraction separation of the fixed narrow size class in relation to the number of purification stages at constant regime parameters of the operation of the classifier depends on the number of stages.

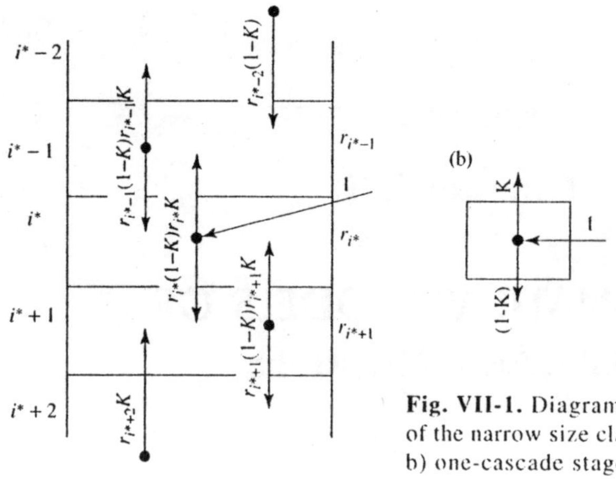

Fig. VII-1. Diagram of separation of particles of the narrow size class: a) multi-stage cascade; b) one-cascade stage.

The fraction $F_f(x)=\dfrac{r_f}{r_s}$ will be referred to as the degree of fractional extraction of the fine product for a narrow size class, where r_f and r_s is the amount of the narrow size class in the fine product and in the initial material, respectively.

For a single stage, the pattern of the process is very simple (see Fig. VII-1,b).

The degree of fractional extraction of a fine product in this case corresponds to the distribution coefficient $F_f(x)_{(1)} = K$.

In the case of two stages of purification, the distribution pattern has the form shown in Fig. VII-2.

Thus, the fractional extraction for two stages represents the sum of an infinite series

$$F_f(x)_{(2)} = \lim_{n\to\infty}\left[K + K^2(1-K) + K^3(1-K)^2 + ... + K^n(1-K)^{n-1}\right] \quad \text{(VII-1)}$$

In the general form

$$F_f(x)_{(2)} = \lim \sum_{n=1}^{\infty}(1-K)^{n-1}K^n \qquad \text{(VII-2)}$$

For three stages of purification, the distribution pattern has the form shown in Fig. VII-3.

Consequently,

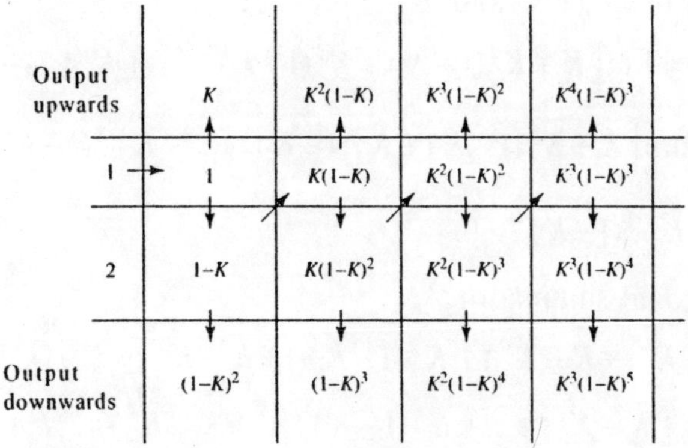

Fig. VII-2 Diagram of separation for two purification stages.

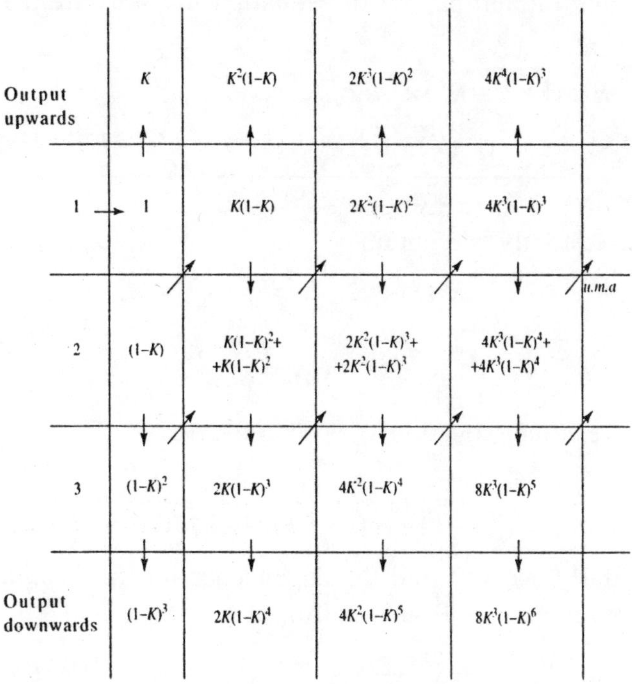

Fig. VII-3 Diagram of separation for three purification stages.

$$F_f(x)_{(3)} = \lim_{n\to\infty}$$

$$\left[K + K^2(1-K) + 2K^3(1-K)^2 + \ldots + \underbrace{2^{n-1}(1-K)^n K^{n+1}}_{a_n} \right] \quad \text{(VII-3)}$$

we determine $F_f(x)_{(2)}$ and $F_f(x)_{(3)}$:

$$F_f(x)_{(2)} = \lim\left[K + K^2(1-K) + K^3(1-K)^2 + ... + K^n \times (1-K)^{n-1}\right] =$$

$$= \lim_{n\to\infty}\left[K + K^2(1-K) + K^3(1-K)^2 + ... + K^{n+1}(1-K)^n\right],$$

since $\lim\limits_{n\to\infty} K^{n+1}(1-K)^n = 0$.

After diagonal summation:

$$K + K(1+K-K^2) + K^2(1-K)(1+K-K^2) + K^3(1-K)^2 \times$$

$$\times(1+K-K^2) + ... + K^n(1-K)^{n-1}(1+K-K^2) = 2F_f(x)_{(2)}$$

From the left part of the equation, we separate the common co-factor:

$$K + (1+K-K^2)\times$$

$$\times\underbrace{\left[K + K^2(1-K) + K^3(1-K)^2 + ... + K^n(1-K)^{n-1}\right]}_{F_f(x)_{(2)}} = 2F_f(x)_{(2)}$$

Consequently, we obtain

$$K + (1+K-K^2)F_f(x)_{(2)} = 2F_f(x)_{(2)},$$

$$F_f(x)_{(2)} = \frac{K}{1+K^2-K}$$

The resultant equation will be converted:

$$\frac{K}{1+K^2-K} = \frac{K}{1-K(1-K)} = \frac{K}{1-F_f(x)_{(1)}(1-K)} = F_f(x)_{(2)}$$

Similarly, we also find the dependence for the fractional extraction for three stages of purification

$$\lim_{n\to\infty}\left[K + K^2(1-K) + 2K^3(1-K)^2 + 4K^4(1-K)^3 + ...\right.$$

$$\left. ... + 2^{n-1}\cdot K^{n+1}(1-K)^n\right] = F_f(x)_{(3)}$$

$$\lim_{n\to\infty}\left[K + K^2(1-K) + 2K^3(1-K)^2 + 4K^4(1-K)^3 + ...\right.$$

$$\left. ... + 2^n K^{n+2}(1-K)^{n+1}\right] = F_f(x)_{(3)}$$

Adding up these two equations, we obtain

$$K-K^2(1-K)+K(1+2K-2K^2)+K^2(1-K)(1+2K-2K^2)+$$
$$+2K^3(1-K)^2(1+2K-2K^2)+...+2^{n-1}K^{n+1}(1-K)^n\times$$
$$\times(1+2K-2K^2)=2F_f(x)_{(3)}$$

Or

$$K-K^2(1-K)+(1+2K-2K^2)F_f(x)_{(3)}=2F_f(x)_{(3)};$$

$$F_f(x)_{(3)}=\frac{K-K^2(1-K)}{1+2K^2-2K}=K\frac{1+K^2-K}{1+2K^2-2K} \qquad \text{(VII-4)}$$

The equation for the fractional extraction in three stages will be converted

$$K\frac{1+K^2-K}{1+2K^2-2K}=\frac{K}{\frac{1+2K^2-2K}{1+K^2-2K}}=\frac{K}{\frac{1+K^2-K+K^2-K}{1+K^2-K}}=$$

$$=\frac{K}{1-\frac{K(1-K)}{1+K^2-K}}=\frac{K}{1-F_f(x)_{(2)}(1-K)}$$

It may be assumed that in a general case, the fractional extraction for the apparatus with n stages has the form

$$F_f(x)_{(n)}=K\frac{1}{1-F_f(x)_{(n-1)}(1-K)} \qquad \text{(VII-5)}$$

Examination of distribution of the material over the height of apparatus at four purification steps results in an infinite converging series where any term represents an odd number from the Fibonacci series. In this case, the fractional extraction has the form which is difficult to express by a finite result:

$$F_f(x)_{(4)}=K+\sum_{n=1}^{\infty}a_{2n-1}^{f}K^{n+1}(1-K)$$

Similarly, it is possible to obtain fractional extractions at 5, 6, 7 or any other finite number of the stages of the classifier which are difficult to express in the final form.

For example,

$$F_{(x)_{(5)}} = K + K^2(1-K) + \sum_{n=1}^{\infty} A_n K^{n+2}(1-K)^{n+1},$$

$$A_n = A_{n-1} + 3^{n-1}; A_0 = 2;$$

$$F_f(x)_{(6)} = K + K^2(1-K) + 2K^3(1-K)^2 + 5K^4(1-K)^3 + 14K^5(1-K)^4 +$$

$$+42K^6(1-K)^5 + 131K^7(1-K)^6 + 420K^8(1-K)^7 + ...;$$

$$F_f(x)_{(7)} = K + K^2(1-K) + 2K^3(1-K) + 5K^4(1-K)^3 + 14K^5(1-K)^4 +$$

$$+42K^6(1-K)^5 + 132K^7(1-K)^6 + 428K^8(1-K) + ...;$$

$$F_f(x)_{(8)} = K + K^2(1-K) + 2K^3(1-K) + 5K^4(1-K)^3 + 14K^5(1-K)^4 +$$

$$42K^6(1-K)^5 + 132K^7(1-K)^6 + 429K^8(1-K)^7 + ...$$

In a general case, it is not always possible to express in some manner the n-th member of these and subsequent series, not to speak about the transition to the sum.

At the same time, for a cascade consisting of eight stages, this formula was derived in the form of the dependence

$$A_{n(8)} = \frac{2n!}{n!(n+1)!}$$

It may be seen that the proportional model illustrates efficiently the mechanism of the process, but is very cumbersome and not suitable for calculations. Consequently, other models will be investigated.

2. DISCRETE MODEL

It will be attempted to prove strictly the main relationships for the model of a regular cascade. For this purpose, we examine the cascade distribution of the material at discrete moments of time (acts) with equal breaks. It is assumed that the material is supplied to the equipment with the same time period and in identical portions.

It is well known that at specific concentration of the solid phase $\mu(\leq 2 \text{ kg/m}^3)$ any size class is distributed as if there were no other particles in the process.

Taking into account previous assumptions, it will be proved that there is the limit $\lim_{m\to\infty} r_{ij}^m$, i.e. in the course of the process of separation in each step the amount of the material tends to a constant amount, where m is the number of the acts of distribution of the material; r_{ij}^m is the amount of the material of the size class j on a stage with the number i in the distribution act m.

It is assumed that $r_{ij} = (r_{1j}^m, r_{2j}^m, .., r_{zj}^m)$. According to definition:

$$r_{i,j}^{m+1} = k_j r_{i+1,j}^m + (1-k_j) r_{i-1,j}^m \quad r_{i,j}^{m+1} = k_j r_{i+1,j}^m + (1-k_j) r_{i-1,j}^m$$

when $i \neq 1$, z, i^* (i^* is the place of introduction of material into the apparatus).

The material is introduced into the apparatus in equal portions. It is assumed that the amount of a single portion is γ. Consequently, for stage i^* the following condition is fulfilled:

$$r_{i^*,j}^{m+1} = k_j r_{i^*+1,j}^m + (1-k_j) r_{i^*-1,j}^m + \gamma$$

For a stage number 1

$$r_{1,j}^{m+1} = k_j r_{2,j}^m$$

For a stage number z

$$r_{z,j}^{m+1} = (1-k_j) r_{z-1,j}^m$$

Thus, we obtain the following matrix iterative relationship

$$r_j^{m+1} = A r_j^m + b$$

$$A = \begin{vmatrix} 0 & k_j & 0 & \dots & 0 & 0 \\ 1-k_j & 0 & k_j & \dots & 0 & 0 \\ 0 & 1-k_j & 0 & k_j & \dots & 0 \\ 0 & 0 & . & . & . & . \\ . & . & . & . & . & k_j \\ 0 & 0 & 0 & 0 & 1-k_j & 0 \end{vmatrix}$$

$$b = \begin{vmatrix} 0 \\ 0 \\ . \\ \gamma \\ . \\ 0 \end{vmatrix} \underline{\text{place}*}$$

It is well known that $r_j^m \rightarrow r_j$ (where r_j is the solution of the iterative equation $r_j = Ar_j + b$) when and only when for all eigenvalues of the matrix A it is fulfilled at: $|\lambda| < 1$. According to the Frobenius theorem, if a matrix consists of non-negative real elements and it cannot be reduced to the block diagonal form by simultaneous permutations, then the following will be fulfilled for its maximum (with respect to modulus) value: $s \le |\lambda| \le S$ where $S = \max \sum a_{ij}, s = \min_i \sum_{ij}$, and if $s \ne S$, then $s < |\lambda| < S$.

For the matrix A, $S = 1$, $s = \min(k_j, 1-k)$, i.e. for any eigenvalue of the matrix A, $|\lambda| < 1$ is fulfilled, then $r_j^m \to r_j$, where r_j is the solution of the equation $r_j = Ar_j + b$.

We examine the process in the limiting (stationary) state. r_i is the amount of the size class j in the separation stage i in the stationary state (index j will be omitted in order to simplify considerations). The resultant conclusion is valid for any size class and, consequently, this also applies to all size classes on the given stage.

We determine the calculation equation for the degree of fractional extraction of the size class j into fine product. The definition of $F_f(x)$ shows that

$$F_f(x) = \frac{r_f}{r_s} = \lim_{t \to \infty} \frac{\sum_{l=1}^{t} r_{f_1}{}^l}{\sum_{l=1}^{t} r_{s_1}}$$

where $r_{f_1}^i$ and r_{s_1} is the amount of the material of size class j in a single portion of the material leaving the apparatus (from the first stage upwards) and entering the apparatus (into stage i^*), respectively, in the separation act l. In this case, it was established that r_{s_1} is constant and independent of l. It will be proved that:

$$F_f(x) = \frac{r_f}{r_s} = \lim_{l \to \infty} \frac{r_{f_1}{}^l}{r_{s_1}}$$

Previous considerations show that $\lim_{l \to \infty} \frac{r_{f_1}^l}{r_{s_1}}$ exists (because $r_{f_1}^l = kr_1^l$) and it will be referred to as α. Consequently, $r_{f_1}^l = r_{s_1}\alpha + o(l)(l \to \infty)$.

We substitute this into the equality $\frac{r_f}{r_s} = \lim_{l\to\infty} \frac{\sum_{l=1}^{t} r_{f_1}^l}{\sum_{l=1}^{t} r_{s_1}}$ and obtain that

$\frac{r_f}{r_s} = \lim_{l\to\infty} \frac{r_{f_1}^l}{r_{s_1}} = \alpha$. Therefore, for the size class j $F_f(x) = \frac{r_f}{r_s} = \lim_{l\to\infty} \frac{r_{f_1}^l}{r_{s_1}}$.

If k is the coefficient of distribution of the size class j, then the amount of the size class, lifted from stage i to stage $i-1$ is $k \cdot r_i$, and the amount lowered to the stage $i+1$ is $(1-k) \cdot r_i$.

On the basis of Fig. VII-1a we normalise the amount of the size class j in each separation stage (in the stationary stage of the process) in relation to this class in every portion of the material, supplied into apparatus, i.e. $R_i = \frac{r_i}{r_{s_1}}$. R_i will be referred to as the normalised amount of the size class j in the stationary state in relation to the amount of this class in each portion of the material supplied into apparatus. Consequently, $R_{i-1} = \frac{r_i}{r_{s_1}} \cdot k = R_i \cdot k$. Therefore, the normalised amount, transferred into the fine product is $R_1 \cdot k$. On the other hand, it is equal to $\frac{r_{f_1}}{r_{s_1}} \left(r_{f_1} = \lim_{l\to\infty} r_{f_1}^l \right)$, i.e. $R_1 \cdot k = \frac{r_{f_1}}{r_{s_1}}$. Consequently, $F_f = R_1 \cdot k$. After determining the calculation equation for $F_f(x)$, it is possible to determine the amount of the size class j, transferred into the fine product in accordance with equation $r_f = F_f \cdot r_s$ (r_s is known initially).

It will be proved that for the size class j the degree of fractional extraction into the fine product is calculated from equation

$$F_f(x) = \begin{cases} \dfrac{1-\chi^{z+1-i^*}}{1-\chi^{z+1}} & k \neq 0.5 \\ \dfrac{z+1-i^*}{z+1} & k = 0.5 \\ 0 & k = 0 \end{cases}$$

normalised amount leaving stage i | normalised amount entering stage i

Where $\chi = \frac{1-k}{k}$, i^* is the number of the stage of introduction of the material into apparatus. To determine F_f, initially we find R_1. It has been proved that the normalised amount of the material of each size class in each step of apparatus during the process remains constant, i.e. the amount of the material entering and leaving any stage of apparatus in a single redistribution act is identical. The following balance equation may be written:

$$(1-k)R_i + kR_i = (1-k)R_{i-1} + kR_{i+1}$$

($i \neq 1, z$, because from these stages the material goes outside, and also $i \neq i^*-1$, i^*, i^*+1, because these steps are directly affected by the material entering the stage i^*. All these cases will be examined later).

The degree of fractional extraction into a coarse product is $\frac{r_c}{r_s}$, where r_c is the amount of the size class j in the yield of the coarse product, therefore $\frac{r_f}{r_s} + \frac{r_c}{r_s} = \frac{r_f + r_c}{r_s} = 1$ and, consequently, the degree of fractional extraction of the size class j into the coarse product is $1-F_f(x)$. Previously, it was proved that $F_f(x) = kR_1$, similarly, it may be determined that $1-F_f(x) = (1-k)R_z$ and consequently

$$kR_1 + (1-k)R_z = F_f + (1-F_f) = 1 \tag{VII-6}$$

we denote $\chi = \frac{1-k}{k}$ and consequently (VII-6) is expressed in the following form

$$R_1 + \chi R_z = 1 + \chi \tag{VII-7}$$

The normalised amount of each size class in each step of separation is constant during the process, consequently, the following relationships can be written:

For the second stage:

$$R_2 = R_1 + \Delta R_2$$

For the third stage:

$$R_3 = R_2 + \Delta R_3 = R_1 + \Delta R_2 + \Delta R_3$$

...... For stage i^*-1

$$R_{i^*-1} = R_{i^*-2} + \Delta R_{i^*-1} = R_1 + \Delta R_2 + \Delta R_3 + \cdots + \Delta R_{i^*-1}$$

The normalised amount of the material of the size class j entering apparatus is $\frac{r_{s_1}}{r_{s_1}}=1$. Consequently, for stage i^*:

$$R_{i^*}=R_{i^*-1}+\Delta R_{i^*}+1=R_1+\Delta R_2+\Delta R_3+\cdots+\Delta R_{i^*}+1$$

For stage i^*+1

$$R_{i^*+1}=R_1+\Delta R_2+\Delta R_3+\cdots+\Delta R_{i^*+1}$$

...... for stage z

$$R_z=R_1+\Delta R_2+\Delta R_3+\cdots+\Delta R_z$$

For stage 1 we shall write the following relationship

$$R_1=R_2k=(R_1+\Delta R_2)k$$

Consequently

$$R_1=\frac{1}{\chi}\Delta R_2 \qquad \text{(VII-8)}$$

For the stage z:

$$R_z=(1-k)R_{z-1} \text{ and } R_z=R_{z-1}+\Delta R_z$$

From these two equalities we obtain $R_{z-1}=-\frac{1}{k}\Delta R_z$. This will be substituted into equation

$$R_z=(1-k)R_{z-1}$$

and obtain

$$R_z=-\chi\Delta R_z \qquad \text{(VII-9)}$$

The examined balance equation for the stage $i\neq 1,z,\ i^*-1,\ i^*,\ i^*+1$.

$$(1-k)R_i+kR_i=(1-k)R_{i-1}+kR_{i+1}$$

It is well known that

$$R_i=R_{i-1}+\Delta R_i \text{ and } R_{i+1}=R_{i-1}+\Delta R_i+\Delta R_{i+1}$$

We substitute these two equalities into the balancing equation and obtain

$$(1-k)(R_{i-1}+\Delta R_i)+k(R_{i-1}+\Delta R_i)=(1-k)R_{i-1}+k(R_{i-1}+\Delta R_i+\Delta R_{i+1})$$

Consequently, we obtain

$$\Delta R_i(1-k)=k\Delta R_{i+1} \qquad \text{(VII-10)}$$

and, consequently

$$\Delta R_i=\frac{1}{\chi}\Delta R_{i+1} \qquad \text{(VII-11)}$$

For the stages with numbers 1 and z, identical relationships have already been obtained, (VII-8) and (VII-9). Now we examine the stage of separation i^*-1. For this stage:

$$(1-k)R_{i^*-1}+kR_{i^*-1}=(1-k)R_{i^*-2}+kR_{i^*}+k$$

k was added since the normalised amount of the material of class j, entering the apparatus is equal to 1 and consequently, the part of it which is transferred to stage i^*-1 is equal to $1 \cdot k = k$. Consequently, in accordance with (VII-10) we obtain

$$\Delta R_{i^*-1}(1-k)=k\Delta R_{i^*}+k$$

and consequently

$$\Delta R_{i^*-1}=\frac{1}{\chi}(\Delta R_{i^*}+1) \quad \text{(VII-12)}$$

We examine the stage i^*+1

$$(1-k)R_{i^*+1}+kR_{i^*+1}=(1-k)R_{i^*}+kR_{*+2}+(1-k)$$

Adding $1-k$ for the same reason as the addition of k to the stage i^*-1, in accordance with (VII-10) we obtain

$$\Delta R_{i^*+1}(1-k)=k\Delta R_{i^*+2}+(1-k)$$

and consequently

$$\Delta R_{i^*+1}=\frac{1}{\chi}\Delta R_{i^*+2}+1 \quad \text{(VII-13)}$$

The balance equation for stage i^* is the same as for the conventional i because we have already examined the normalised amount of the material of the size class j in the balance equations for the stages i^*-1 and i^*+1 and consequently

$$(1-k)R_{i^*}+kR_{i^*}=(1-k)R_{i^*-1}+kR_{i^*+1}$$

which gives

$$\Delta R_{i^*}=\frac{1}{\chi}\Delta R_{i^*+1}$$

....

Consequently, the following sequence of relationships can be presented

$$R_1=\frac{1}{\chi}\Delta R_2$$

$$\Delta R_2=\frac{1}{\chi}\Delta R_3$$

$$\Delta R_3 = \frac{1}{\chi}\Delta R_4$$

$$\Delta R_{i^*-2} = \frac{1}{\chi}\Delta R_{i^*-1}$$

$$\Delta R_{i^*-1} = \frac{1}{\chi}(\Delta R_{i^*} + 1)$$

$$\Delta R_{i^*} = \frac{1}{\chi}\Delta R_{i^*+1}$$

$$\Delta R_{i^*+1} = \frac{1}{\chi}\Delta R_{i^*+2} + 1$$

$$\Delta R_{i^*+2} = \frac{1}{\chi}\Delta R_{i^*+1}$$

$$\Delta R_{z-1} = \frac{1}{\chi}\Delta R_z$$

The second equation is substituted into the first one

$$R_1 = \frac{1}{\chi^2}\Delta R_3$$

The third equation is substituted into this equation

$$R_1 = \frac{1}{\chi^3}\Delta R_4$$

And we continue ....

$$R_1 = \frac{1}{\chi^{i-1}}\Delta R_i$$

$$R_1 = \frac{1}{\chi^{i^*-2}}\Delta R_{i^*-1}$$

$$R_1 = \frac{1}{\chi^{i^*-1}}(\Delta R_{i^*} + 1)$$

$$R_1 = \frac{1}{\chi^{i^*}}\Delta R_{i^*+1} + \frac{1}{\chi^{i^*-1}}$$

$$R_1 = \frac{1}{\chi^{i*+1}} \Delta R_{i*+2} + \frac{1+\chi}{\chi^{i*}}$$

$$R_1 = \frac{1}{\chi^{z-1}} \Delta R_z + \frac{1+\chi}{\chi^{i*}} \qquad \text{(VII-14)}$$

From equation (VII-9) $\Delta R_z = -\frac{1}{\chi} R_z$. Substituting this expression into this equation (VII-14), we obtain

$$R_1 = -\frac{1}{\chi^z} R_z + \frac{1+\chi}{\chi^{i*}}$$

Equation (VII-7) shows that $R_1 = 1 + \chi - \chi R_z$. Consequently, we obtain the following system of equations

$$R_1 = -\frac{1}{\chi^z} R_z + \frac{1+\chi}{\chi^{i*}}$$

$$R_1 = 1 + \chi - \chi R_z \qquad \text{(VII-15)}$$

We determine R_1 from this system

$$-\frac{1}{\chi^z} R_z + \frac{1+\chi}{\chi^{I*}} = 1 + \chi - \chi R_z$$

$$\frac{1+\chi}{\chi^{I*}} - (1+\chi) = \frac{1}{\chi^z} R_z - \chi R_z$$

$$(1+\chi)(\frac{1-\chi^{I*}}{\chi^{I*}}) = R_z(\frac{1-\chi^{z+1}}{\chi^z})$$

Consequently

$$R_z = \frac{(1+\chi)(1-\chi^{I*})\chi^{z-I*}}{(1-\chi^{z+1})}$$

This is substituted into (VII-15)

$$R_1 = (1+x)\frac{(1+\chi)(1-\chi^{I*})\chi^{z+1-I*}}{(1-\chi^{z+1})} =$$

$$= (1+\chi)(\frac{1-\chi^{z+1} - \chi^{z+1-I*} + \chi^{z+1}}{1-\chi^{z+1}}) =$$

$$(1+\chi)(\frac{1-\chi^{z+1-I*}}{1-\chi^{z+1}})$$

Now, substituting R_1 into equation $F_f(x) = kR_1$:

$$F_f(x)=kR_1=\frac{1}{1+\chi}R_1=\frac{1-\chi^{z+1-i*}}{1-\chi^{z+1}}$$

i.e.

$$F_f(x)=\frac{1-\chi^{z+1-i*}}{1-\chi^{z+1}} \qquad \text{(VII-16)}$$

This is the relationship for calculating the degree of fractional extraction of the size class j into the fine product (where $\chi=\frac{1-k}{k}$).

The function $F_f(x)$ is not determined for k = 0.5 and k = 0. We examine two cases:

1. At k = 0.5, χ = 1. According to l'Hopital's rule: $\lim_{\chi\to 1}=\frac{1-\chi^{z+1-i*}}{1-\chi^{z+1}}$

$$=\lim_{\chi\to 1}\frac{(1-\chi^{z+1-i*})'}{(1-\chi^{z+1})'}=\frac{z+1-i*}{z+1}$$

2. At k = 0, $\chi=\frac{1-0}{0}$ this means that $\lim_{\chi\to 1}\frac{1-\chi^{z+1-i*}}{1-\chi^{z+1}}=0$, because $z + 1 - i^* < z + 1$ is always satisfied.

Consequently

$$F_f(x)=\begin{cases}\dfrac{1-\chi^{z+1-i^*}}{1-\chi^{z+1}} & k\neq 0{,}05\\[2ex] \dfrac{z+1-i^*}{z+1} & k=0{,}5\\[2ex] 0 & k=0\end{cases} \qquad \text{(VII-17)}$$

where $\chi=\frac{1-k}{k}$, i^* is the number of stage of introduction of the material.

In order to determine the amount of the material of size class

j, entering the fine product, it is necessary to use the equation $F_f(x) = \frac{r_f}{r_s}$,

i.e. $r_f = F_f r_s \cdot r_s$, is available from the composition of the initial material.

Also, it is possible to calculate the yield of the narrow class into the coarse product:

$$r_{c,j} = r_{s,j} - r_{f,j}$$

3. ANALYSIS OF THE MATHEMATICAL MODEL OF A REGULAR CASCADE

The main properties of the resultant function $F_f(\chi)$ can be formulated as follows:

1. According to the dependence (VII-17) with the variation of χ from 0 to ∞, $F_f(\chi)$ is always greater than zero.
2. The limits of variation of $F_f(\chi)$ is the range of the values from 0 to 1. At $\chi = 0$, $F_f(\chi) = 1$, and at $\chi \to \infty$ $F_f(\chi) = 0$.
3. Function $F_f(\chi)$ is continuous and differentiable in the entire range of variation of the distribution parameter.
4. According to equation (VII-17), $F_f(\chi)$ is an unambiguous function.
5. It may be shown that the relationship (VII-17) is reduced to the equation

$$F_f(\chi)^{z+1} - \chi^{z+1-i^*} + \left[1 - F_f(\chi)\right] = 0 \qquad \text{(VII-18)}$$

with real coefficients. Consequently, according to the Decartes theorem, equation (VII-18) should have two real roots or generally should not have them. However, since unity is the identity root of equation (VII-18), the latter also has the second eal root in a particular case, which can also be equal to unity. Since equation (VII-17) is used at $\chi \neq 1$, then, consequently, only one real value $\chi \neq 1$ will correspond to any fixed value of $F_f(\chi)$ of equation (VII-18).

6. The items 4 and 5 show that $F_f(\chi)$ is a monotonic function. This is in good agreement with the physical meaning of the process. Taking into account item 2, we have a function $F_f(\chi)$ which is monotonically decreasing. Consequently, $\partial F_f(\chi)/\partial\chi \leq 0$ in the entire range of variation of the distribution parameter. Transferring to the distribution coefficient K, we consequently obtain:

$$\frac{\partial F_f(\chi)}{\partial K} = \frac{\partial F_f(\chi)}{\partial \chi}\frac{d\chi}{dK} = -\frac{1}{K^2}\frac{\partial F_f(\chi)}{\partial \chi} \geq 0$$

in the entire range of variation of this coefficient. The equality to zero of $\partial F_f(\chi)/\partial\chi$ or $\partial F_f(\chi)/\partial K$ is observed only in extreme cases at $\chi = 0$ or ∞ and $K = 1$ or 0, respectively. Both these values for χ or K are also extreme values for the classification process.

The curve of fractional separation in the optimum regime fully characterises the quality of the classification process, i.e. as the steepness of the curve $F_f(x)$ increases, the quality of separation improves.

The limit to which the fractional extraction tends at a relatively large number of the section of apparatus will now be determined. For this purpose, equation (VII-17) will be reduced to the following expressions:

$$F_f(x)^{(i^*)}_{(z)/K\neq 0.5} = \frac{1-\left(\frac{1-K}{K}\right)^{z+1-i^*}}{1-\left(\frac{1-K}{K}\right)^{z+1}}$$

$$F_f(x)^{(i^*)}_{(z)/K\neq 0.5} = \frac{\left(\frac{K}{1-K}\right)^{z+1}-\left(\frac{K}{1-K}\right)^{i^*}}{\left(\frac{K}{1-K}\right)^{z+1}-1},$$

$$F_f(x)^{(i^*)}_{(z)/K=0.5} = \frac{z+1-i^*}{z+1}1-\frac{i}{z+1}$$

We examine possible variants:

1). $K > 0{,}5; \frac{1-K}{K} < 1$

In this case

$$F_f(x)^{(i^*)}_{(z)/z\to\infty} = \lim_{z\to\infty}\frac{1-\left(\frac{1-K}{K}\right)^{z+1-i^*}}{1-\left(\frac{1-K}{K}\right)^{z+1}} =$$

$$\lim_{z\to\infty}\left[1-\left(\frac{1-K}{K}\right)^{z+1-i^*}\right] = 1-\lim_{z\to\infty}\left(\frac{1-K}{K}\right)^{z+1-i^*}$$

2) $K = 0.5$

For this case

$$F_f(x)^{(i^*)}_{(z)/z\to\infty} = 1 - \lim_{z\to\infty}\frac{i^*(z)}{z+1};$$

$$3).\quad K < 0,5; \frac{K}{1-K} < 1$$

$$F_f(x)^{(i^*)}_{(z)z\to\infty} = \lim_{z\to\infty}\frac{\left(\frac{K}{1-K}\right)^{z+1} - \left(\frac{K}{1-K}\right)^{i^*}}{\left(\frac{K}{1-K}\right)^{z+1} - 1} = \lim_{z\to\infty}\left(\frac{K}{1-K}\right)^{i^*}$$

For example, when the material is introduced into the top part of apparatus ($i^* = 1$) at $K \geq 0.5$ we have

$$F_f(K)_{z\to\infty} = 1 - \lim_{z\to\infty}\left(\frac{1-K}{K}\right)^z = 1;$$

At $K \leq 0.5$

$$F_f(K)_{z\to\infty} = \lim_{z\to\infty}\left(\frac{K}{1-K}\right) = \frac{K}{1-K}$$

In the case of symmetric central introduction of the material into the apparatus $\left(i^* = \frac{z+1}{2}\right)$ at $K > 0.5$

$$F_f(K)_{z\to\infty} = 1 - \lim_{z\to\infty}\left(\frac{1-K}{K}\right)^{(z+1)/2} = 1;$$

At $K = 0.5$

$$F_f(K)_{z\to\infty} = 1 - \lim_{z\to\infty}\frac{z+1}{2(z+1)} = 0.5 \quad \text{invariant to } z$$

At $K \leq 0.5$

$$F_f(K)_{z\to\infty} = \lim_{z\to\infty}\left(\frac{K}{1-K}\right)^{(z+1)/2} = 0$$

When the material is introduced into the lower part of apparatus ($i^* = z$) at $K = \geq 0.5$

$$F_f(K)_{z\to\infty} = 1 - \lim_{z\to\infty}\left(\frac{1-K}{K}\right) = 1 - \frac{1-K}{K} = \frac{2K-1}{K};$$

At $K \le 0.5$

$$F_f(K)_{z\to\infty} = \lim_{z\to\infty}\left(\frac{K}{1-K}\right)^z = 0$$

Thus, the boundary grain size can be shown in advance only for the introduction into the central part of the apparatus, irrespective of the number of sections of the apparatus. In this case, it should be that $K = 0.5$ because in accordance with the optimum of the process, the boundary grain is extracted at the top or at the bottom by 50% (for apparatus with an arbitrary number of sections).

For any other area of introduction of the material, to determine the boundary grain specified by the distribution coefficient $K \neq 1(w,d)$, it is necessary to solve the equation in relation to K_m

$$\frac{1-\left(\dfrac{1-K_m}{K_m}\right)^{z+1-i^*}}{1-\left(\dfrac{1-K_m}{K_m}\right)^{z+1}} = 0.5$$

This equation shows the dependence of the boundary coefficient of distribution on the number of sections and the area of introduction of the material into apparatus $K_m = f(z, \iota^*)$, and the optimum boundary grain in the given regime depends on the number of sections, and so on:

$$d_{m.opt} = f(w, z, i^*)$$

We determine the general equation for the first derivative of fractional extraction with respect to the distribution coefficient:

$$\left[F_f(K)\right]' = \left(\frac{1-\chi^n}{1-\chi^m}\right)' \frac{dx}{dK},$$

where

$$\frac{1-K}{K} = \chi; z+1-i^* = n; z+1 = m;$$

$$\frac{d\chi}{dK} = \left(\frac{1-K}{K}\right)' = \frac{-K-(1-K)}{K^2} = -\frac{1}{K^2};$$

$$\left[F_f(K)\right]' = \frac{-nx^{n-1}(1-x^m)+mx^{m-1}(1-x^n)}{(1-x^m)^2}\left(-\frac{1}{K^2}\right),$$

at $\chi = 1$, $K = 0.5$, $F_m = 0/0$.

To solve the indeterminacy, we use the l'Hopital's rule:

$$\left[F_f(K)'_{\chi=1}\right] = \lim_{\chi\to1;K\to0.5}\frac{n-\chi^{n-1}(1-\chi^m)-mx^{m-1}(1-\chi^n)}{(1-x^m)^2}\frac{1}{K^2} =$$

$$= 4\lim_{x\to1}\frac{\left[n\chi^{n-m}(1-\chi^m)-m(1-\chi^n)\right]'}{\left[(1-\chi^m)^2\right]'} = 4\lim_{x\to1}\frac{n(n-m)\chi^{n-m-1}}{-2m\chi^{m-1}} =$$

$$= -\frac{2n(n-m)}{m} = -\frac{2(z+1-i^*)(-i^*)}{z+1} = \frac{2i^*(z+1-i^*)}{z+1}$$

This is the maximum value of tgα – i.e. the angle of inclination of the curve of fractional extraction for the i^*-introduction. The highest value of the angle of inclination of the curve for different areas of introduction of the material into apparatus should be determined from the condition:

$$\frac{d}{di^*}\left[F_f'(x)\right]_{K=0,5} = 0;$$

$$\frac{d}{di^*}\left[\frac{\partial}{\partial}\frac{F_f(x)}{K}\right]_{K=0,5} = \frac{2(z+1-i^*)-2i^*}{z+1} = 0$$

Consequently,

$$2(z+1)-4i^* = 0,$$

where

$$i^* = \frac{z+1}{2}$$

This shows that the largest angle of inclination of the curve is typical of the curve of fractional extraction F_f for the symmetric central input of the material into the apparatus.

Thus, the highest separation capacity should be characteristic of a cascade apparatus with symmetric central introduction of the material.

It should be mentioned that the results can be used with equal efficiency for cascade separating processes of different nature, such as adsorption, rectification, extraction, separation of isotopes, etc. The processes of different nature have different mechanisms, forming the basis of formulation of the distribution coefficients of monocomponents.

4. SEPARATION IN CYCLIC FEED OF BULK MATERIAL INTO CASCADE APPARATUS

The process of separation of bulk material in a cascade apparatus and the supply of material have been examined so far for the same periods of time $\Delta\tau$. We examine a case in which the material is supplied into apparatus every $p\Delta\tau$ step, where p is the non-negative integer.

For the case in which separation and feed of the material into the apparatus are carried out in the same period of time, we have obtained the following matrix iteration relationship:

$$r_j^m = Ar_j^{m-1} + b$$

$$A = \begin{vmatrix} 0 & k_j & 0 & \dots & 0 & 0 \\ 1-k_j & 0 & k_j & \dots & 0 & 0 \\ 0 & 1-k_j & 0 & k_j & \dots & 0 \\ 0 & 0 & . & . & . & . \\ . & . & . & . & . & k_j \\ 0 & 0 & 0 & 0 & 1-k_j & 0 \end{vmatrix} \qquad b = \begin{vmatrix} 0 \\ 0 \\ . \\ \gamma \\ . \\ 0 \end{vmatrix} \underline{\text{coordinate i*}}$$

where γ is the amount of material in a single portion supplied into the apparatus; $r_j^m = (r_{1j}^m, r_{2j}^m, r3_{zj}^m, \dots, r_{zj}^m)$, where $r_{i,j}^m$ is the amount of material of size class j in the stage of apparatus i at the m-th act of separation.

When examining the process in which the material is supplied into apparatus every p acts of separation, the matrix iteration relationship has the following form:

$$r_j^m = Ar_j^{m-1} + b$$

$$r_j^{m+1} = Ar_j^m$$

$$r_j^{m+2} = Ar_j^{m+1}$$

$$r_j^{m+p-1} = Ar_j^{m+p-2}$$

$$r_j^{m+p} = Ar_j^{m+p-1} + b$$

and so on.

We determine a new iteration process which reflects the state of the initial process every p steps:

$$r_j^0 = u_j^0,\ r_j^p = u_j^1,\ r_j^{2p} = u_j^2,\ \ldots,\ r_j^{np} = u_j^n,\ \ldots$$

Consequently,

$$u_j^1 = A^p u_j^0 + b$$

$$u_j^2 = A^p u_j^1 + b$$

....

$$u_j^n = A^p u_j^{n-1} + b$$

It has been proved that in matrix A all eigenvalues with respect to the modulus are lower than 1, and, consequently, A^p has the same property. Consequently, the process $u_j^n \underset{n\to\infty}{\to} u_j$ where u_j is the solution of the matrix iteration equation $u_j = A^p u_j + b$.

On the basis of these considerations, it may be concluded that the initial process r_j^n is cyclic and that every sub-sequence $r_j^{n_k}$ converges (where $n_k - n_{k-1} = p$).

The examined process in its limiting state will be referred to as a stationary-cyclic process and will be examined. The amount of material of size class j in the separation stage i during p acts of separation in the stationary-cyclic state is denoted as follows: (index j is omitted to simplify considerations): $r_i^{(1)}$, $r_i^{(2)}$, .., $r_i^{(p)}$. It should be mentioned that $r_i^{(m)} = r_i^{(m+p)}$ for any integer non-negative m.

We denote $r_i = r_i^{(1)} + r_i^{(2)} + \ldots + r_i^{(p)}$, i.e r_i is the total amount of the material of size class j which passes through stage i during p acts of separation. It is clear that r_i is constant.

It is assumed that R_i is a normalised amount of the material of size class j which passes through stage i during p acts of separation in relation to the amount of material of size class j in a single portion supplied into the apparatus.

It is also evident that the degree of fractional extraction of size class j in p acts of separation into the fine product will be:

$$F_f = kR_1$$

and into the coarse product:

$$1 - F_f = (1-k)R_z$$

The amounts of the material leaving stage i and entering this stage during p acts of separation, respectively, are equal to each other and, consequently, the balance conditions, obtained for the stationary process will be fulfilled, i.e.

$$(1-k)R_i + kR_i = (1-k)R_{i-1} + kR_{i+1} \qquad i \neq 1, z,, i^* - 1, i^* + 1$$

$$(1-k)R_{i^*-1} + kR_{i^*-1} = (1-k)R_{i^*-2} + kR_{i^*} + k$$

$$(1-k)R_{i^*+1} + kR_{i^*+1} = (1-k)R_{i^*} + kR_{*+2} + (1-k)$$

$$R_1 = kR_2$$

$$R_z = (1-k)R_{z-1}$$

Consequently, the equation for calculating the degree of fractional extractiɔn in p separation acts will be the same as $k \neq 0.5$ for the stationary process, namely:

$$F_f = \left\| \begin{array}{ll} \dfrac{1-\chi^{z+1-i^*}}{1-\chi^{z+1}} & k = 0.5 \\ & k = 0 \\ \dfrac{z+1-i^*}{z+1} & \\ 0 & \end{array} \right.$$

Similarly, the degree of fractional extraction of the material of size class j during p separation acts may be regarded as the ratio of the amount of this size class, transferred into the fine product during p acts of separation, to its amount in a single portion of the material supplied into the apparatus.

The resultant dependence for the degree of fractional extraction remains constant throughout the entire process, as indicated by the balance equations.

In the stationary-cyclic process, the only difference from the stationary process is that this process is regarded for the number of separation cycles multiple to p if it is finite. All calculations should be carried out in a corresponding manner.

5. ABSORBING MARKOV CHAINS IN THE CASCADE SEPARATION OF BULK MATERIALS

The method of redistribution of a narrow size class with the distribution coefficient k in a cascade apparatus is shown in Fig.VII-

1. The examined process is similar to the 'random wandering' upwards and downwards with transition upwards with the probability k and downwards with $1-k$ and with two adsorbing states. This shows that the redistribution of a particle of a fixed class in the apparatus with z stages, may be represented by the Markov adsorbing chains, having the transition matrix of the following type:

$$\begin{vmatrix} 1 & 0 & 0 & 0 & 0 & 0 & \dots & 0 \\ k & 0 & 1-k & 0 & 0 & 0 & \dots & 0 \\ 0 & k & 0 & 1-k & 0 & 0 & \dots & 0 \\ . & . & . & . & . & . & . & . \\ . & . & . & . & . & . & . & . \\ 0 & 0 & \dots & 0 & k & 0 & 1-k & 0 \\ 0 & 0 & \dots & 0 & 0 & k & 0 & 1-k \\ 0 & 0 & 0 & 0 & 0 & \dots & 0 & 1 \end{vmatrix}$$

In this matrix there are $z + 2$ lines and columns, the first and last states are the yields into the coarse and fine products, and all remaining states are the probabilities of transition of the particle between the stages of the apparatus.

This matrix will be modified to a more suitable, canonic form by combining all ergodic (adsorbing) states into a single group and all non-returnable states into another group. In this case, there are z non-returnable states (corresponding to the number of stages of the apparatus) and two ergodic states (corresponding to the coarse and fine product). Consequently, the canonic form will be:

$$\begin{array}{c} \quad I \qquad\qquad O \\ R \rightarrow \left|\begin{array}{cc|ccccccc} 1 & 0 & 0 & 0 & . & . & \dots & 0 & 0 \\ 0 & 1 & 0 & 0 & . & . & \dots & 0 & 0 \\ \hline k & 0 & 0 & 1-k & 0 & . & . & \dots & 0 \\ 0 & 0 & k & 0 & 1-k & 0 & . & \dots & 0 \\ 0 & 0 & 0 & k & 0 & 1-k & 0 & \dots & 0 \\ . & . & . & . & . & . & . & . & . \\ 0 & 0 & 0 & \dots & 0 & k & 0 & 1-k & 0 \\ 0 & 0 & 0 & 0 & \dots & 0 & k & 0 & 1-k \\ 0 & 1-k & 0 & 0 & 0 & \dots & 0 & k & 0 \end{array}\right| \leftarrow Q \end{array}$$

Here region O consists totally of zeros, the submatrix Q (dimension $z \times z$) describes the behaviour of the particle prior to exit from the apparatus (from the set of the non-returnable states), the submatrix r with the dimension $z \times 2$ corresponds to transitions from the apparatus to the coarse and fine product (from non-returnable to ergodic states), the submatrix I (dimension 2 × 2) relates to the process after exit of the particle from apparatus (after the particle has reached the ergodic state). In the adsorbing chains, the probability of reaching one of such states tends to unity. Consequently, it should be noted that with the probability of 1 the particle reaches some adsorbing state earlier or later, i.e. it exits in one of the two products of separation.

For any adsorbing chain Q^n tends to zero and $I–Q$ is reversible, and $(I-Q)^{-1}=\sum_{k=0}^{\infty}Q^k$

For any adsorbing Markov chain, the fundamental matrix is the matrix $N = (I - Q)^{-1}$.

n_j denotes a function equal to the total number of redistribution acts, carried out by the particle on stage j, i.e. in the non-returnable state j. Consequently, it may be stated that the mathematical expectation of the particle, which is in stage i at the initial moment of time, to be on stage j of n_j acts of redistribution, is the coordinate ij of the matrix N, i.e.

$$\{E_i(n_j)\}=N \tag{VII-19}$$

This shows that the mean time spent by the particle in the given stage is always finite, and that these mean times are simply the matrix N.

Each particle travels into the apparatus through the stage i^*, and equation (VII-19) can be used to calculate the mean time spent by the particle on each stage.

We introduce the following notations:

$$N_2 = N(2N_{dg}-I)-N_{sq}$$

The matrix of the dimension $z \times z$, where N_{dg} corresponds to the matrix, whose diagonal is equal to the diagonal of the matrix N, and all other elements are equal to zero, and N_{sq} corresponds to the matrix whose elements are the squares of the elements of the matrix N. It is clear that in a general case for any matrix A we have $A^2 \neq A_{sq}$, and the equality will be valid only for diagonal matrices.

$B = NR$ is the matrix with size $z \times 2$.

$$\tau = N\xi$$

where ξ is the column vector, and all elements of the vector are equal to 1:

$$\tau_2=(2N-I)\tau-\tau_{sq}$$

For the adsorbing Markov chains, it may be asserted that the dispersion of the particle, situated at the initial moment of time on stage i, to be found on the stage j of n_j acts of redistribution, is the co-ordinate ij of matrix N_2, i.e.

$$\{D_i(n_j)\} = N_2 \qquad \text{(VII-20)}$$

It is also assumed that function T is equal to the total time (the number of redistribution acts), including the initial position, spent by the particle inside the apparatus prior to its exit from the apparatus. The value of T shows how many steps the Markov process should take prior to arriving to the ergodic set.

If at the initial moment of time the particle is in the stage i, the mathematical expectation of the number of redistribution acts up to the exit of the particle into the coarse of fine product, is the i-th co-ordinate of the vector τ, and the dispersion of the former is the i-th co-ordinate of the vector τ_2, i.e.

$$\{E_i(T)\} = \tau$$

$$\{D_i(T)\} = \tau_2$$

In the process of separation, the particle is introduced into the apparatus through stage i^* and, consequently, we can determine the mean time (the number of redistribution acts) of stay of the particle in the apparatus and deviation from this time.

It is assumed that b_{ij} is the probability of the particle being in stage i to penetrate into the coarse or fine product (j = 1 for the fine product and j =2 for the coarse product).
Consequently,

$$\{b_{ij}\} = B = NR$$

This shows that it is possible to control the probability of exit of the particle into the required product, by changing the area of introduction of the particle into the apparatus.

References

1. Fadeev D.K. and Fadeev V.H., Computational Methods of Linear Algebra, WH Freeman and Company, San Francisco and London (1959).
2. Gantmacher F.R., Applications of the Theory of Matrices, Interscience Publishers, Division of John Wiley & Sons, New York, London, Sydney (1959).
3. Shishkin S., Dissertation for the title of Candidated of Technical Sciences, Ural Polytechnic, Sverdlovsk, 1983.
4. Barsky E. and Buikis M., A Mathematical Model for the Cascade Separation at Identical Stages of the Separator, Latvian Journal of Physics and Technical Science, N5, p.22–32 (2001).

5. Barsky M.D., Optimisation of the Process of Separation of Granular Materials, Nedra, Moscow (1978).
6. Kemeny J. and Snell J., Finite Markov Chains, Princeton, New Jersey (1967).
7. Kemeny J., Snell J. and Knapp A., Denumerable Markov Chains, Springer-Verlag, Berlin (1976).
8. Govorov A.V., Cascade and Combined Process of Fractionation of Bulk Materials. Dissertation for the title of the candidate of technical sciences, Ural Polytechnic Institute, Sverdlovsk (1986).
9. Barsky E., Mathematical Models of Separation Process and Optimisation of these Processes, M.Sc.Thesis, Ben-Gurion University of Negev (1998).

Chapter VIII

STRUCTURAL MODEL OF THE PROCESS

1. MAIN PROBLEMS OF THEORY

The current theoretical considerations regarding the fundamentals of the process are based on the accepted classic representations, created by Rittinger, Richards and Finkel. These representations are based on the following assumptions:

1. The velocity of the flow is assumed to be uniform in the cross section of the apparatus.

2. The mass nature of the process is restricted only by empirical consideration of constricted conditions in the determination of the finite velocities of deposition of particles of fixed narrow size classes. In this case, the hovering velocity is assumed to be equal to the consolidated velocity of settling. This should reflect in an implicit manner the effects of the mechanical interaction of the particles with each other and with the walls of the apparatus, and also the effect of the solid phase on the carrying capacity of the flow as a whole. Otherwise, the mechanism of the process of classification is examined on the basis of the behaviour of a separate isolated particle.

3. It is assumed that the distribution of the solid phase in the cross section of the apparatus is uniform.

The determination of the relationship between the hovering velocity and the settling velocity of the same particles in a moving medium indicates that it is necessary to take into account the structure of the flow when developing a model of cascade separation. Our investigations of mass phenomena in the separation processes have contributed this factor to the account when developing the process theory. However, at present, this account is highly superficial and, consequently, it is not possible to explain completely the accumulated experimental data.

Consequently, within the framework of the currently available classification theories it is not possible to explain many relationships of the process determined by experiments.

Using the new approach, the following facts should be combined in a new common concept:

1. In practice, the fine product is always contaminated with coarse particles, and vice versa.

2. The experiments show that for pneumatic classification, the mean flow velocity w_0 at which the narrow fixed size class of the particles starts to be extracted into the fine product, is approximately 2.5 times smaller than the finite velocity of settling of a single particle of a given size in the medium.

3. The empirical dependence of the degree of fractional extraction on the Froude criterion, general for all cases, has not as yet been explained:

$$F_f(d) = F_f(Fr),$$

where $F_f(d)$ is the fractional extraction into the fine products of particles of a fixed size class d.

Consequently, the curves of the fractional extraction of the particles of a fixed size class into the fine product in relation to the relative size of the particles and the relative velocity are universal:

$F_f = F_f\left(\dfrac{d}{d_x}\right)$ – the curve is universal in all regimes;

$F_f = F_f\left(\dfrac{w}{w_x}\right)$ – the curve is universal for particles of all sizes,

where w_x is the velocity at which the particles of the fixed narrow size class of the particles are extracted by $x\%$; d_x is the size of the particles extracted into the fine product by $x\%$ at an arbitrary fixed velocity.

4. In the light of the existing theories, the relationship between the parameters Fr_x and the relative density of material cannot be efficiently explained.

To explain the above and a number of other experimental data, we have used a new approach to interpreting the mechanism of gravitational cascade classification. This approach is based on the non-uniform kinematic structure of a two-phase flow.

Previously, we used the experimental dependence of the distribution coefficient on the Froude criterion:

$$K = f(Fr)$$

for a cascade classifier. However, when estimating the distribution coefficient, it is also possible to use an analytical approach from the position of the structure of the moving flow.

2. Generalised coefficient of distribution based on the structure of the flow

The structural model of the formation of the distribution coefficient, like any other model, uses a number of assumptions. The main of these are as follows:

1. Particles are spherical.
2. The distribution of particles of any narrow size class in the cross section of apparatus is uniform because of intensive interaction between the particles and also with the wall of the apparatus and the internal devices.
3. The rising two-phase flow should be regarded as a continuum with increased density. As established in our studies, the carrying capacity of the 'dust-laden' flow is higher than that of a clean medium. Conventionally, this may be taken into account by increasing the effective density of the flow. The distribution of the local speeds of the solid phase is some function of the geometrical characteristics of the effective cross section of the channel.

In a general case:

$$u_r = w \cdot f\left(\frac{r}{R}\right), \qquad \text{(VIII-1)}$$

where r is the characteristic co-ordinate of some point of the cross section of apparatus; R is the characteristic boundary size of the effective cross section of apparatus; u_r is the local velocity of the solid phase at the point with the co-ordinate r; w is the mean velocity of the flow.

Thus, dependence (VIII-1) takes into account the form of the cross section of the channel.

According to the Newton–Rittinger law, the dynamic effect of the flow on a single particle is determined by the following dependence:

$$F_r = \lambda \frac{\pi d^2}{4} \rho_n \frac{(u_r - v_r)^2}{2},$$

where λ is the drag coefficient of the particle; $\frac{\pi d^2}{4}$ is the area of the middle section of the particle; ρ_n is the density of the flow; v_r is the local absolute velocity of movement of the particle; $(u_r - v_r)$ is the velocity of particles in relation to the flow.

The difference of the absolute velocities is algebraic. The positive direction of the velocities u_r and v_r is represented by the direction of movement of the flow.

If the total number of particles of the given monofraction in the examined cross section is regarded as unity, the distribution coefficient is written in the form $k = n_{v\geq 0}$ – the number of particles of a given narrow size class with the absolute velocity higher than or equal to zero.

Examination of equilibrium of the particle distance r_0 from the axis gives:

$$\frac{\pi d^3}{6}(\rho - \rho_0) = \lambda \frac{\pi d^2}{4}\rho_0 \frac{(u_r - v_r)^2}{2} \tag{VIII-2}$$

consequently

$$u_r - v_r = w\sqrt{\frac{4}{3\lambda}\cdot B}, \tag{VIII-3}$$

where $B = \frac{qd(\rho - \rho_0)}{w^2 \rho_0}$

We examine the regime of turbulent flow around the particle characterised by a constant value of drag coefficient λ. In this case, the Reynolds criterion of flow around the particle is:

$$\text{Re}_r = \frac{(u_r - v_r) d\rho_0}{\mu} \geq 500$$

where μ is the dynamic viscosity of the medium.

Taking into account (VIII-3) gives:

$$\frac{w\sqrt{\frac{4}{3\lambda}\cdot B}\cdot d\rho_0}{\mu} \geq 500$$

This condition corresponds to the expression (at $\lambda = 0.5$)

$$\sqrt{\frac{8}{3}Ar} \geq 500,$$

where Ar is the Archimedes criterion

$$Ar = \frac{qd^3\rho\,\rho_0}{\mu^2}$$

It is thus possible to determine the limiting size of the particles above which the flow around other particles is definitely turbulent.

Special features of the laminar flow around the particles will now be examined. In this case:

$$\text{Re}_p = \frac{(u_r - v_r)d\rho_0}{\mu} \leq 1$$

The equilibrium conditions give

$$\frac{\pi d^3}{6}q\rho = 3\pi\mu(u_r - v_r)d$$

Taking into account the previous expression for this case one can write:

$$Ar = 18\text{Re}$$

It is well known that in laminar flow-around the drag coefficient of the particles is

$$\lambda = \frac{24}{\text{Re}} = \frac{24\mu}{(u_r - v_r)d\rho_0}$$

Deriving the relationship

$$\frac{u_r - v_r}{w} = \sqrt{\frac{4}{3}B\frac{(u_r - v_r)d\rho_0}{24\mu}},$$

gives $\dfrac{u_r - v_r}{w} = \dfrac{1}{18}\text{Re}_w \cdot B,$

where Re_w is the Reynolds number calculated from the mean flow velocities.

We now return to relationship (VIII-3). For a particle of a narrow size class with absolute velocity $v_r \geq 0$, we can arrive at the following equation

$$u_r \geq w_r\sqrt{\frac{4B}{3\lambda}}$$

Substituting (VIII-4) into (VIII-1) gives:

$$f\left(\frac{r}{R}\right) \geq \sqrt{\frac{4}{3\lambda} \cdot B}; \qquad \text{(VIII-5)}$$

Similarly, for particles with $v_r \leq 0$:

$$f\left(\frac{r}{R}\right) \leq \sqrt{\frac{4B}{3\lambda}}; \qquad \text{(VIII-6)}$$

The inequalities (VIII-5) and (VIII-6) include the following limiting cases:

1. For any co-ordinate:

$$f\left(\frac{r}{R}\right) > \sqrt{\frac{4B}{3}};$$

In this case the distribution coefficient is $K = 1$;

2. Correspondingly, at $f\left(\frac{r}{R}\right) < \sqrt{\frac{4B}{3}}$ for any co-ordinate r we have $K = 0$.

An intermediate case is characterised by the fact that some co-ordinates form the lines of the level in accordance with the equality:

$$f\left(\frac{r_0}{R}\right) = \sqrt{\frac{4B}{3}}; \qquad \text{(VIII-7)}$$

It is assumed that equation (VIII-7) has one real root:

$$\frac{r_0}{R} = f^{-1}\left(\frac{4B}{3\lambda}\right); \qquad \text{(VIII-8)}$$

Taking this into account, we determine the corresponding area ω_{r_0} for which:

$$f\left(\frac{r_i}{R}\right) \geq f\left(\frac{r_0}{R}\right);$$

Consequently, the distribution coefficient is written in the form:

$$K = \frac{\omega_{r_0}}{\omega_R} = c\left(\frac{r_0}{R}\right) \quad \text{– for the convex profile } f\left(\frac{r}{R}\right)$$

$$K = C\left[1 - \left(\frac{r_0}{R}\right)^2\right] \quad \text{– for the concave profile } f\left(\frac{r}{R}\right)$$

Coefficient C characterises the form of the level lines and of the effective cross section. For example, for a circle $C = 1$. Substituting the dependence (VIII-5) into the resultant equation we finally obtain:

$$K = \varphi\left(\frac{B}{\lambda}\right);$$

An identical dependence holds for two and more roots of equation (VIII-6). This is the case for complex profiles $f\left(\frac{r}{R}\right)$. For example in the case of the profile shown in Fig. VIII-1 for some monofraction we have three real roots of the equations VIII-6: $r_{0_1}; r_{0_2}; r_{0_3}$. The latter form isotachs taking into account the shape of the effective section of the apparatus. The isotachs determine the corresponding total area $\sum \omega_{r_{0i}}$ for which:

$$f\left(\frac{r}{R}\right) \geq f\left(\frac{r_{0i}}{R}\right)$$

Thus, for the given case:

$$\sum \omega_{r_{0i}} = \omega_{r_{01}} + \omega_{r_{02-3}}$$

and for the distribution coefficient we may write:

$$K = \frac{\sum \omega_{r_{0i}}}{\omega_R}$$

Since $\sum \omega_{r_{0i}}$ is expressed unambiguously through r_{0i}, being roots of equation (VIII-6), the final expression for the distribution coefficients will have the form:

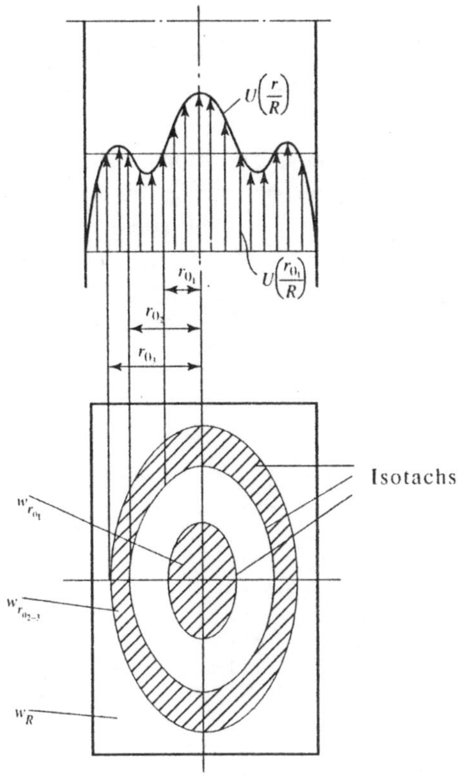

Fig.VIII-I Formation of the distribution coefficient for a complicated profile of the structure of the solid phase.

$$K = \varphi\left(\frac{B}{\lambda}\right) \tag{VIII-9}$$

The specific expression for the distribution coefficient may be obtained by transferring to the specific profile of the curve of the solid medium in the cross section of the apparatus. We examine consecutively the cases of interaction between the particles and the flow.

1. Turbulent flow around particles and the turbulent regime of movement of the medium in the apparatus

In this case, for an equilibrium apparatus with a circular cross section, the distribution of the velocities of the solid medium along the radius is usually expressed by an empirical dependence:

$$u_r = w\frac{(m+1)(m+2)}{2}\left\langle 1-\frac{r}{R}\right\rangle^m = wf\left(\frac{r}{R}\right) \qquad \text{(VIII-10)}$$

where m is an exponent which depends on the regime of movement of the medium, roughness of the pipe walls ($m < 1$).

According to (VIII-7) we determine the co-ordinate of the isotach at which the absolute velocity of the fixed monofraction is equal to zero:

$$\frac{(m+1)(m+2)}{2}\left\langle 1-\frac{r_0}{R}\right\rangle^m = \sqrt{\frac{4B}{3\lambda}}$$

Consequently

$$\frac{r_0}{R} = 1 - \left[\frac{2}{(m+1)(m+2)}\sqrt{\frac{4B}{3\lambda}}\right]^{\frac{1}{m}} \qquad \text{(VIII-11)}$$

The area of the cross section for which

$$u\left(\frac{r}{R}\right) \geq f\left(\frac{r_0}{R}\right)$$

is $\omega_{r_0} = \pi\, r_0^2$

Consequently, the distribution coefficient is expressed by the relationship:

$$K = \left(\frac{r_0}{R}\right)^2$$

Therefore, taking into account (VIII-11) we have:

$$K = \left\{1 - \left[\frac{2\cdot\sqrt{\frac{4B}{3\lambda}}}{(m+1)(m+2)}\right]^{\frac{1}{m}}\right\}^2$$

Instead of the dependence (VIII-10) we can examine a different profile of the distribution of the velocities of the solid medium along the radius:

$$u_r = \frac{n+2}{n} \cdot w\left[1-\left(\frac{r}{R}\right)^n\right] \tag{VIII-12}$$

where n is the degree of turbulisation of the flow ($n = 2 \div \infty$).

The following cases will now be examined:

1. The gradient of velocity along the axis of the flow.

For the dependence (VIII-12):

$$\frac{du}{dr} = -(n+2)\frac{w}{R}\left(\frac{r}{R}\right)^{n-1} \tag{VIII-13}$$

for the dependence (VIII-11):

$$\frac{du}{dr} = -\frac{w}{R} \cdot \frac{n(n+1)(n+2)}{2\left(1-\frac{r}{R}\right)^{1-n}} \tag{VIII-14}$$

Consequently, for equation (VIII-12) in accordance with (VIII-13) we have:

$$\left(\frac{du}{dr}\right)_{r=0} = 0$$

Correspondingly, for (VIII-10) from (VIII-14):

$$\left(\frac{du}{dr}\right)_{r=0} = -\frac{f(0) \cdot n}{R}$$

Thus, the relationship (VIII-10) in contrast to (VIII-12) describes a discontinuous function along the flow axis.

2. The gradient of velocity in the wall of the pipe.

For function (VIII-12) from (VIII-13) we obtain:

$$\left(\frac{du}{dr}\right)_{r=R} = -(n+2)\frac{w}{R} = -\frac{f(0) \cdot n}{R}$$

which shows that the gradient increases with an increase of the degree of turbulisation of the flow and mean velocity.

Taking into account the (VIII-14), for the dependence (VIII-10) we obtain:

$$\left(\frac{du}{dr}\right)_{r=R} = \infty$$

which indicates to the transfer of an infinite momentum (corresponds to an infinite friction force).

3. Expression (VIII-12) in contrast to (VIII-10) combines all movement regimes up to laminar.

Taking into account (VIII-8), the radius forming the distribution coefficient is:

$$\frac{r_0}{R}=\left[1-\frac{n}{n+2}\cdot\sqrt{\frac{4B}{3\lambda}}\right]^{\frac{1}{n}}$$

Consequently, we obtain a relationship for the distribution coefficient:

$$K=\left[1-\frac{n}{n+2}\sqrt{\frac{4B}{3\lambda}}\right]^{\frac{2}{n}}$$

The regime of laminar flow around particles

Because in this case the regime of movement of medium in the channel spreads from laminar to turbulent, we shall use dependence (VIII-12) for the structure of the flow. For example, at $n = 2$ we have a parabolic profile of the velocity profile, and at $2 < n < 8$ a transition regime, and at $n > 8$ turbulent movement. Therefore, according to (VIII-8) we have:

$$\frac{n+2}{n}\left[1-\left(\frac{r_0}{R}\right)^n\right]=\sqrt{\frac{4B}{3\lambda}} \qquad \text{(VIII-15)}$$

Using this equation the expression for the drag coefficient we obtain:

$$\frac{n+2}{n}\left[1-\left(\frac{r_0}{R}\right)^n\right]=\sqrt{\frac{4}{3}\cdot\frac{u_{r_0}d\rho_0 B}{24\mu}}$$

or

$$\frac{n+2}{n}\left[1-\left(\frac{r_0}{R}\right)^n\right]=\sqrt{\frac{\frac{n+2}{n}w\left[1-\left(\frac{r_0}{R}\right)^n\right]d\rho_0}{18\mu}\cdot B}$$

Simplifying, we obtain:

$$\frac{n+2}{n}\left[1-\left(\frac{r_0}{R}\right)^n\right]=\frac{\mathrm{Re}_w B}{18}$$

Taking into account that $\mathrm{Re}_w^2 \cdot B = Ar$, the resultant equation has the form:

$$\frac{n+2}{n}\left[1-\left(\frac{r_0}{R}\right)^n\right]=\frac{\sqrt{Ar \cdot B}}{18}$$

Taking into account that the distribution coefficient $K=\left(\frac{r_0}{R}\right)^2$, we obtain the final equation:

$$K=\left[1-\frac{\sqrt{Ar \cdot B}}{18}\cdot\frac{n}{n+2}\right]^{\frac{2}{n}}$$

For the parabolic profile (n = 2):

$$K=1-\frac{\sqrt{Ar \cdot B}}{36};$$

Intermediate regime of flow around the particles

Using the well-known dependence of the drag coefficient of criteria Re and Ar:

$$\lambda=\frac{4}{3}\cdot\frac{Ar}{\mathrm{Re}^2}$$

and the interpolation formula proposed by R.B. Rozenbaum and O.M. Todes, holding in all flow regimes:

$$\mathrm{Re}=\frac{Ar}{18+0,61\sqrt{Ar}}$$

we obtain

$$\lambda=\frac{4}{3}\cdot\frac{(18+0,61\sqrt{Ar})^2}{Ar}$$

Substituting the last equation into (VIII-15) and passing to the distribution coefficient, we obtain:

$$\frac{n+2}{n}\left(1-K^{\frac{n}{2}}\right)=\frac{\sqrt{Ar\cdot B}}{18+0,61\sqrt{Ar}}$$

The generalised dependence of the distribution coefficient in an arbitrary regime of movement of the medium and for an arbitrary regime of flow around other particles has the form:

$$K_{\Sigma}=\left[1-\frac{n}{n+2}\cdot\frac{\sqrt{Ar\cdot B}}{(18+0,61\sqrt{Ar})}\right]^{\frac{2}{n}} \qquad \text{(VIII-16)}$$

3. ANALYSIS OF THE GENERALISED DISTRIBUTION COEFFICIENT

We analyse (VIII-16) for turbulent regimes ($Ar \geq 10^5$). In these conditions, the term '18' of the denominator can be ignored and the equation is reduced to:

$$K=\left(1-\frac{n}{n+2}\sqrt{\frac{8}{3}\cdot B}\right)^{\frac{2}{n}};$$

Taking into account the model of the regular cascade (VII-17), this expression is in good qualitative agreement with the experimentally determined dependences.

Two approaches are used in the examination of suspension-bearing flows. The first approach examines the two-phase flow as some continuous medium (continuum) with averaged properties. The characteristics of the dispersoid are some mean velocity, density, etc.

In the pure form, this approach is not suitable for the classification process, because the dispersoid must be divided into individual phases since the result of the process is the separation of each monofraction which form together the discrete phase. In addition to this, the accuracy of this approach increases with a decrease in the slip of the phases. Therefore, this approach can be used efficiently in, for example, describing the processes similar to pneumatic transport and not classification processes characterised by the counterflow movement of dispersed particles in relation to a continuous medium (in fact, for any fixed narrow size class).

In the second approach, the behaviour of every phase is examined separately. In this case, when examining the classification process, it is important to take into account a coarse number of random factors.

This determines the insurmountable difficulties for the quantitative description of the results of the process in the explicit form. Therefore, this approach cannot be used only in a limited number of cases, solving only the simplest problems of the behaviour of two-phase flows and is completely unsuitable for describing the classification process as a whole.

For the classification process, it is sufficient to use a combined approach. In this approach, on the basis of the evaluation of the effect of a continuous medium on a discrete phase, considering the behaviour and interaction of individual monofractions, a transition is made to the dispersoid with the effective carrying capacity. Thus, the continuous medium and every individual monofraction take part in the formation of the dispersoid whereas the dispersoid affects the behaviour of the particles of each fixed narrow size class. This reflects implicitly the intraphase and interphase interaction, but on the basis of the continuum.

To justify the transition to the separated dispersoid, we evaluate the density of the flow of monofractions through the cross section of a classifier. The number of the particles of a fixed monofraction, passing through the cross section of apparatus per unit time, can be expressed from the equation

$$Q_{\Sigma} = \frac{G \cdot r_s \cdot r_i}{P} \qquad \text{(VIII-17)}$$

where G is the mass of the monofractions; P is the mass of a single particle of the given narrow size class.

In this case, the rising flow of the particles of the fixed monofraction is expressed in the form

$$Q_a = \frac{G_f \cdot r_s \cdot r_i K}{P} \qquad \text{(VIII-18)}$$

where the co-factor $r_i K$ is the relative flow of the examined particles from section i into the section positioned above it.

The resultant rising flow of the monofraction is:

$$Q_r = \frac{G_f r_s F_f}{P}; \qquad \text{(VIII-19)}$$

According to the model of the regular cascade, equations (VIII-17), (VIII-18) depend on K, z, i^*, i, whereas (VIII-19) does not depend on the examined section (for the upper branch of the apparatus). Therefore, the density of the flow of the particles will be estimated for sections i for which the relative flows are equal also on the basis of equation (VIII-19).

Assuming that the distribution of particles in the cross section of apparatus is uniform, we obtain equations for the density of the flow of the particles of a fixed narrow size class:

$$q_1 = \frac{G_f \cdot r_s}{F \cdot P}$$

for a unit relative flow, and

$$q = q_1 \cdot F_f \qquad \text{(VIII-20)}$$

for the resultant rising flow.

The resultant equations will be transformed and, for this purpose, we shall multiply the numerator and the denominator of the right hand part by the volume flow rate of the continuous phase V:

$$q = \frac{G_f \cdot r_s \cdot V}{V \cdot F \cdot P} = \frac{\mu \cdot r_s \cdot V}{F \cdot P} = \frac{\mu \cdot r_s \cdot w}{P}; \qquad \text{(VIII-21)}$$

We examine the dependences (VIII-20) and (VIII-21) on a specific example. For example, in separation of periclase (ρ = 3600 kg/m^3) in equilibrium apparatus in the regime w = 2.83 m/s and at a consumption concentration of μ = 1.5 kg/m^3, the yield of the fine product is approximately 20%.

The grain size composition of the initial material, the degree of fractional extraction and the density of particle flows, calculated from equations (VIII-20) and (VIII-21) are presented in Table VIII-1.

The data in the Table indicate that the density of the flow of the

Table VIII-1 The density of flows of particles of different monofractions calculated from equations (VIII-20 and (VIII-21)

Narrow size class (mm)	–0.14	–0.2+0.14	–0.3+0.2	–0.5+0.3
Mean size d (mm)	0.07	0.17	0.25	0.40
r_s,%	10.93	13.51	15.75	26.59
F_f, %	93	45.5	8.0	2.0
q (cm^2·s)$^{-1}$	71.8·10^3	6.2·10^3	2.3·10^3	0.94·10^3
q_Σ (cm^2·s)$^{-1}$	66.8·10^3	2.8·10^3	184	19

particles, especially the fine ones, in the apparatus are relatively high, regardless of the fact that the yield of the fine product is low. This may be interpreted by assuming that the fine particles, catching up with the coarse ones, exert an additional effect on them, in comparison with the continuous medium. In this case, the higher density of particles averages out and equalises this effect with respect to time. Consequently, it is possible to transfer to the carrying capacity of the flow as a whole – the dispersoid – and examine its effect separately on particles of each narrow size class ('separated' dispersoid).

It is interesting to compare the quantitative value of the density of the particle flow with the experimental data. For example, I.M. Razumov presented the experimental data on the number of collisions (in the conditions of a vertical pneumatic transport system) of a suspension-bearing flow, containing a monofraction with the size d = 2.3mm on a stationary surface with an area of 1 cm^2. In this case, the parameters, characterising the experimental conditions, were as follows:

- the density of the medium ρ_0 = 1.29 kg/m^3;
- the density of the material of particles ρ = 1200 kg/m^3;
- the initial mass concentration $3.5\dfrac{\text{kg/h}}{\text{kg/h}}$, corresponding to μ = 4.515 kg/m^3;
- r_s = 100%, since the tests were carried out with the monofraction;
- the mean flow rate of the medium was varied in the range of 10÷17.5 m/s.

In the experiments, 300 to 1300 impacts per second were recorded per 1 cm^2 of the surface situated in the rising flow. It may easily be seen that, according to the experimental conditions, the number of collisions is nothing else but the density of the flow of particles determined by equation (VIII-20). Assuming on average that w = 14 m/s, we find $N = q = 827 = 827\dfrac{1}{\text{cm}^2\times\text{s}}$, which is close to mean N recorded in the experiment. For velocities w = 10 m/s and w = 15 m/s, the number of collisions, determined by equation (VIII-20), is respectively 590 and 1033 1/cm^{2}s. Evidently, as estimates, these results are fully satisfactory.

For transition to the 'separated' dispersoid, differing for particles of each narrow size class, it is essential to estimate an important parameter such as density ρ_n (the density of the dispersoid for the particles of the j-th narrow size class).

We examine two monofractions with a mass of individual particles

'm' and 'M'. It is assumed that N fine particles collide in inelastic manner with a single coarse particle and transfer their momentum to it. To a first approximation, number N may be evaluated through the ratio of the densities of the flow of the examined monofractions:

$$N = \frac{q_m}{q_M} = \frac{r_m}{r_M} \cdot \left(\frac{d_m}{d_M} \right)^3; \tag{VIII-22}$$

As a result of inelastic collision, the ensemble of the fine and coarse particles have the same velocity $v_{Km} = v_{KM}$. In this case, the speed of fine particles decreases from v_{Hm} to v_{KM}, and the velocity of coarse particles increases from v_{HM} to v_{KM}. The change of the momentum of the ensemble of the fine particles is:

$$\Delta L_m = N \cdot m(v_{Hm} - v_{Km});$$

For the coarse particle

$$\Delta L_M = M(v_{KM} - v_{HM})$$

It is evident that $\Delta L_m = \Delta L_M$. This leads to:

$$v_{Km} = v_{KM} = \frac{N \cdot m v_{Hm} + M v_{HM}}{N \cdot m + M} \tag{VIII-23}$$

The conditions of uniform movement of the fine and coarse particles with the initial velocities have the form:

$$(u - v_{Hm})^2 = \frac{4}{3\lambda} \cdot \frac{\rho}{\rho_0} \cdot g d_m \tag{VIII-24}$$

$$(u - v_{HM})^2 = \frac{4}{3\lambda} \cdot \frac{\rho}{\rho_0} \cdot g d_M \tag{VIII-25}$$

The condition of the uniform movement of the coarse particle with a finite velocity in a dispersoid is:

$$(u - v_{KM})^2 = \frac{4}{3\lambda} \cdot \frac{\rho}{\rho_0} g d_M \tag{VIII-26}$$

Dividing the equations (VIII-25) by (VIII-26), we obtain:

$$\frac{\rho_n}{\rho_0} = \left(\frac{u - v_{HM}}{u - v_{KM}} \right)^2 \tag{VIII-27}$$

From equation (VIII-23) we obtain:

$$u - v_{KM} = u - \frac{N \cdot m v_{Hm} + m v_{HM}}{N \cdot m + M} \tag{VIII-28}$$

After quite simple transformations the equation has the form:

$$u - v_{KM} = \frac{N \cdot m (u - v_{Hm}) + M (u - v_{HM})}{N \cdot m + M} \tag{VIII-29}$$

Using the results from (VIII-27) gives:

$$\frac{\rho_n}{\rho_0} = \frac{(N \cdot m + M)^2}{\left(N \cdot m \dfrac{u - v_{Hm}}{u - v_{HM}} + M \right)^2} \tag{VIII-30}$$

or taking into account (VIII-24) and (VIII-25)

$$\frac{\rho_n}{\rho_0} = \frac{(N \cdot m + M)^2}{\left(N \cdot m \sqrt{\dfrac{d_m}{d_M}} + M \right)^2} = \left(\frac{N \dfrac{m}{M} + 1}{N \dfrac{m}{M} \sqrt{\dfrac{d_m}{d_M}} + 1} \right)^2 \tag{VIII-31}$$

Using (VIII-22) and taking into account $\dfrac{m}{M} = \left(\dfrac{d_m}{d_M} \right)^3$, we obtain:

$$\frac{\rho_n}{\rho_0} = \left(\frac{\dfrac{r_m}{r_M} + 1}{\dfrac{r_m}{r_M} \sqrt{\dfrac{d_m}{d_M}} + 1} \right)^2 \tag{VIII-32}$$

Since $\rho_n = \rho_0 + \Delta\rho_n$, the relative increase of the density of the dispersoid is:

$$\frac{\Delta\rho_n}{\rho_0} = \frac{\rho_n}{\rho_0} - 1$$

Taking into account the n-th amount of the examined monofractions, transferring the momentum to a coarser particle, the relative increase of the density of the dispersoid has the form:

$$\frac{\Delta\rho_n}{\rho_0} = \sum_{j=1}^{n}\left(\frac{\rho_n}{\rho_0}\right)_j - n + 1$$

Taking into account (VIII-32), the final equation for the density of the dispersoid is:

$$\rho_n = \rho_0\left[\sum_{j=1}^{n}\left(\frac{\frac{r_m}{r_M}+1}{\frac{r_m}{r_M}\sqrt{\frac{d_m}{d_M}}+1}\right)_j^2 + 1 - n\right] \qquad \text{(VIII-33)}$$

In order to carry out the numerical verification using equation (VII-33) for the previously examined experiment with the classification of periclase, the density of the flow of the dispersoids for the given monofractions was calculated:

d_m = 0.40 mm $\qquad \rho_n$ = 2.29 kg/m^3

d_m = 0.25 mm $\qquad \rho_n$ = 2.06 kg/m^3

d_m = 0.17 mm $\qquad \rho_n$ = 1.70 kg/m^3

The results indicate a small change in the density of the flow of the dispersoid acting on the individual narrow classes of the particles. On average, for the given case as the first approximation it may be accepted that ρ_n = 2.0 kg/m^3 = const for all monofractions. It should be mentioned that for the materials which do not differ greatly in the grain size distribution, the difference in the mean density of the flow will be small. For more accurate calculations (or for materials greatly differing in the composition), it is recommended to use equation (VIII-33) for each monofraction.

The second important problem in transition to the dispersoid is the evaluation of its structure, the profile of distribution of its velocities in the cross section of apparatus. In equations (VIII-24) and (VIII-25) we considered a dispersoid with the local effective speeds 'u'. The transition to ρ_n takes place on the basis of the transfer of the momentum by small particles to coarse ones. In this case, the density of the flow ρ_n was assumed to be constant in the cross section of apparatus (as a result of the assumption of the uniform distribution of the particles). Since the local carrying capacity, related to the unit surface, is characterised by the product $\rho_n u^2$, since the profile of its curve should be affine to the distribution of the square of the velocities of the dispersoid in the cross section. In our investigations, presented in the chapter concerned with the examination of the force effect of the two-phase flow, we found a sharp-tip of the profile of the curve for the upper part of the apparatus. Taking into account its affine transformation with a scale factor of 0.5 and the non-linear transformation according to the square root, we obtain the profile of the curve of the velocities of the dispersoid, close to parabolic. Since the volume flow rate of the dispersoid must be equal to the volume flow rate of the continuous medium, the equation of distribution of the speed of the dispersoid should have the form:

$$f\left(\frac{r}{R}\right)=2w\left[1-\left(\frac{r}{R}\right)^2\right] \qquad \text{(VIII-34)}$$

which corresponds to the coefficient n = 2.

It is evident that the above considerations should be regarded only as an approximate estimate because it is based on a number of assumptions.

Substituting n = 2 into equation (VIII-17) and replacing ρ_0 by ρ_n, we have

$$K=1-\sqrt{0,4\cdot B}; \qquad \text{(VIII-35)}$$

The resultant expression describes quite adequately B_{max} = 2.5 and is in quite satisfactory agreement with the description of experimental distribution curves $F_f(d)$ on the basis of a cascade model (for turbulent conditions). In the case of an arbitrary regime of flow around the particles from (VIII-17), we obtain

$$K=1-\frac{\sqrt{Ar\cdot B}}{36+1,575\sqrt{Ar}}; \qquad \text{(VIII-36)}$$

Equations (VIII-35) and (VIII-36) hold for ρ_0 = 1.2 kg/m^3 and ρ_n = 2.0 kg/m^3, which are included in the coefficients, and the criteria Ar and B are expressed as previously by means of ρ_0.

4. ANALYSIS OF THE MAIN EXPERIMENTAL DEPENDENCES FROM THE VIEWPOINT OF THE STRUCTURAL MODEL

Experiments were carried out to determine the main relationships governing the gravitational classification in cascade apparatus of different design. It has become possible to explain the results from the viewpoint of structural and cascade models:

1. From the equations (VIII-35) and (VIII-17) we obtain directly

the possibility of constructing the separation curve $F_f(x)$ in any regime. This was not possible on the basis of the currently available theory.

2. The same equations make it possible to consider the results of separation in relation to the number of sections of the cascade apparatus (the height of the classifier) and the area of supply of the initial material into the separating column.

3. The effect of the design special features of different apparatuses on the fractionation process is considered using different types of cascade models.

4. In order to verify the form of the curves of fractional separation for particles of different narrow size classes in relation to the classification regime, and also other relationships, the calculations were carried out for a tray cascade apparatus consisting of four sections (z = 4) with upper supply of the initial material (i^* =1). The experimental data for separation of quartzite in this apparatus (ρ = 2650 kg/m^3) are shown in Fig.VIII-2. Comparison of results of calculations of $F_f[d_j, w]$ using equations (VIII-35) and (VIII-17) with the experimental data are given in the same graph.

5. According to equation (VIII-35), velocity w_0 of the start of extraction of the fixed monofraction (intersection of the curve $F_f[d_i,w]$ with the abscissa) is determined from the condition:

$$K = 0 = 1 - \sqrt{0.4 \cdot B};$$

and consequently, ignoring the value of ρ_0 in comparison with ρ_f, we obtain:

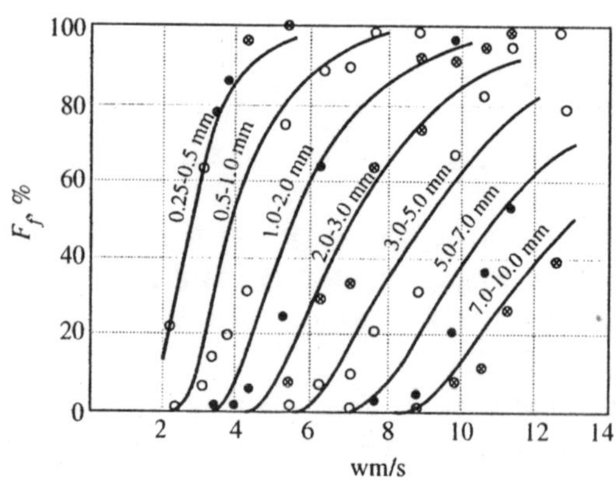

Fig. VIII-2 Dependence of the degree of fractional extraction into the fine product on the velocities of the airflow: ○, ⊗, ● – experimental points, ——— – calculated curve.

$$w_0 = \sqrt{0.4 \cdot \frac{\rho}{\rho_0} gd}$$

Consequently, we can predict the ratio of the velocity w_0 to the final rate of settling v_0 of a single particle of the given size class in air. It is well known that:

$$v_0 = \sqrt{\frac{4}{3\lambda} \cdot \frac{\rho}{\rho_0} \cdot gd} \qquad \text{(VIII-37)}$$

At an aerodynamic drag coefficient $\lambda = 0.5$, we obtain:

$$\frac{v_0}{w_0} = \sqrt{\frac{4}{3 \cdot 0.5 \cdot 0.4}} = 2.59 \qquad \text{(VIII-38)}$$

In order to verify this relationship, we calculated velocity v_0 from equation (VIII-37) for particles of all narrow size classes examined in the previous example, and experimental values of w_0 were taken from Fig.VIII-3. In this case, the coefficient of aerodynamic drag in equation (VIII-37) was determined from the more accurate dependence:

$$\lambda = 0.5 + \frac{29.2}{\sqrt{Ar}} + \frac{430}{Ar}$$

Comparison of the calculated dependence (VIII-38) with the experimental data is shown in Fig.VIII-3.

6. The characteristic properties of the separation curves are their affine properties, in particular, in relation to regime. This results in

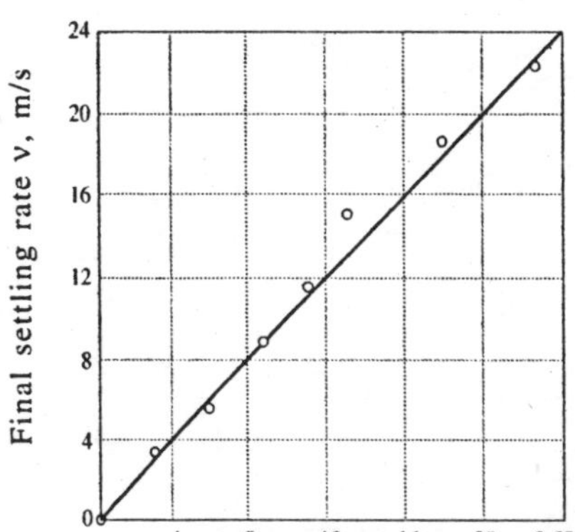

Fig. VIII-3 Relationship between the final rate of settling of the particles of different monofractions and the maximum velocity of the airflow resulting in $F_f(x) = 0$.

the unitary form of the curve $F_f\left(\frac{w}{w_{50}}\right)$ for all monofractions. Combined analysis of the structural and cascade models confirms this fact. Thus, the distribution coefficient K_{50} for any monofraction, extracted by 50%, is determined from the equation:

$$\frac{1-\left(\frac{1-K_{50}}{K_{50}}\right)^{z+1-i^*}}{\left(\frac{1-K_{50}}{K_{50}}\right)^{z+1}}=0,5 \qquad \text{(VIII-39)}$$

The value of K_{50}, determined from (VIII-39), is substituted into (VIII-35):

$$K_{50}=1-\sqrt{0.4\frac{\rho}{\rho_0}\cdot\frac{gd}{w_{50}^2}}$$

and consequently

$$0.4\frac{\rho}{\rho_0}gd=(1-K_{50})^2 w_{50}^2$$

The resultant value of $0.4\frac{\rho}{\rho_0}gd$ is substituted into (VIII-35) for the expression of an arbitrary distribution coefficient:

$$K=1-\frac{w_{50}}{w}(1-K_{50})$$

The latter is used in equation (VIII-17):

$$F_f=\frac{1-\left[\frac{1}{\left(\frac{1}{1-K_{50}}\right)\frac{w}{w_{50}}-1}\right]^{z+1-i^*}}{1-\left[\frac{1}{\left(\frac{1}{1-K_{50}}\right)\frac{w}{w_{50}}-1}\right]^{z+1}} \qquad \text{(VIII-40)}$$

The resultant equation (VIII-40) satisfies the unitary nature of the curve $F_f\left(\frac{w}{w_{50}}\right)$; to plot this curve, it is necessary to solve initially equation (VIII-39) and then (VIII-40). In particular, for the examined example (z = 4, i^* = 1), from equation (VIII-39) we have:

$$K_{50} = 0.342; \qquad \frac{1}{1-K_{50}} = 1.519$$

In this case, the dependence (VIII-40) has the form:

$$F_f\left(\frac{w}{w_{50}}\right) = \frac{1-\left(\dfrac{1}{1.519\dfrac{w}{w_{50}}-1}\right)^4}{1-\left(\dfrac{1}{1.519}\dfrac{w}{w_{50}}-1\right)^5} \qquad \text{(VIII-41)}$$

Comparison of the results of calculations using equation (VIII-41) and the experimental data is presented in Fig.VIII-4.

7. According to previous considerations, the consequence of the affinity of the separation curves $F_f(x)$ is the unitary form of the curve $F_f\left(\frac{d}{d_{50}}\right)$ for all velocities. Using the solution of equation (VIII-39)

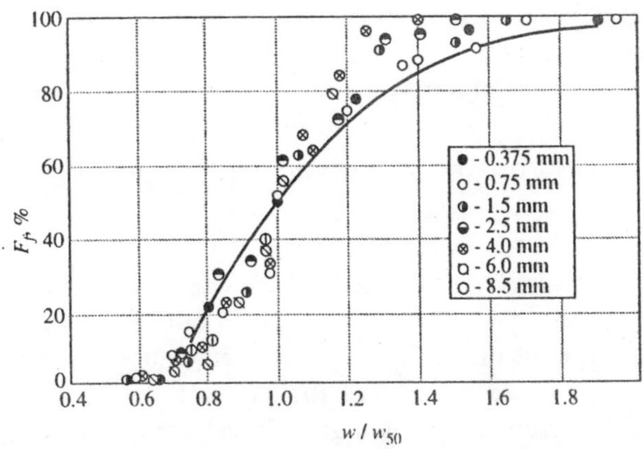

Fig.VIII-4 Dependence of the fractional extraction of the relative speed: circles – experimental points, —— – calculated curve.

in relation to K_{50} in (VIII-35), for an arbitrary regime:

$$K_{50} = 1 - \sqrt{0.4 \cdot \frac{\rho}{\rho_0} \cdot \frac{g d_{50}}{w^2}} \qquad \text{(VIII-42)}$$

$$\frac{(1-K_{50})^2}{d_{50}} = 0.4 \cdot \frac{\rho}{\rho_0} \cdot \frac{g}{w^2} \qquad \text{(VIII-43)}$$

Passing to an arbitrary distribution coefficient, we obtain:

$$K = 1 - (1-K_{50}) \cdot \sqrt{\frac{d}{d_{50}}} \qquad \text{(VIII-44)}$$

Consequently, (VIII-17) assumes the form:

$$F_f\left(\frac{d}{d_{50}}\right) = \frac{1 - \left[\dfrac{(1-K_{50})\sqrt{\dfrac{d}{d_{50}}}}{1-(1-K_{50})\sqrt{\dfrac{d}{d_{50}}}}\right]^{z+1-i^*}}{1 - \left[\dfrac{(1-K_{50})\sqrt{\dfrac{d}{d_{50}}}}{1-(1-K_{50})\sqrt{\dfrac{d}{d_{50}}}}\right]^{z+1}} \qquad \text{(III-45)}$$

Experimental verification of the dependence (VIII-45) for z = 4, i^* = 1 is in Fig.VIII-5.

It should be mentioned that the same procedure can be used to show the unitary form of the curves $F_f\left(\frac{d}{d_s}\right)$ and $F_f\left(\frac{w}{w_s}\right)$ for an arbitrary value of fractional extraction.

8. The universal form of the curve $F_f(Fr)$ determined by experiments for different conditions and different monofractions is revealed directly from examination of the structural and cascade models. For a specific apparatus (z; i^*), the density of the particles of the material (ρ), the fractional extraction is unambiguously determined by the parameter gd/w^2. Comparison of the calculated curve using equations (VIII-35) and (VIII-17) with experimental data in the examined example (z = 4; i^* = 1; 2500 kg/m^3) is shown in Fig.VIII-6.

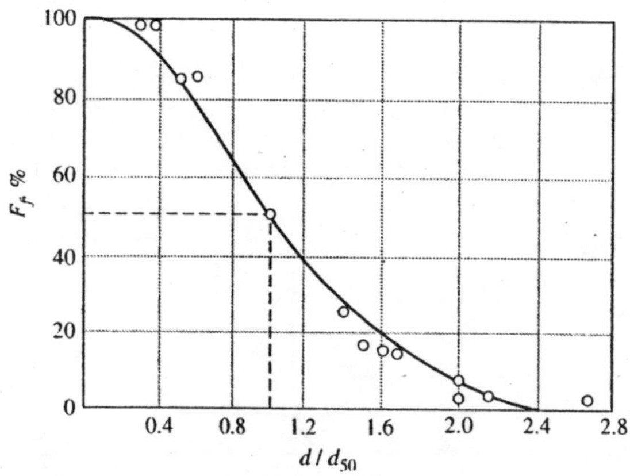

Fig.VIII-5 Dependence of the fractional extraction on the relative size: o – experimental points, —— – calculated curve.

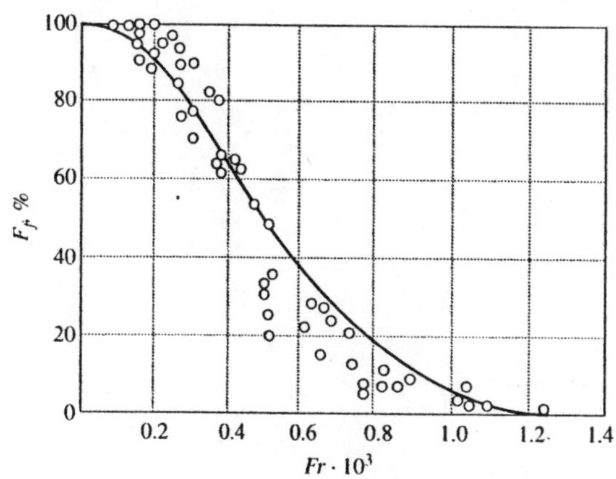

Fig.VIII-6 Dependence of the fractional extraction on the Froude criterion (Fr); o – experimental points, —— – calculated curve.

9. In a more general case, when the density of the classified material differs, the universal dependence is a function of the generalised parameter of classification $F_f(B)$. This fact also follows directly from expressions (VIII-35) and (VIII-17). In particular, for the separation of different materials in an equilibrium apparatus with a circular cross section ($D_{ann.}$ = 100 mm; z_{cond} = 9; i^* = 6; μ = 1.5 kg/m^3) the results of calculations of the values $F_f(B)$ and experimental data are presented in Fig.VIII-7.

10. In order to verify the correspondence of the structural model to the empirical dependence, from expression (VIII-17) we determine the value of the parameter Fr_{50}:

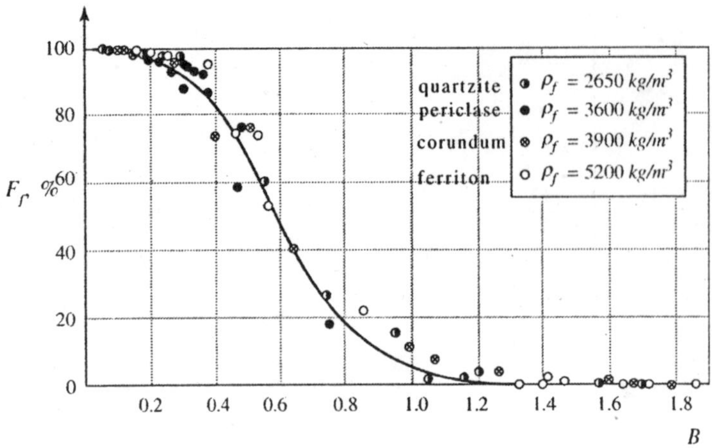

Fig. VIII-7 Dependence of the fractional extraction of the generalised classification parameter (B) for different materials: circles – experimental points; _____ solid line – calculated curve.

$$K_{50} = 1 - \sqrt{0.4 \frac{(\rho - \rho_0)}{\rho_0} \cdot Fr_{50}}$$

Consequently

$$Fr_{50} = \frac{(1 - K_{50})^2}{0.4} \cdot \frac{\rho_0}{(\rho - \rho_0)}$$

Taking into account that in the given example K_{50} = 0.342, we obtain:

$$Fr_{50} = 1.08 \cdot \frac{\rho_0}{(\rho - \rho_0)}$$

This is close to experimental correlation.

On the whole, all examined examples indicate convincingly that the structure of the flow prevails in the process of gravitational classification. An advantage of the method is that it is simple and the results are in good agreement with the experimental values and fundamental experimental dependences for the process of gravitational classification obtained up to now.

5. VERIFICATION OF THE ADEQUACY OF THE STRUCTURAL MODEL

In Chapter VIII, a number of models of the cascade organisation of the process were proposed. On the basis of these models, and taking into account the structure of the flow, it is possible to use a more differentiated approach to predicting the results of fractionation in individual systems differing in design features. Without discussing the extremely complicated pattern of the formation of the flow in actual systems, in the roughest examination, it is possible to separate three special features characteristic of the moving flow which can be associated with the design of apparatus:

– the nature of change of the velocity field of the solid phase along the height of the classification column;

– the presence of stagnant zones and the degree of filling of the cross section of the apparatus with the moving flow;

– the nature of movement of the material in the first phase of the process (in the feed section), the intensity of its interaction with the flow and internal elements, leading to the equalisation of concentration and a decrease in the size of jumps and dips.

For example, the velocity field of the solid medium may be both uniform along the height of the apparatus and non-uniform. The first case evidently includes hollow (equilibrium) apparatuses with a constant cross section whose operating characteristics are very close to this case. The uniform velocity field determines the same regime of interaction of the flow with particles of any level of the section of the separating column and predetermines the distribution coefficient constant in the height of the apparatus. The model describing most efficiently these conditions of organisation of the process is the model of the regular cascade (MRC). Thus, the MRC should be a basis for calculating the equilibrium apparatus with circular, square or rectangular sections and also zigzag-type apparatus.

Conical apparatuses with expanding or decreasing cross section and also with a complex configuration of the walls are characterised by the distribution coefficient changing along the height. A basis for calculating these systems for the known variation of K (cross section) may be the model of a completely non-uniform cascade.

For circular apparatus, apparatus with conical built-in elements, and a number of other systems, a characteristic feature is the variable narrowing and expansion of the flow of the medium. Each section of these systems may be regarded as consisting of two separating elements with the distribution coefficients of the monofraction K_1 and K_2. Therefore, the most suitable model for predicting these results for classification under these conditions should be the duplex cascade model.

The intensity of the interaction of the medium with the particles depends not only on the non-uniform velocity field along the height of the apparatus but also on the non-uniformity of the field in the cross section. For example, for apparatus with a circular cross section, the effect of transverse non-uniformity of the flow is taken into account by the structure model. In this case, it is assumed evident that the degree of filling of the cross section of the apparatus by the moving solid phase is 100%. A different situation exists in apparatus with square and right-angled cross sections characterised by the formation of dead zones in the corners. These dead zones reduce to zero the effect of the medium on the yield of particles. If in the rough estimates the line of the zero velocity of the flow in the square cross section apparatus is represented by a circle inscribed into the square, the degree of filling of this cross section by the uprising flow is:

$$C_{sq} = \frac{F_{cir}}{F_{sq}} = \frac{\pi}{4}$$

Since the distribution coefficient (taking into account the structural model) is determined on the basis of the ratio of the areas, in calculating the coefficient for the square section it is necessary to introduce a correction coefficient: $C_{sq} = \frac{\pi}{4}$.

Taking this coefficient into account:

$$k_{sq} = \frac{\pi}{4} \cdot K_0 \qquad \text{(VIII-46)}$$

where K_0 is the coefficient of distribution for apparatus with a circular cross section.

If the line of zero velocity of the apparatus with a rectangular section is represented by an inscribed ellipse, then K_{rec} can also be determined using equation (VIII-46), since

$$C_{rec} = \frac{F_{el}}{F_{rec}} = \frac{\pi}{4} \qquad \text{(VIII-47)}$$

Equation (VIII-47) for the square and right-angled cross sections is recommended only as the first approximation, because the actual degrees of filling will be slightly higher.

Finally, the nature of movement of the material in the first phase of the process in the absence of intensive interaction of the particles with the internal elements of the apparatus and in the presence of the dead zones may lead to a large dip in the main bulk of the particles in relation to the initial level of introduction. The most favourable conditions for passage are obviously characteristic of the hollow apparatus with the square or right-angled section.

Taking this into account, it was attempted to carry out the quantitative prediction of the results of the process of fractionation for a number of systems for different designs. The results of calculations compared with the experimental data are presented in appropriate figures.

Figure VII-8 shows a calculated curve and experimental values for classification of periclase with a density of ρ = 3600 kg/m^3 in an equilibrium apparatus with a circular cross section (D_{an} = 100 mm). The number of conventional cross sections Z = 9, the feed section i^* = 6. The consumption concentration of the material is μ = 1.5 kg/m^3. Calculations were carried out using the model of the regular cascade and the structural model in accordance with the equations (VIII-17) and (VIII-35). For the dependence $F_f(B)$, the mean deviation of the calculated curve from experimental points in Fig.VIII-8 is +1.8% ÷ 2.1%.

Figure VIII-9 shows the dependence $F_f(B)$ for the classification of quartzite with a density of ρ = 2650 kg/m^3 in equilibrium apparatus with a square cross section, size 100 × 100 mm^2. The number of conventional sections z = 6, the feed section i^* = 3. The consumption of concentration of the material μ = 2 kg/m^3. Calculations were carried out using the model of a regular cascade with a skip of 1.5 conventional sections

$$F_f = \frac{1-\chi^{2.5}}{1-\chi^{7}}$$

The distribution coefficient is determined with a correction for the square section in accordance with (VIII-46):

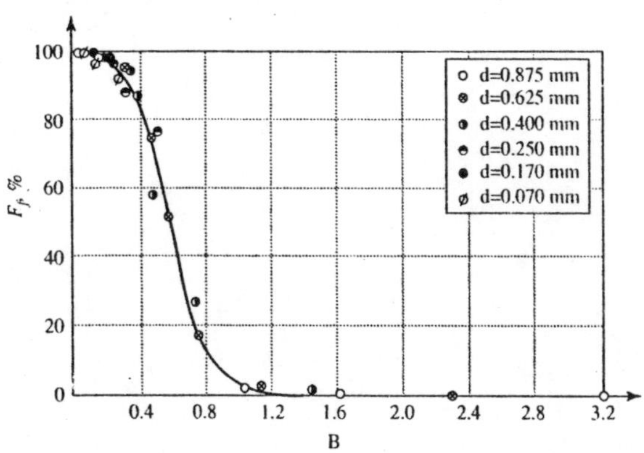

Fig.VIII-8 Dependence of $F_f(x) = f(B)$ for circular cross-section apparatus: circles – experimental points; —— – calculated curve.

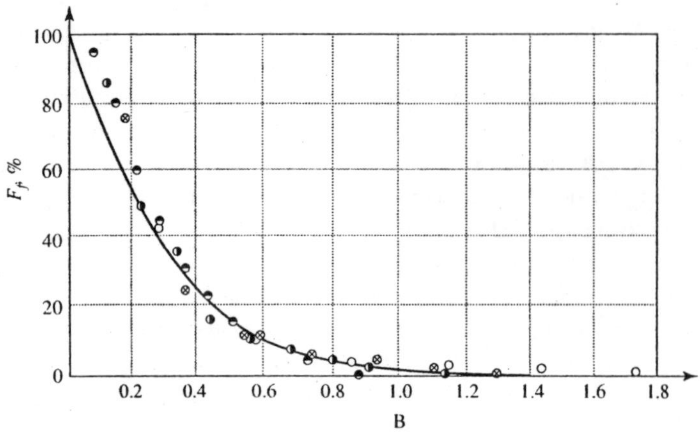

Fig.VIII-9 Dependence $F_f(x) = f(B)$ for square section apparatus: circles - experimental points; —— - calculated curve.

$$k_{sq} = \frac{\pi}{4}\left[1-\sqrt{0.4\cdot B}\right]$$

The maximum deviation of the calculated curve from the experimental point does not exceed 15%.

Figure VIII-10 shows the dependence $F_f(B)$ in separation of quartzite (ρ = 2650 kg/m^3) in apparatus with a rectangular cross section of the zigzag type. The number of sections z = 6, i^* = 3. The consumption concentration is μ = 2.0 kg/m^3. Calculations were carried out using the model of the regular cascade:

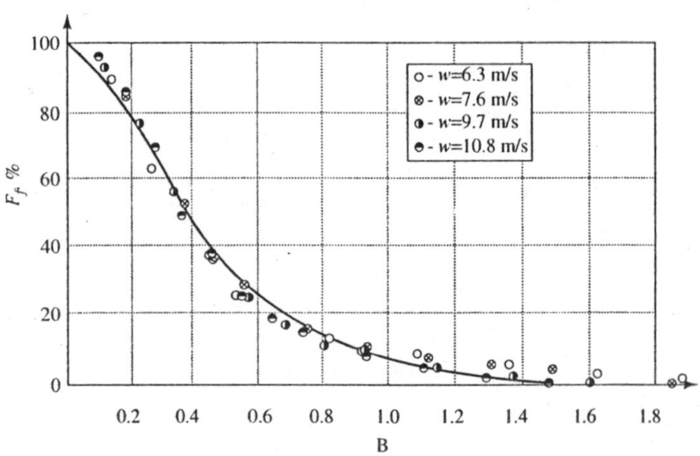

Fig.VIII-10 Dependence $F_f(x) = f(B)$ for "zigzag" type apparatus: circless - experimental points; —— - calculated curve.

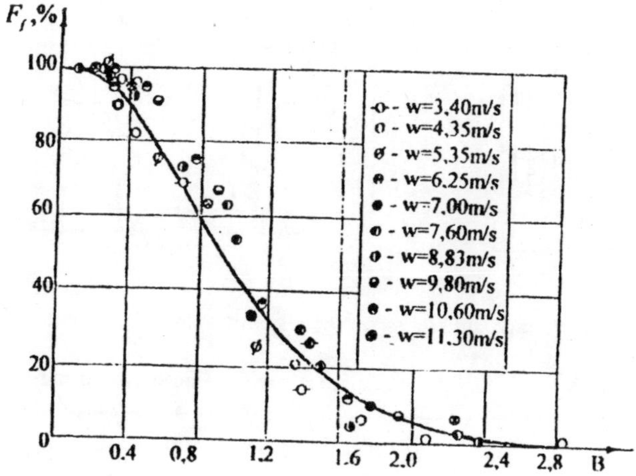

Fig.VIII-11 Dependence $F_f(x) = f(B)$ for tray apparatus: symbols – experimental points; —— – calculated curve.

$$F_f = \frac{1-\chi^4}{1-\chi^7}$$

The distribution coefficient was determined with a correction for the square section:

$$k_{sq} = \frac{\pi}{4}\left[1-\sqrt{0.4 \cdot B}\right]$$

The maximum deviation of the calculated curve from the experimental point does not exceed 7%.

6. MULTIROW CLASSIFIER

We have developed a multirow cascade classifier with a productivity of 60–70 t/h consisting of 7 rows of cascade purification (Fig.VIII-12). The apparatus is designed for fractionation of potassium chloride containing at least 80% of material with the size greater than 0.1mm.

In accordance with the technical conditions in a product from which dust was removed, the content of dust fractions (0.1mm) should not exceed 2%. For calculation of the fractional schema we use the resultant dependence.

Calculations were carried out for the optimum speed of the flow which according to industrial tests is 4.1 m/s (Table VIII-2). Initially, we examine the operation of apparatus without re-circulation (Fig.VIII-12). The experimental separation curves are shown in a

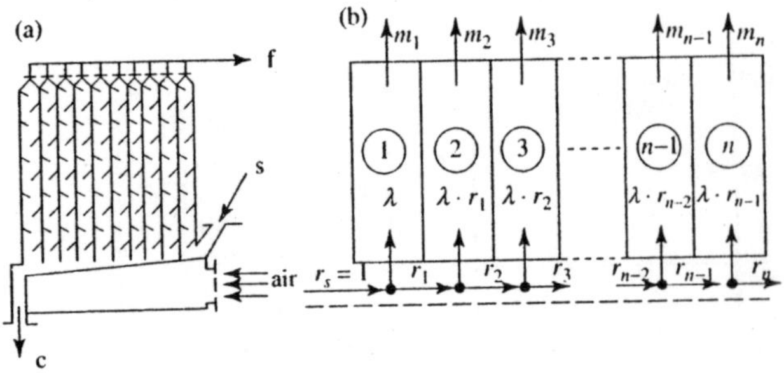

Fig. VIII-12 Multistage apparatus: a) principal diagram; b) calculation scheme.

Table VIII-2. Results of calculations of fractionation methods

Variant	γ_c,%	$D_c^{(-0.1)}$,%	γ_{nn},%	$R_{nn}^{(+0.1)}$	$R_{(s)}^{(+0.1)}$
I	66.610	0.434	33.390	40.069	13.3
II	66.001	0.374	33.999	41.011	13.94
III	66.682	0.449	30.318	34.087	10.33

dependence on the number of rows in Fig.VIII-13.

The yield of products from the section of the apparatus is determined in accordance with equations:

1st section

$$\gamma_1 = \sum r_s F_4(x);$$

2nd section

$$\gamma_2 = \sum r_s \left[F_8(x) - F_4(x)\right];$$

3rd section

$$\gamma_3 = \sum r_s \left[F_{12}(x) - F_8(x)\right]$$

The composition of the products is determined as follows:

1st section

$$r_1 = \frac{r_s F_4(x)}{\gamma_1};$$

2nd section

$$r_2 = \frac{r_s \left[F_8(x) - F_4(x)\right]}{\gamma_2};$$

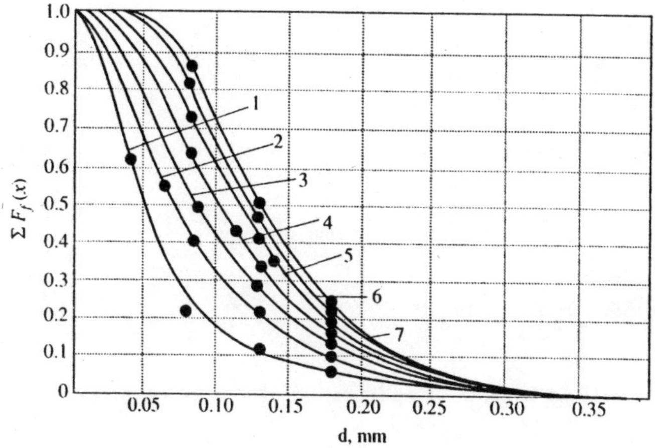

Fig.VIII-13 Fractional separation curves for a different number of rows (1–7) in a multirow cascade apparatus.

3rd section

$$r_3 = \frac{r_s\left[F_{12}(x) - F_8(x)\right]}{\gamma_3}$$

The coarse product, calculated using this method, has a contamination equal to 0.4%.

For the given organisation of the process where all yields of the final product are combined and represent the dust fraction we obtain the following parameter of the process:

$$\gamma_{int} = 64.3\%; \qquad D_{-0.1} = 0.4$$

In a dust-free product $D_{-0.1}$ = 0.4%, which corresponds to technical conditions. However, this results in a relatively low yield (64.3%) of the dust-free product and, in addition to this, a large part (up to 15.62% of the initial product) of the class +0.1mm is lost in the fine product. In this case, the losses of the class +0.1mm for the fraction of the 2nd and 3rd sections with their small yields (7.08 and 4.28% respectively) equal 8.71% of the initial value, i.e. 50% of total losses.

We examine a sectioned multirow apparatus consisting of three sections (Fig.VIII-14). These sections may consist of one or several rows.

Evidently, it is efficient not to mix the fine products of the 2nd and 3rd sections with the dusty product of the 1st section, and classification should be carried out in the same apparatus.

We determine the optimum schema of fractionation of recirculation.

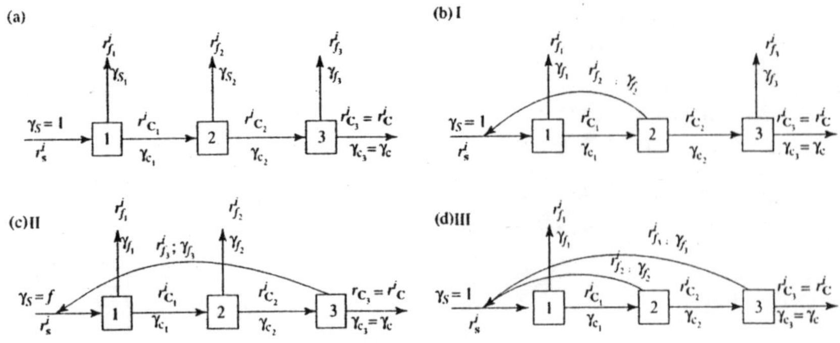

Fig.VIII-14 Diagram of optimisation of multirow separation.

The fractional extraction of the narrow size classes into the fine product will be calculated. Variant I (see Fig.VIII-12) will be examined.

The following rotations were used: $r_{f_1}, r_{f_2}, r_{f_3}$ – the content of some narrow size class in the fine product of the sections 1, 2 and 3, respectively; $r_{c_1}, r_{c_2}, r_{c_3}$ – is the content of the same size class in the coarse product of the sections 1, 2 and 3, respectively.

The equations used in the calculations

$$F_f(x) = \frac{r_f}{r_s}\gamma_f$$

or

$$F_f(x) = \frac{r_{f_1}}{r_s}\gamma_{f_1} + \frac{r_{f_3}}{r_s}\gamma_{f_3};$$

$$F_{f_1} = \frac{r_{f_1}}{r_s + r_{f_2}\gamma_{f_2}}\gamma_{f_1};$$

$$F_{f_2} = \frac{r_{f_2}}{r_{c_1}}\frac{\gamma_{f_2}}{\gamma_{c_1}};$$

$$F_{f3} = \frac{r_{f_3}}{r_{c_2}}\frac{\gamma_{f_3}}{\gamma_{c_2}};$$

$$r_s = r_{f_2}\gamma_{f_2} = r_{f_1}\gamma_{f_1} + r_{c_1}\gamma_{c_1}$$

$$r_{c_1}\gamma_{c_1} = r_{f_2}\gamma_{f_2} + r_{c_2}\gamma_{c_2}$$

$$r_{c_2}\gamma_{c_2} = r_{f_3}\gamma_{f_3} + r_{c_3}\gamma_{c_3}$$

$$r_s = r_{f_1}\gamma_{f_1} + r_{f_3}\gamma_{f_3} + r_{c_3}\gamma_{c_3} \qquad \text{(VIII-48)}$$

where $F_f(x)$ is the fractional extraction of the fixed fraction into the fine product for the entire system; $F_{f_1}; F_{f_2}; F_{f_3}$ is the fractional extraction of the fixed fraction into the fine product in the sections 1, 2 and 3, respectively.

For the examined variation,

$$F_f = \frac{r_{f_1}}{r_s}\gamma_{f_1} + \frac{r_{f_3}}{r_s}\gamma_{f_3} \qquad \text{(VIII-49)}$$

Consequently

$$F_f = F_{f_1}\left[1 + F_{f_2}\frac{1}{(1-F_{f_2})(1-F_{f_3})}\frac{r_{c3}}{r_s}\gamma_c\right] +$$

$$+F_{f_3}\frac{1}{1-F_{f_3}}\frac{r_{c3}}{r_s}\gamma_3 \qquad \text{(VIII-50)}$$

Finally, we obtain:

$$F_f = \frac{F_{f_1} + F_{f_3} - F_{f_1}F_{f_2} - F_{f_2}F_{f_3} + F_{f_1}F_{f_2}}{1 - F_{f_1} + F_{f_1}F_{f_2}} \qquad \text{(VIII-51)}$$

On the condition that each section has four separation columns:

$$F_{f_1} = F_{f_2} = F_{f_3} = \left[1-(1-F_{f_1})^4\right] \qquad \text{(VIII-52)}$$

In transformation of the previous dependence, we obtain:

$$F_f = \frac{1 - F_{c_1}^1 + F_{c_1}^8 F_{c_1}^{12}}{1 - F_{c_1}^4 + F_{c_1}^8} \qquad \text{(VIII-53)}$$

where $F_{c_1} = 1 - F_{f_1}$ is the extraction of some narrow size class in the first separation column into the coarse product.

Consequently,

$$F_c = \frac{F_{c_1}^{12}}{1 - F_{c_1}^4 + F_{c_1}^8} \qquad \text{(VIII-54)}$$

for variant II (see Fig.VIII-12)

$$F_c = \frac{F_{c_1}^{12}}{1 - F_{c_1}^{8} + F_{c_1}^{12}} \qquad \text{(VIII-55)}$$

for variant III (see Fig.VIII-12)

$$F_c = \frac{F_{c_1}^{12}}{1 - F_{c}^{4} - F_{c}^{12}} \qquad \text{(VIII-56)}$$

Thus, depending on the method of recirculation for the fine products, it is possible to calculate and select the optimum organisation of the process.

The method was used to calculate the results of all three variants and for comparison the results are presented in Table VIII-2.

Thus, variant III (the method of recirculation of the product of the 2nd and 3rd sections) is most efficient for removing the dust from potassium chloride. It increases by 5% the yield of the complete product and decreases to 10.33% (in comparison to 15.62% in the open cycle) the loss of target classes into the dust fraction.

An 11% increase in the product to apparatus with respect to initial feed does not change the regime of operation of the classification in the region of self-modelling with respect to concentration.

This example illustrates convincingly the possibilities of combined principle of organisation of cascade separation.

References

1. Barsky M.D., Fractionation of powders, Nedra, Moscow (1980).
2. Ushakov S.G. and Zverev N.I., Inertia of separation of dust, Energiya, Moscow (1984).
3. Shraiber A.A., Milyutin N.I. and Yatsenko V.P., Hydromechanics of two-component flows with a solid polydispersed substance, Naukova Dumka, Moscow (1980).
4. Razumov I.M., Pneumatic and hydraulic transport in chemical industry, Khimiya, Moscow (1979).
5. Kanusik Yu.P. and Barsky M.D., Effect of the height of the counterflow air classifier on the efficiency of the process, Izv. VUZ, Gornyi Zhurnal, No.8, 153–154 (1969).
6. Boothroyd R., The flow of gas with suspended particles, Russian translation, Mir, Moscow (1975).
7. Govorov A.V., Cascade and combined processes of fractionation of bulk materials. Dissertation for the title of Candidate of technical sciences, Sverdlovsk (1986).

Chapter IX

IRREGULAR CASCADES

1. COMPLEX CASCADES

When organising a separation cascade, there may be deviations from the operating principle forming the base of the regular cascade model. For example, investigations of the carrying capacity of a two-phase flow in the conditions of classification showed different force effects of a dispersoid for the upper (above the feed section) and lower branches of the cascade (sections from $i^* + 1$ to z). As a result, the separation processes in these branches is characterised by different distribution coefficients. For the upper branch it is k_1, for the lower branch k_2 with $k_1 > k_2$. Thus, we have an interlinked complex consisting of two regular cascades with their parameters (Fig.IX-1a,b).

The second possible case of organisation of a complex cascade is that a classifier is produced from two parts (Fig.IX-2a,b), characterised by different flow rates of the continuum (different distribution coefficients). However, a regular cascade is realised within the limits of each part. The principal difference of the examined case from the previous one is that the relationship between the upper and lower columns is realised through the lower section of the upper cascade and the upper section of the lower cascade, whereas feeding is carried out through an arbitrary section of the upper or lower cascades. In the previous case, the feed section was included in the number of linking sections.

Thus, in both examined cases, the complex cascade represents a partially non-uniform cascade.

The calculation methods for the first case are shown in Fig.IX-1a,d.

For the variant in Fig.IX-1a,c, the work of the upper branch of the cascade is characterised by the equation:

$$F_f = F_1(1+x) \qquad \text{(IX-1)}$$

From the material balance from the flows in the upper branch we have:

$$y = 1 + x - F_f$$

Taking into account (XI-1)

$$y = (1+x)(1-F_1) \qquad \text{(IX-2)}$$

For operation of the lower branch of the cascade:

$F_2 = \dfrac{x}{y}$ or taking (IX-2) into account $F_2 = \dfrac{x}{(1+x)(1-F_1)}$

and consequently

$$x = \frac{F_2(1-F_1)}{1-F_2(1-F_1)}$$

Substituting the last equation into (IX-1) we finally obtain

$$F_f = \frac{F_1}{1+F_1F_2-F_2} \qquad \text{(IX-3)}$$

According to Fig.IX-1a, in equation (IX-3):

$$F_1 = \frac{1-\chi_1}{1-\chi_1^{i^*+1}}; \qquad F_2 = \frac{1-\chi_2^{z-i^*}}{1-\chi_2^{z+1-i^*}}$$

For the variant in Fig.IX-1b we obtain

$$F_f = x \cdot F_1 \qquad \text{(IX-4)}$$

$y = x - F_f$ considering (IX-4) $\qquad y = x(1-F_1) \qquad$ (IX-5)

$F_2 = \dfrac{x}{1+y}$ and therefore $y = \dfrac{x}{F_2} - 1$

Equating the results in equation (IX-5) we have:

$$\frac{x}{F_2} - 1 = x(1-F_1)$$

Consequently

$$x = \frac{F_2}{1-F_2(1-F_1)}$$

Finally, from (IX-4) we have

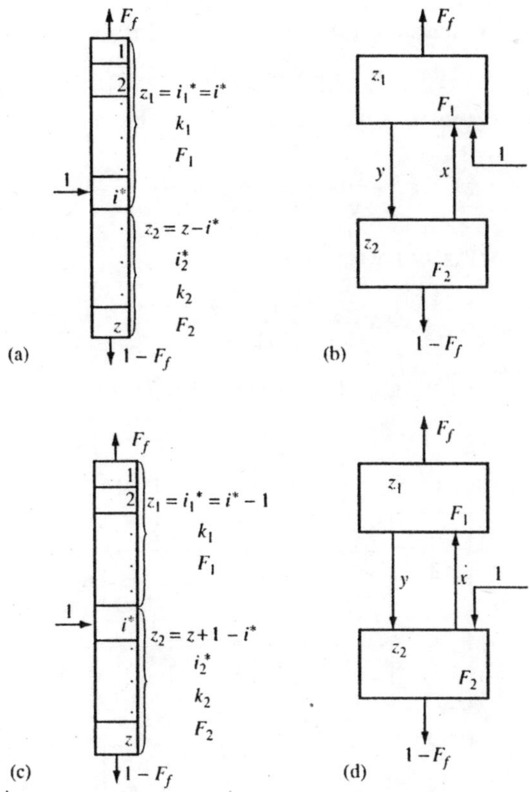

Fig.IX-1 Diagram of complex regular cascades produced from sections of the same type: a) feed section in the upper branch; b) feed section in the lower branch; c,d) calculation schemes.

$$F_f = \frac{F_1 F_2}{1 + F_1 F_2 - F_2} \qquad \text{(IX-6)}$$

According to Fig.IX-1c in this equation:

$$F_1 = \frac{1-\chi_1}{1-\chi_1^{i^*}}; \qquad F_2 = \frac{1-\chi_2^{z+1-i^*}}{1-\chi_2^{z+2-i^*}}$$

For the variant Fig.IX-2, the upper regular cascade has two feed sections i^* and z_1 with appropriate input flows of the fixed monofraction in the amount of 1 and x and the appropriate degrees of the fractional extraction into the fine product F_1 and F_0.

Therefore, for the operation of the upper cascade:

$$F_f = F_1 + xF_0 \qquad \text{(IX-7)}$$

From the material balance taking into account (IX-7) we get:

$$y = 1 + x - F_1 - xF_0$$

For the lower regular cascade it is evident that:

$$F_2 = \frac{x}{y} = \frac{x}{1 + x - F_1 - xF_0}$$

From the latter equation we obtain:

$$x = \frac{F_2(1 - F_1)}{1 - F_2(1 - F_0)}$$

Substituting the resulting equation into (IX-7) we finally have:

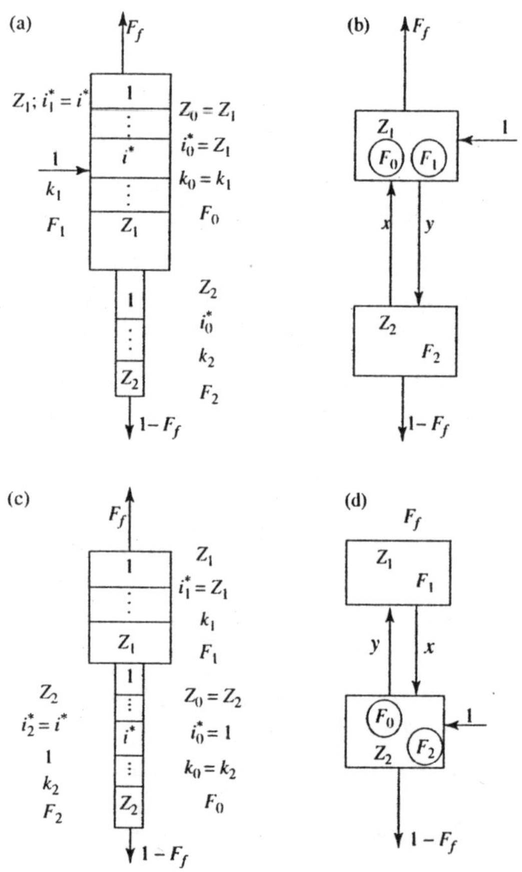

Fig.IX-2 Diagram of complex regular cascades produced from sections of different type: a) feed section in the upper branch, b) feed section in the lower branch, c,d) calculation schemes.

$$F_f = \frac{F_1 - F_2(F_1 - F_0)}{1 - F_2(1 - F_0)} \qquad \text{(IX-8)}$$

For (IX-8) the following is valid (Fig.IX-2a):

$$F_1 = \frac{1 - \chi_1^{z_1 + 1 - i_1^*}}{1 - \chi_1^{z_1 + 1}};$$

$$F_0 = \frac{1 - \chi_1}{1 - \chi_1^{z_1 + 1}};$$

$$F_2 = \frac{1 - \chi_2^{z_2}}{1 - \chi_2^{z_2 + 1}}$$

Similarly for the variant in Fig.IX-2c,d:

$$F_f = F_1 \cdot y;$$

$$x = y - F_f = y(1 - F_1) \qquad \text{(IX-9)}$$

For the lower cascade it holds that:

$$y = F_2 + xF_0$$

or

$$y = F_2 + y(1 - F_1)F_0$$

Consequently

$$y = \frac{F_2}{1 + F_0F_1 - F_0}$$

Substituting the resultant equation into (IX-9) we finally obtain:

$$F_f = \frac{F_1F_2}{1 - F_0(1 - F_1)} \qquad \text{(IX-10)}$$

From (XI-10) it holds that (Fig.IX-2b):

$$F_1 = \frac{1 - \chi_1}{1 - \chi_1^{z_1 + 1}}; \qquad F_2 = \frac{1 - \chi_2^{z_2 + 1 - i_2^*}}{1 - \chi_2^{z_2 + 1}}; \qquad F_0 = \frac{1 - \chi_2^{z_2}}{1 - \chi_2^{z_2 + 1}}$$

2. UNBALANCED CASCADE

An unbalanced cascade is the one in which the work of the sections is characterised by the fact that the flow of the fixed monofraction in the section is not equal to the flow leaving the section into the adjacent sections. The stationary nature of the process remains unchanged. The operating diagram of such a cascade is in Fig.IX-3a.

Similar organisation of the process may have two realisations:
1. The flow in the section is smaller than that leaving into adjacent sections

$$r_i < r_i(k+\lambda);$$

2. The flow in the section is larger than the flow leaving the section

$$r_i > r_i(k+\lambda).$$

The first case is characterised by the distribution coefficient $k + \lambda > 1$ and corresponds to operations of the cascade with a 'skip' of the monofraction into the sections. The size of the 'skip' is $r_i(k + \lambda - 1)$. The model with a skip can be realised in accordance with the variant in Fig.IX-3b, i.e. the skip of the monofraction in the direct direction, and variant Fig.IX-3c – the skip in the reverse direction.

The second case is characterised by the distribution coefficient $k + \lambda < 1$ and corresponds to the operation of an unbalanced cascade with a circulation (dead) zone (Fig.IX-3c). In this case, the circulation flow in the section:

$$r_{cir} = r_i(1-\lambda-k).$$

From the condition of the stationary nature of the process, for mathematical description of the operation of the unbalanced cascade we have a recurrent equation (Fig.IX-3a):

$$r_i(k+\lambda) = r_{i-1}\lambda + r_{i+1}k \qquad \text{(IX-11)}$$

and the boundary conditions:

– for the upper branch of the cascade $r_i(k + \lambda) = r_2 k$ (IX-12)

– for the lower branch of the cascade $r_z(k + \lambda) = r_z - 1$

It may easily be seen that the boundary condition (IX-12) are in the framework of the recurrent equation (IX-11) transformed to the form:

$$r_0 = 0; r_{z+1} = 0 \qquad \text{(IX-13)}$$

The solution of equations (IX-11) and (IX-13) on the basis of calculations of finite differences leads to the result:

In the case $\lambda \neq k$

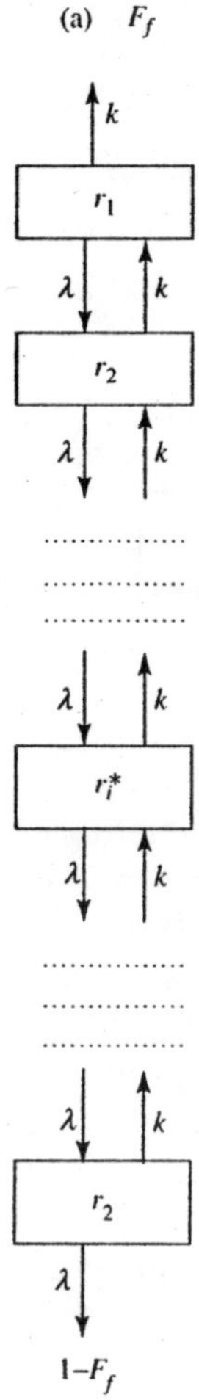

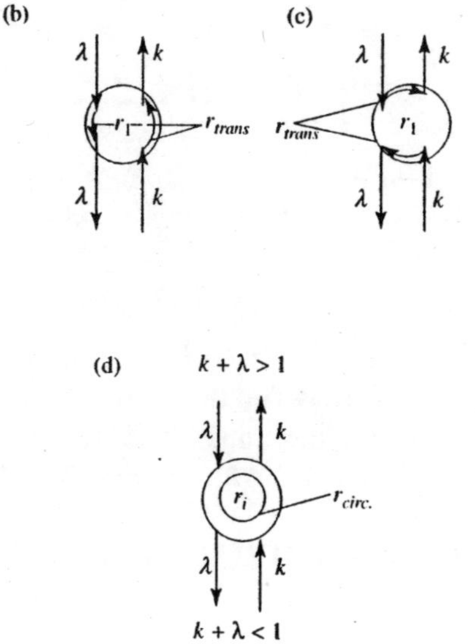

Fig.IX-3 Diagram of a homogeneous non-balanced cascade: a) diagram of distribution of the monofraction over the section; b) diagram with a skip of the monofraction in the direct direction; c) diagram with a skip of the monofraction in the reverse direction; d) diagram of operation of the section with the circular flow.

$$r_i = \frac{F_f}{(\lambda-k)}\cdot\left[\left(\frac{\lambda}{k}\right)^i - 1\right]; \qquad (1 \le i \le i^*)$$

$$r_i = \frac{1-F_f}{(\lambda-k)}\cdot\left[1-\left(\frac{k}{\lambda}\right)^{z+1-i}\right]; \qquad (i^* \le i \le z)$$

$$F_f = \frac{\left[1-\left(\frac{\lambda}{k}\right)^{z+1-i^*}\right]}{\left[1-\left(\frac{\lambda}{k}\right)^{z+1}\right]};$$

and for $\lambda = k$

$$r_i = \frac{F_f}{k} \cdot i; \qquad (1 \le i \le i^*)$$

$$r_i = \frac{1-F_f}{\lambda}(z+1-i); \qquad (i^* \le i \le z)$$

$$F_f = \frac{z+1-i^*}{z+1}$$

It may easily be seen that for the case of an equilibrium cascade where $\lambda = 1 - k$, all the above equations coincide with the previously derived equations holding for the model of the regular cascade.

3. UNIFORM EQUILIBRIUM CASCADE WITH ADDITIONAL FLOWS

In accordance with the previous assumptions, the model of the uniform cascade assumes that the distribution coefficients are constant in sections. Within the framework of the examined model, this concept is also applied to additional flows. Thus, the fraction of a flow, removed (supplied) from (to) an arbitrary section is the same. Without discussing the methods of organisation of a similar process, one can mention its possible applications for multi-product separation. The diagram of operation of a cascade with additional flows is shown in Fig.IX-4.

From the condition of equilibrium operation of each section of the cascade we have

$$r_i = r_i(k + \lambda + \delta)$$

Consequently, the fraction of the removed flow:

$$\delta = 1 - (k + \lambda)$$

From the condition of stationary operation of sections we have a recurrent equation characterising the operation of the cascade:

$$r_i = r_{i-1}\lambda + r_{i+1}k \qquad \text{(IX-14)}$$

The dependence (IX-14) is a homogeneous linear finite-difference equation with constant coefficients of the second order. Therefore, we have two boundary conditions for this equation:

– for the upper branch of the cascade $r_1 = r_2 k$

– for the lower branch of the cascade $r_z = r_{z-1}$ (IX-15)

The solution of equation (IX-14) with the boundary conditions (IX-15) has the form:

at $\lambda k \neq \frac{1}{4}$;

$$r_i = \frac{F_f(u_1^i - u_2^i)}{k(u_1 - u_2)}; \qquad (1 \le i \le i^*)$$

$$r_i = \frac{F_c}{\lambda\left(1 - \frac{u_1}{u_2}\right)u_2^{z-i}}\left[1 - \left(\frac{u_2}{u_1}\right)^{z+1-i}\right]; \qquad (i^* \le i \le z)$$

$$F_f = \frac{\left[1 - \left(\frac{u_1}{u_2}\right)^{z+1-i^*}\right]}{u_2^{i^*}\left[1 - \left(\frac{u_1}{u_2}\right)^{z+1}\right]};$$

where

$$u_1 = \frac{1 + \sqrt{1 - 4\lambda k}}{2k}; \qquad u_2 = \frac{1 - \sqrt{1 - 4\lambda k}}{2k};$$

for $\lambda \cdot k = \frac{1}{4}$;

$$r_i = 2iF_f\left(\frac{1}{2k}\right)^i; \qquad (1 \le i \le i^*)$$

$$r_i = \frac{2i^*}{(z+1)} \cdot \frac{\left[z + 1 - i(2k)^{z+1-i}\right]}{(i^* + 1 - i^* 2k)}; \qquad (i^* \le i \le z)$$

$$F_f = \frac{(2k)^{i^*}}{(z+1)} \cdot \frac{\left[z + 1 - i^*(2k)^{z+1-i^*}\right]}{(i^* + 1 - i^* 2k)};$$

$$F_c = \left(1 - \frac{2kz}{z+1}\right) \cdot \frac{i^*}{(i^* + 1 - i^* 2k)}.$$

In a particular case, where there are no additional flows, i.e. $k + \lambda = 1$, we obtain:

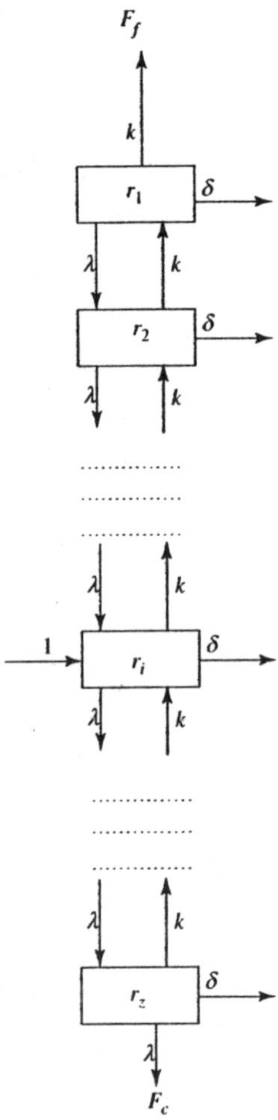

Fig. IX-4 Diagram of distribution of the monofraction in sections of an equilibrium cascade with additional flows.

$\delta = 1 - (k + \delta) > 0$ fraction of the flow supplied to each section.

$\delta = k + \lambda - 1 > 0$ fraction of the flow removed from each section;

$$u_1 = \frac{1+\sqrt{1-4k(1-k)}}{2k} = \chi;$$

$$u_2 = \frac{1-\sqrt{1-4k(1-k)}}{2k} = 1$$

It may easily be seen that all these equations are transferred into equations valid for the model of the regular cascade.

4. MATHEMATICAL MODEL OF A DUPLEX CASCADE

The model of the regular cascade may be used most efficiently for equipment with a uniform velocity field along the height of the column (for example, zigzag classifier, equilibrium apparatus). However, the majority of systems can be regarded as consisting of sections operating in two different conditions (tray classifier, apparatus with conical inserts, poly-cascade, etc.). Each section of such a cascade may be regarded as consisting of two separating elements A and B which have their own coefficients of distribution for a fixed monofraction: k_1 and k_2. The duplex model assumes a non-uniform (in the sense of k_1 and k_2) cascade of alternating equilibrium separating elements A and B.

Four possible variants of organisation of the duplex cascade are shown schematically in Fig.IX-5 and IX-6. The conventional notations:

I – the unit flow of a fixed monofraction;

R_{iA}, r_{iB} – the total flow of the given monofraction in the element A, B of the i-th section of the cascade;

i^* – the section of input of the unit flow;

k_1, k_2 – coefficients of distribution of the monofraction in element A,B.

We shall examine the cascade A–B with supply of a unit flow of a fixed monofraction in element A (Fig.IX-5a). From the conditions of the material balance or the stationary process for each element we have a system of canonic equations:

$$r_{1A} = r_{1B}k_2;$$

$$r_{1B} = r_{1A}(1-k_1) + r_{2A}k_1;$$

$$r_{2A} = r_{1B}(1-k_2) + r_{2B}k_2;$$

....

....

$$r_{i-1,B} = r_{i-1,A}(1-k_1) + r_{iA}k_1; (a)$$

$$r_{iA} = r_{i-1,B}(1-k_2) + r_{iB}k_2; (b)$$

$$r_{iB} = r_{iA}(1-k_1) + r_{i+1,A}k_1;(c)$$
$$r_{i+1,A} = r_{iB}(1-k_2) + r_{i+1,B}k_2;(d)$$

....

....

$$r_{i^*A} = r_{z-1,B}(1-k_2) + r_{i^*B}k_2 + 1;$$

....

....

$$r_{zA} = r_{z-1,B}(1-k_2) + r_{zB}k_{2;}$$
$$r_{zB} = r_{zA}(1-k_1); \qquad \text{(IX-16)}$$

From (IX-16a,b,c) we have:

$$r_{iA} = r_{i-1,A}(1-k_1)(1-k_2) + r_{iA}k_1(1-k_2) + r_{iA}k_2(1-k_1) + r_{i+1,A}k_1k_2;$$

Consequently

$$r_{iA}\left(\frac{1}{k_1k_2} - \chi_1 - \chi_2\right) = r_{i-1,A}\chi_1\chi_2 + r_{i+1,A};$$

or

$$r_{iA}(1+\chi_1\chi_2) = r_{i-1,A}\chi_1\chi_2 + r_{i+1,A} \qquad \text{(IX-17)}$$

We denote

$$\frac{1}{1+\chi_1\chi_2} = q \qquad \text{(IX-18)}$$

Taking into account (IX-18), equation (IX-17) is transformed to the form

$$r_{iA} = r_{i-1,A}(1-q) + r_{i+1,A}q \qquad \text{(IX-19)}$$

We have obtained a recurrent equation for the flow of the monofraction in element A. It has the form of a recurrent equation of the model of the regular cascade in which the coefficient of distribution of the monofraction k is replaced by parameter q, determined in accordance with (IX-19). From (Fig.IX-6a,b,c) we have an identical recurrent equation for element B:

$$r_{iB} = r_{i-1,B}(1-q) + r_{i+1,B}q \qquad \text{(IX-20)}$$

The boundary conditions are determined from the appropriate canonic equations of the system (IX-16)

$$r_{0,B}=0=r_{0,A}(1-k_1)+r_{1A}k_1;$$

And consequently

$$r_{0,A}=-\frac{1}{\chi_1}r_{1A};$$

$$r_{z+1,A}=0=r_{zB}(1-k_2)+r_{z+1,B}k_2;$$

and

$$r_{z+1,B}=-\chi_2 r_{zB}$$

Thus, we obtain the following boundary condition for the cascade A÷B:

For element A

$$r_{0A}=-\frac{1}{\chi_1}r_{1A}$$

$$r_{z+1,A}=0 \tag{IX-21}$$

For element B

$$r_{0B}=0$$

$$r_{z+1,B}=-\chi_2 r_{zB} \tag{IX-22}$$

The boundary conditions (IX-21) and (IX-22) are valid only in the case of the cascade A-B and do not depend on the element into which the unit flow of the monofraction is supplied.

In the case of a separating cascade A-A (Fig.IX-6) we have constant recurrent equations (IX-19), (IX-20) with different boundary conditions:

For element A

$$r_{0A}=-\frac{1}{\chi_1}r_{1A}$$

$$r_{z+1,A}=-\chi_1 r_{zA} \tag{IX-23}$$

For element B

$$r_{0B}=0$$

$$r_{zB}=0 \tag{IX-24}$$

The solution of the recurrent equations of the model of the duplex cascade is expressed by a dependence identical with the dependence valid for the model of regular cascade. For example, for the flow in section i for the elements A and B we have:

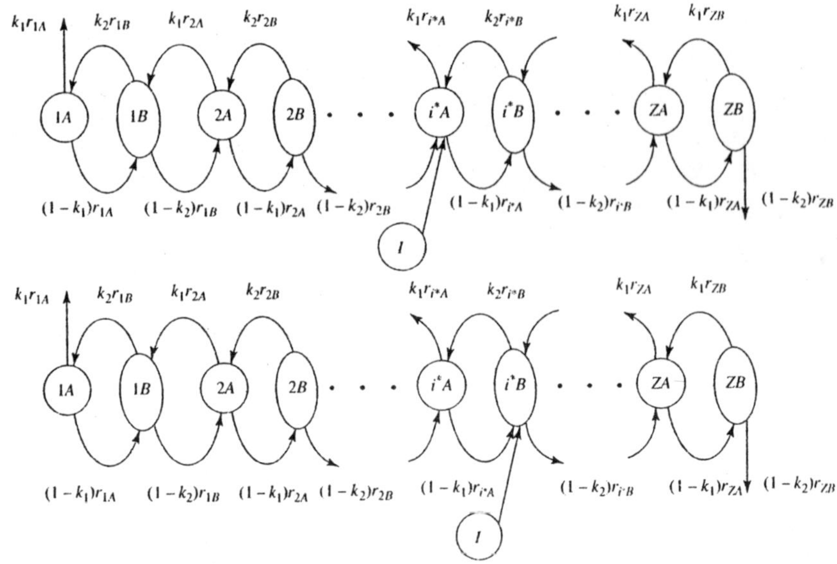

Fig IX-5 Diagram showing streams in an A-B duplex cascade: a) input of material into element A; b) input of material into element B.

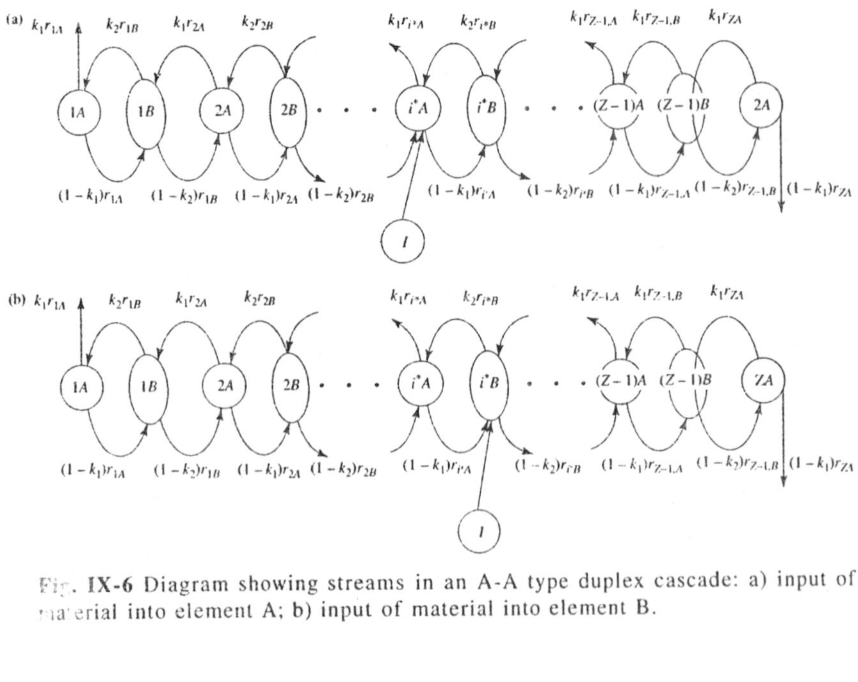

Fig. IX-6 Diagram showing streams in an A-A type duplex cascade: a) input of material into element A; b) input of material into element B.

$$r_{i(A,B)} = c_1 + c_2 Q^i; (Q \neq 1)$$

$$r_{i(A,B)} = c_1 + c_2 i; \qquad (Q = 1), \qquad \text{(IX-25)}$$

where $Q = \dfrac{1-q}{q} = \chi_1 \chi_2$ is the duplex parameter.

It should be mentioned that the coefficients c_1 and c_2, included in equation (IX-25) are valid for one of the branches of the cascade (for example, the upper one). Consequently, for the lower branch of the cascade they have different values, c_3 and c_4, respectively, as a result of the effect of the feeding section, where the material balance of the flows differs by unity from the flows in an arbitrary section of both branches of the cascade. Thus, the following equations hold for the lower branch of the cascaded:

$$r_i = c_3 + c_4 Q^i; (Q \neq 1) \qquad i^* \leq i \leq z$$

$$r_i = c_3 + c_4 i; \quad (Q = 1) \qquad \text{(IX-26)}$$

The specific forms of relationships (IX-26) will depend on specific boundary conditions (the type of duplex cascade). We examine gradually all possible cases.

a. Cascade A–B

For the examined case, in accordance with (IX-21) the boundary conditions have the form

$$r_{0A} = \frac{1}{\chi_1} r_{1A}$$

$$r_{z+1,A} = 0 \qquad \text{(IX-27)}$$

$$r_{0B} = 0$$

$$r_{z+1,B} = -\chi_2 r_z B \qquad \text{(IX-28)}$$

a. We examine the flows in the element $A (Q \neq 1)$.

Equations (IX-27) and (IX-28) are not sufficient for determining the coefficient c_1 and c_2 because they include also unknown coefficients c_3 and c_4. Therefore, we use an additional condition – the material balance for the entire cascade – the sum of flows, leaving elements I and the last element, is equal to unity:

$$r_{1A}k_1 = F_f$$
$$r_{zB}(1-k_2) = 1-F_f \qquad \text{(IX-29)}$$

According to the scheme of flows (Fig.IX-5)

$$r_{zB} = r_{zA}(1-k_1)$$

Consequently, the system of equations (IX-29) has the form:

$$r_{1A}k_1 = F_f$$
$$r_{zA}(1-k_1)(1-k_2) = 1-F_f \qquad \text{(IX-30)}$$

For the first equations of the system (IX-27) and (IX-30) taking into account (IX-25) we have:

$$c_1 + c_2 = -\frac{1}{\chi_1}(c_1 + c_2 Q)$$
$$(c_1 + c_2 Q)k_1 = F_f \qquad \text{(IX-31)}$$

Solving (IX-31) we obtain:

$$c_1 = -\frac{F_f}{(Q-1)k_1k_2}; \qquad c_2 = \frac{F_f}{(Q-1)k_1k_2} \cdot \frac{k_2}{(1-k_1)}$$

Consequently, taking into account (IX-25):

$$r_{iA} = \frac{\left[1 - Q^i \cdot \dfrac{k_2}{(1-k_1)}\right]}{(1-Q)k_1k_2} \cdot F_f$$

Thus, taking into account the relationships for Q

$$(1-Q)k_1k_2 = k_1 + k_2 - 1$$

Finally we obtain:

$$r_{iA} = \frac{\left(1 - Q^i \cdot \dfrac{k_2}{1-k_1}\right)}{(k_1 + k_2 - 1)} \cdot F_f; (1 \le i \le i^*) \qquad \text{(IX-32)}$$

It should be mentioned that (IX-32) is valid both when supplying the monofraction into element A and into element B.

For the second equation of the system (IX-27) and (IX-29) taking into account (IX-25) we have

$$c_3 + c_4 Q^{z+1} = 0$$

$$(c_3 + c_4 Q^z)(1-k_1)(1-k_2) = 1 - F_f \qquad \text{(IX-33)}$$

Solving (IX-33), we obtain:

$$c_3 = \frac{1-F_f}{(Q-1)k_1k_2}; \qquad c_4 = -\frac{1-F_f}{(Q-1)k_1k_2Q^{z+1}}$$

Consequently, taking into account (IX-25)

$$r_{iA} = \frac{(Q^{i-z-1}-1)}{(k_1+k_2-1)} \cdot (1-F_f) \qquad \text{(IX-34)}$$

It should be mentioned that (IX-34) is valid when supplying the monofraction both into the element i^*A and the element i^*B.

For the second elements of the system (IX-27) and (IX-30), taking into account (IX-25) we have

$$c_3 + c_4 Q^{z+1} = 0$$

$$(c_3 + c_4 Q^z)(1-k_1)(1-k_2) = 1 - F_f \qquad \text{(IX-35)}$$

Solving this system, we have

$$c_3 = \frac{1-F_f}{(Q-1)k_1k_2}; \qquad c_4 = -\frac{1-F_f}{(Q-1)k_1k_2Q^{z+1}}$$

Taking into account (IX-26)

$$z_{iA} = \frac{(Q^{i-z-1}-1)}{(k_1+k_2-1)}(1-F_f) \qquad \text{(IX-36)}$$

It should be mentioned that when supplying a single monofraction into element i^*A, equation (IX-36) is valid for $i^* \le i \le z$, but when supplying into element i^*B, the number of the examined section should be in the range $i^* \le i \le z$, because element i^*A in these conditions is included in the upper branch of the cascade.

b. We examine the flows in the element B($Q \neq 1$).In this case, the systems of equations (IX-28) and (IX-29) taking into account (IX-25) and (IX-27) are reduced to the form:

$$c_1 + c_2 = 0$$

$$(c_1 + c_2 Q)k_1k_2 = F_f \qquad \text{(IX-37)}$$

$$c_3 + c_4 Q^{z+1} = -\chi_2 (c_3 + c_4 Q^z)$$
$$(c_3 + c_4 Q^z)(1-k_2) = 1 - F_f \qquad \text{(IX-38)}$$

The solution of the system (IX-38) is:

$$c_1 = -\frac{F_f}{(Q-1)k_1 k_2}; \qquad c_2 = \frac{F_f}{(Q-1)k_1 k_2}$$

Consequently, taking into account (IX-25) we obtain:

$$r_{iB} = \frac{(1-Q^i)}{(k_1+k_2-1)} \cdot F_f; \qquad \left.\begin{array}{l}(I \to i^* A; 1 \le i \le i^* - 1) \\ (I \to i^* B; 1 \le i \le i^*)\end{array}\right\} \qquad \text{(IX-39)}$$

The solution of the system (IX-38) is:

$$c_3 = \frac{(1-F_f)}{(Q-1)k_1 k_2}; \qquad c_4 = -\frac{(1-F_f)}{(Q-1)k_1 k_2} \cdot \frac{k_1}{(1-k_2)} \cdot \frac{1}{Q^z}$$

Consequently, using (IX-27) we obtain:

$$r_{iB} = \frac{\left(Q^{i-z} \cdot \dfrac{k_1}{1-k_2} - 1\right)}{(k_1+k_2-1)} \cdot (1-F_f); (I \to i^* A, B; i^* \le i \le z) \qquad \text{(IX-40)}$$

To determine the degree of fractional extraction when supplying a unit flow of the monofraction into element A, we use the condition of unambiguity of determination of r_{i*A} from the expression (IX-34)

$$\frac{\left[Q^{i^*} \cdot \dfrac{k_2}{(1-k_1)} - 1\right]}{(k_1+k_2-1)} \cdot F_f = \frac{(1-Q^{i^*-z-1}}{(k_1+k_2-1)} \cdot (1-F_f)$$

Consequently, we easily obtain that:

$$F_f = \frac{1-Q^{z+1-i^*}}{1-Q^{z+1} \cdot \dfrac{k_2}{(1-k_1)}}; \qquad (A \div B; I \to i^* A)$$

Determination of F_f when supplying into i^*B is determined from the expression (IX-38), (IX-40):

$$\frac{(1-Q^{i^*})}{(k_1+k_2-1)}\cdot F_f = \frac{\left(Q^{i^*-z}\cdot\frac{k_1}{(k_1+k_2-1)}-1\right)}{(k_1+k_2-1)}\cdot(1-F_f)$$

Consequently

$$F_f = \frac{1-Q^{z+1-i^*}\cdot\frac{k_2}{(1-k_1)}}{1-Q^{z+1}\cdot\frac{k_2}{(1-k_1)}}; \qquad (A\div B; I\rightarrow i^*B)$$

c. Separation cascade A÷B operates in the regime when the parameter $Q = \chi_1 = \chi_2$, which is equivalent to $k_1 + k_2 = 1$.

Using for element A the boundary conditions (IX-27) and (IX-30), taking into account (IX-23), we have:

$$c_1 = -\frac{1}{\chi_1}(c_1+c_2)$$
$$(c_1+c_2)k_1 = F_f$$
$$c_3+c_4(z+1)=0$$
$$(c_3+c_4 z)(1-k_1)(1-k_2)=1-F_f$$

Solving the given system of equations, we obtain:

$$c_1 = -\frac{F_f}{(1-k_1)};\quad c_2=\frac{F_f}{k_1(1-k_1)};\quad c_3=\frac{(1-F_f)(z+1)}{(1-k_1)(1-k_2)};\quad c_4=-\frac{(1-F_f)}{(1-k_1)(1-k_2)}$$

Consequently, taking into account (IX-23) and (IX-25) gives:

$$r_{iA}=\frac{(1-k_1)F_f}{k_1k_2};\quad (I\rightarrow i^*A,B);\quad 1\le i\le i^* \qquad \text{(IX-41)}$$

$$r_{iA}=\frac{(z+1-i)}{k_1k_2}(1-F_f);\quad (I\rightarrow i^*A; i^*\le i\le z;)$$
$$(I\rightarrow i^*B; i^*+1\le i\le z) \qquad \text{(IX-42)}$$

To determine the flow in the element B(Q = 1) we use the expressions:

$$c_1 = 0$$
$$(c_1 + c_2)k_1k_2 = F_f$$
$$c_3 + c_4(z+1) = -\chi_2(c_3 + c_4 z)$$
$$(c_3 + c_4 z)(1 - k_2) = -1 - F_f$$

Solving the given system of equations leads to:

$$c_1 = 0; \quad c_2 \frac{F_f}{k_1k_2}; \quad c_3 = \frac{(1-F_f)}{k_1k_2}(z+k_2); \quad c_4 = -\frac{(1-F_f)}{k_1k_2}$$

On the basis of equations (IX-23) and (IX-25) we have:

$$r_{iB} = \frac{iF_f}{k_1k_2}; \qquad (I \to i^*A)0; 1 \le i \le i^* - 1);$$
$$(I \to i^*B; 1 \le i \le i^*) \qquad \text{(IX-43)}$$

$$r_{iB} = \frac{(z + k_2 - i)}{k_1k_2}(1 - F_f); \qquad (I \to i^*A, B; i^* \le i \le z) \quad \text{(IX-44)}$$

When supplying a unit flow of the monofraction into element i^*A, the degree of fractional extraction is determined from the unambiguity condition, using expressions (IX-41) and (IX-42):

$$\frac{(i^* - k_1)}{k_1k_2}F_f = \frac{(z+1-i^*)}{k_1k_2}(1 - F_f)$$

$$F_f = \frac{z+1-i^*}{z+k_2}; (I \to i^*A)$$

Similarly, using (IX-43) and (IX-44), we obtain:

$$\frac{i^* \cdot F_f}{k_1k_2} = \frac{(z+k_2-i^*)}{k_1k_2}(1 - F_f)$$

$$F_f = \frac{z+k_2-i^*}{z+k_2}; (I \to i^*B)$$

d. Duplex cascade of the type A÷A

In this case, we have different boundary conditions:

$$r_{0A} = -\frac{1}{\chi_1} r_{1A} \tag{IX-45}$$

$$\begin{aligned} r_{z+1,A} &= -\chi_1 r_{zA} \\ r_{0B} &= 0 \\ r_{zB} &= 0 \end{aligned} \tag{IX-46}$$

and the conditions of the material balance for the entire column are:

$$\begin{aligned} r_{1A}k_1 &= F_f \\ r_{zA}(1-k_1) &= 1-F_f \end{aligned} \tag{IX-47}$$

Equations (IX-47), for element B have the form:

$$\begin{aligned} r_{1B}k_1k_2 &= F_f \\ r_{z-1,B}(1-k_1)(1-k_2) &= 1-F_f \end{aligned} \tag{IX-48}$$

For operation of the cascade in the regime $Q \neq 1$ for element A from previously obtained relationships we have system of equations:

$$\begin{aligned} c_1 + c_2 &= -\frac{1}{\chi_1}(c_1 + c_2 Q) \\ (c_1 + c_2 Q)k_1 &= F_f \\ c_3 + c_4 Q^{z+1} &= -\chi_1 (c_3 + c_4 Q^z) \\ (c_3 + c_4 Q^z)(1-k_1) &= 1-F_f \end{aligned} \tag{IX-49}$$

For element B using (IX-46) and (IX-48), a similar system is even simpler:

$$\begin{aligned} c_1 + c_2 &= 0 \\ (c_1 + c_2 Q)k_1k_2 &= F_f \\ c_3 + c_4 Q^z &= 0 \\ (c_3 + c_4 Q^{z-1})(1-k_1)(1-k_2) &= 1-F_f \end{aligned}$$

Its solution has the form:

$$c_1 = -\frac{F_f}{(Q-1)k_1k_2}; c_2 = \frac{F_f}{(Q-1)k_1k_2}; c_3 = \frac{1-F_f}{(Q-1)k_1k_2}; c_4 = -\frac{1-F_f}{(Q-1)k_1k_2Q^z}$$

Consequently:

$$r_{iB} = \frac{(1-Q^{i})}{(k_1+k_2-1)}\cdot F_f;(I\rightarrow i^{*}A;1\le i\le i^{*}-1);(I\rightarrow i^{*}B;1\le i\le i^{*})$$

(IX-50)

$$r_{iB} = \frac{(Q^{i-z}-1)}{(k_1+k_2-1)}\cdot(1-F_f);(I\rightarrow i^{*}AB;i^{*}\le i\le z-1)$$

(IX-51)

$$F_f = \frac{1-Q^{z-i^{*}}}{1-Q^{z}};(A\div A;I\rightarrow iB)$$

When using these expressions, it is not necessary to solve the system of equations (IX-49) because from the canonic system of equations, we obtain:

$$r_{iA} = r_{i-1,B}(1-k_2)+r_{iB}k_2$$

Using in the above equation (IX-50) and (IX-51), leads to:

$$r_{iA} = \frac{(1-Q^{i-1})}{(k_1+k_2-1)}\cdot F_f(1-k_2)+\frac{(1-Q^{i})}{(k_1+k_2-1)}\cdot F_f k_2 = \frac{\left(1-Q^{i}\cdot\frac{k_2}{1-k_1}\right)}{(k_1+k_2-1)}\cdot F_f;$$

$(A\div A;Q\neq 1);(I\rightarrow i^{*}A,B;1\le i\le i^{*})$

$$r_{iA} = \frac{(Q^{i-z-1}-1)}{(k_1+k_2-1)}(1-F_f)(1-k_2)+\frac{(Q^{i-z}-1)}{(k_1+k_2-1)}(1-F_f)k_2 = \frac{\left(Q^{i-z}\cdot\frac{k_2}{1-k_1}-1\right)}{(k_1+k_2-1)}(1-F_f);$$

at $(A\div A;Q\neq 1);(I\rightarrow i^{*}A;i^{*}\le i\le z);(I\rightarrow i^{*}B;i^{*}+1\le i\le z)$

From the obtained relationship, using the unambiguity condition, we determine the degree of fractional extraction when supplying the monofraction into $i*A$:

$$F_f = \frac{1-Q^{z+1-i*}\cdot\frac{k_1}{1-k_2}}{1-Q^{z}};(A\div A;I\rightarrow i^{*}A)$$

In the regime Q = $\chi_1\chi_2 = k_1 + k_2 = 1$ for element B we correspondingly obtain:

$$\begin{aligned}&c_1=0\\&(c_1+c_2)k_1k_2=F_f\\&(c_3+c_4z)=0\\&\left[c_3+c_4(z-1)\right]k_1k_2=1-F_f\end{aligned} \tag{IX-52}$$

Consequently:

$$r_{iB}=\frac{iF_f}{k_1k_2};(I\to i^*A;1\le i\le i^*-1);(I\to i^*B;1\le i\le i^*) \tag{IX-53}$$

$$r_{iB}=\frac{(z-i)}{k_1k_2}(1-F_f);(I\to i^*A,B;i^*\le i\le z-1) \tag{IX-54}$$

Similarly:

$$r_{iA}=\frac{(i-1)}{k_1k_2}\cdot F_f(1-k_2)+\frac{iF_f}{k_1k_2}\cdot k_2=\frac{(i-k_1)}{k_1k_2}\cdot F_f;\quad (1-i^*\,A,B;\,1\le i\le i^*) \tag{IX-55}$$

$$r_{iA}=\frac{(z+1-i)}{k_1k_2}\cdot(1-F_f)(1-k_2)+\frac{(z-i)}{k_1k_2}\cdot(1-F_f)k_2=\frac{(z-i+k_1)}{k_1k_2}\cdot(1-F_f);$$

$$(I\to i^*A;i^*\le i\le z);(I\to i^*B;i^*+1\le i\le z) \tag{IX-56}$$

From the unambiguity conditions for supplying the monofraction into i^*A the degree of fractional extraction from (IX-55) and (IX-56) is:

$$F_f=\frac{z-i^*+k_1}{z}$$

For supply into i^*B from (IX-53) and (IX-54) respectively:

$$F_f=\frac{z-i^*}{z}$$

It is clear that all equations for the duplex cascade at $k_1 = k_2$ are transformed into appropriate equations of the regular cascade. In this case $Q = \chi^2$. For example, for cascade A÷B when supplying into i^*A we have:

$$z_1=2z;i_1^*=2i^*-1;\Rightarrow i^*=\frac{i_1^*+1}{2}$$

Consequently

$$F_f = \frac{1-(\chi^2)^{\frac{z_1}{2}+1-\frac{i_1^*+1}{2}}}{1-(\chi^2)^{\frac{z_1}{2}+1}\cdot\frac{1}{\chi}} = \frac{1-\chi^{z_1+1-i_1^*}}{1-\chi^{z_1+1}}$$

When supplying the monofraction into i^*B: $z_1 = 2z$; $i_1^* = 2i^*$;

$$F_f = \frac{1-(\chi^2)^{\frac{z_1}{2}+1-\frac{i_1^*}{2}}\cdot\frac{1}{\chi}}{1-\chi^{z_1+1}} = \frac{1-\chi^{z_1+1-i_1^*}}{1-\chi^{z_1+1}}$$

For the section $iA \Rightarrow i_1 = 2i-1$, consequently

$$r_1 = \frac{1-\frac{1}{\chi}(\chi^2)^{\frac{i_1+1}{2}}}{(2k-1)}\cdot F_f = \frac{(1-\chi^{i_1})}{(2k-1)}\cdot F_f;(1 \le i \le i_1^*)$$

For $k_1 = k_2 = 0.5$ when supplying the monofraction into i^*A; $z_1 = 2z$; $i^*_1 = 2i^*-1$

$$F_f = \frac{\frac{z_1}{2}+1-\frac{i_1^*+1}{2}}{\frac{z_1}{2}+\frac{1}{2}} = \frac{z_1+1-i_1^*}{z_1+1}$$

When supplying the monofraction into i^*B; $i_1^* = 2i^*$

$$F_f = \frac{\frac{z_1}{2}+\frac{1}{2}-\frac{i_1^*}{2}}{\frac{z_1}{2}+\frac{1}{2}} = \frac{z_1+1-i_1^*}{z_1+1}$$

Similarly, other equations are transformed in the same manner.

All calculation equations are summarised in Table No. IX-1. In this table, $Q = \chi_1\chi_2$ is the duplex parameter.

5. THE MATHEMATICAL MODEL OF THE PROCESS OF CASCADE EQUILIBRIUM CLASSIFICATION WITH ARBITRARY SEPARATION COEFFICIENTS

The previously developed models of the process of cascade classification have different areas of application. For example, the region of application of the model of the regular cascade are systems with the same velocity field along the area height. The area of application of a non-balanced cascade model are systems consisting of two branches operating in different conditions. The range of the application of the duplex cascade are systems where each section operates in two different conditions. However, there are systems working in a non-uniform regime along the entire height of the separation column: a classifier with a variable cross section in the direction of height (for example, conical classifier), with additional supply or removal of flows of the continuous medium, etc. These systems are characterised by a non-uniform velocity field along the height and, consequently, different coefficients of distribution of monofractions in different sections and may be regarded as a completely non-uniform cascade of separation elements. In the case of the equilibrium principle of operation of the elements, the scheme of functioning of a completely non-uniform cascade is shown in Fig.IX-7.

Let it be that $k_1 \ldots k_z$ – are the coefficients of separation on each stage, respectively.

Examination of the flows of the monofraction in a non-uniform cascade (Fig.IX-7) gives for the material balance the entire column:

$$r_1 k_1 + r_z(1-k_z) = 1 \qquad \text{(IX-57)}$$

The condition of the material balance for the upper part of the column from the first element to the i-th element has the form:

$$r_1 k_1 + r_i(1-k_i) = r_{i+1}k_{i+1}; \quad i \le i^* - 1 \qquad \text{(IX-58)}$$

Consequently:

$$r_{i+1}k_{i+1} - r_i(1-k_i) = F_f; \quad i \le i^* - 1 \qquad \text{(IX-59)}$$

Similarly, for the part of the column of the lower branch of the apparatus:

$$r_z(1-k_z) + r_i k_i = r_{i-1}(1-k_{i-1}); \quad i \geqslant i^* + 1 \qquad \text{(IX-60)}$$

This gives

$$r_{i-1}(1-k_{i-1}) - r_i k_i = 1 - F_f; \quad i \geqslant i^* + 1$$

Changing indices in recurrent equations (IX-59) we obtain:

Table IX-1 – Summary of Formulas for Duplex Cascade

A-B Cascade				
Input into element ⇒			i*A	I*B
	$F_f \Rightarrow$		$\frac{\left(1-Q^{z+1-i^*}\right)}{\left(1-Q^{z+1}\cdot\frac{k_2}{1-k_1}\right)}$	$\frac{\left(1-Q^{z+1-i}\cdot - \right.}{\left(1-Q^{z+1}\cdot - \right.}$
$Q \neq 1$	r_{iA}	$\frac{\left(1-Q^i\cdot\frac{k_2}{1-k_1}\right)}{(k_1+k_2-1)}\cdot F_f$	$1\le i\le i^*$	$1\le i\le i$
		$\frac{(Q^{i-z-1}-1)}{(k_1+k_2-1)}\cdot(1-F_f)$	$i^*\le i\le z$	$i^*\le i\le$
	r_{iB}	$\frac{(1-Q^i)}{(k_1+k_2-1)}\cdot F_f$	$1\le i\le i^*-1$	$1\le i\le i$
		$\frac{\left(Q^{i-z}\cdot\frac{k_1}{1-k_2}-1\right)}{(k_1+k_2-1)}\cdot(1-F_f)$	$i^*\le i\le z$	$i^*\le i\le$
	$F_f \Rightarrow$		$\frac{z+1-i}{z+k_2}$	$\frac{z+k_2-i}{z+k_2}$
$Q=1;\ k_1+k_2=1$	r_{iA}	$\frac{(i-k_1)}{k_1k_2}\cdot F_f$	$1\le i\le i^*$	$1\le i\le i$
		$\frac{(z+1-i)}{k_1k_2}\cdot(1-F_f)$	$i^*\le i\le z$	$i^*\le i\le$
	r_{iB}	$\frac{i}{k_1k_2}\cdot F_f$	$1\le i\le i^*-1$	$1\le i\le i$
		$\frac{(z+k_2-i)}{k_1k_2}\cdot(1-F_f)$	$i^*\le i\le z$	$i^*\le i\le$

$$r_ik_i - r_{i-1}(1-k_{i-1}) = F_f;(i\le i^*);$$
$$r_i(1-k_i) - r_{i+1}k_{i+1} = 1-F_f;(i\ge i^*); \qquad \text{(IX-61)}$$

At $i = i^*$, the dependence (IX-61) reflect the condition of the material balance for an element of the source. Using the relationship (IX-61) and the boundary conditions for the upper part of the column, we obtain a system of equations:

Table IX -1 (Continued)

A-A Cascade				
		Input into element ⇒	i*A	i*B
$Q \neq 1$		$F_f \Rightarrow$	$\dfrac{\left(1-Q^{z+1-i^*}\cdot\dfrac{k_1}{1-k_2}\right)}{(1-Q^z)}$	$\dfrac{(1-Q^{z-i^*})}{(1-Q^z)}$
	r_{iA}	$\dfrac{\left(1-Q^i\cdot\dfrac{k_2}{1-k_1}\right)}{(k_1+k_2-1)}\cdot F_f$	$1\le i\le i^*$	$1\le i\le i^*$
		$\dfrac{(Q^{i-z}\cdot\dfrac{k_2}{1-k_1}-1)}{(k_1+k_2-1)}\cdot(1-F_f)$	$i^*\le i\le z$	$i^*\le i\le z$
	r_{iB}	$\dfrac{(1-Q^i)}{(k_1+k_2-1)}\cdot F_f$	$1\le i\le i^*-1$	$1\le i\le i^*$
		$\dfrac{(Q^{i-z}-1)}{(k_1+k_2-1)}\cdot(1-F_f)$	$i^*\le i\le z-1$	$i^*\le i\le z-$
$Q=1;\ k_1+k_2=1$		$F_f \Rightarrow$	$\dfrac{(z+k_1-i^*)}{z}$	$\dfrac{(z-i^*)}{z}$
	r_{iA}	$\dfrac{(i-k_1)}{k_1k_2}\cdot F_f$	$1\le i\le i^*$	$1\le i\le i^*$
		$\dfrac{(z+1-i)}{k_1k_2}\cdot(1-F_f)$	$i^*\le i\le z$	$i^*\le i\le z$
	r_{iB}	$\dfrac{i}{k_1k_2}\cdot F_f$	$1\le i\le i^*-1$	$1\le i\le i^*$
		$\dfrac{(z-i)}{k_1k_2}\cdot(1-F_f)$	$i^*\le i\le z-1$	$i^*\le i\le z-$

$$
\begin{aligned}
r_1k_1 &= F_f\\
r_2k_2 &= r_1(1-k_1)+F_f\\
r_3k_3 &= r_2(1-k_2)+F_f\\
&\cdots\cdots\cdots\cdots\\
r_ik_i &= r_{i-1}(1-k_{i-1})+F_f\\
&\cdots\cdots\cdots\cdots\\
r_{i^*}k_{i^*} &= r_{i^*-1}(1-k_{i^*-1})+F_f
\end{aligned}
\qquad \text{(IX-62)}
$$

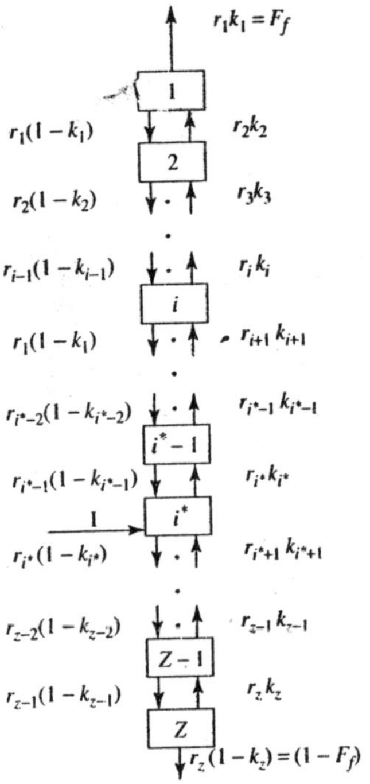

Fig. IX-7 Diagram of fully heterogeneous equilibrium cascasde.

Using the method of successive substitution, we transform the system (IX-62) to the form ($i \le i^*$):

$$r_1k_1 = F_f$$

$$r_2k_2 = F_f\chi_1 + F_f$$

$$r_3k_3 = \chi_2(F_f\chi_1 + F_f) + F_f =$$

$$= \chi_2\chi_1F_f + \chi_2F_f + F_f;$$

$$r_4k_4 = \chi_3\left[\chi_2\left(F_f\chi_1 + F_f\right) + F_f\right] + F_f =$$

$$= \chi_3\chi_2\chi_1F_f + \chi_3\chi_2F_f + \chi_3F_fF_f;$$

$$r_5k_5 = \chi_4\chi_3\chi_2\chi_1F_f + \chi_4\chi_3\chi_2F_f +$$

$$+ \chi_4\chi_3F_f + \chi_4F_f + F_f;$$

....

Consequently, for the i-th element:

$$r_i k_i = F_f\left(\sum_{l=1}^{i-1}\prod_{l}^{i-1}\chi_l + 1\right)$$

or

$$r_i k_i = \frac{F_f}{\chi_i}\sum_{l=1}^{i}\prod_{l}^{i}\chi_l ; (1 \le i \le i^*) \qquad \text{(IX-63)}$$

where $\chi_i = \dfrac{1-k_i}{k_i}$ is the parameter of distribution for the i-th element of the cascade.

Taking into account equation (IX-63) for the element-source, we obtain

$$r_{i^*} k_{i^*} = \frac{F_f}{\chi_{i^*}}\sum_{l=1}^{i^*}\prod_{l}^{i^*}\chi_l \qquad \text{(IX-64)}$$

Using the boundary condition and expression (IX-61), we determine the flows of monofraction upwards for the lower branch of the cascade ($i \ge i^*$):

$$\begin{aligned}
r_{i^*+1}k_{i^*+1} &= r_{i^*}(1-k_{i^*}) - (1-F_f)\\
r_{i^*+2}k_{i^*+2} &= r_{i^*+1}(1-k_{i^*+1}) - (1-F_f)\\
r_{i^*+3}k_{i^*+3} &= r_{i^*+2}(1-k_{i^*+2}) - (1-F_f)\\
&\ldots\ldots\ldots\ldots\ldots\ldots\\
&\ldots\ldots\ldots\ldots\ldots\ldots\\
&\ldots\ldots\ldots\ldots\ldots\ldots\\
r_z k_z &= r_{z-1}(1-k_{z-1}) - (1-F_f)\\
r_z k_z \chi_z &= 1 - F_f
\end{aligned} \qquad \text{(IX-65)}$$

Taking into account the dependences (IX-64), the system of equations (IX-65) is transformed to the form:

$$r_{i^*+1}k_{i^*+1} = F_f\sum_{l=1}^{i^*}\prod_{l}^{i^*}\chi_l - (1-F_f);$$

$$r_{i^*+2}k_{i^*+2} = \chi_{i^*+1}F_f\sum_{l=1}^{i^*}\prod_{l}^{i^*}\chi_l - \chi_{i^*+1}(1-F_f) - (1-F_f);$$

$$r_{i^*+3}k_{i^*+3}=\chi_{i^*+2}\chi_{i^*+1}F_f\sum_{l=1}^{i^*}\prod_{l}^{i^*}\chi_l-\chi_{i^*+2}\chi_{i^*+1}(1-F_f)-$$

$$\chi_{i^*+2}(1-F_f)-(1-F_f);$$

The resultant expressions enable us to add the following equation for an arbitrary section ($i \geq i^*$):

$$r_ik_i=\prod_{i^*+1}^{i-1}\chi_l\cdot F_f\sum_{l=1}^{i^*}\prod_{l}^{i^*}\chi_l-(1-F_f)(1+\sum_{l=i^*+1}^{i-1}\prod_{l}^{i-1}\chi_l)$$

Transforming the last equation:

$$r_ik_i=\frac{F_f}{\chi_i}\sum_{l=1}^{i^*}\prod_{l}^{i}\chi_l-\frac{(1-F_f)}{\chi_i}\sum_{l=i^*+1}^{i}\prod_{l}^{i}\chi_l$$

$$r_ik_i=\frac{F_f}{\chi_i}\sum_{l=1}^{i}\prod_{l}^{i}\chi_l-\frac{1}{\chi_i}\sum_{l=i^*+1}^{i}\prod_{l}^{i}\chi_l;(i^*\leq i\leq z) \qquad \text{(IX-66)}$$

Consequently, for the last element of the cascade we obtain:

$$r_zk_z=\frac{F_f}{\chi_z}\sum_{l=1}^{z}\prod_{l}^{z}\chi_l-\frac{1}{\chi_z}\sum_{l=i^*+1}^{z}\prod_{l}^{z}\chi_l$$

Taking into account (IX-65) we can write that:

$$1-F_f=F_f\sum_{l=1}^{z}\prod_{l}^{z}\chi_l-\sum_{l=i^*+1}^{z}\prod_{l}^{z}\chi_l$$

From the resultant relationship, the degree of fractional extraction is:

$$F_f=\frac{1+\sum_{l=i^*+1}^{z}\prod_{l}^{z}\chi_l}{1+\sum_{l=1}^{z}\prod_{l}^{z}\chi_l} \qquad \text{(IX-67)}$$

The resultant expression may be transformed to a different form. For this purpose, the numerator and denominator will be multiplied by the co-factor $\prod_{l=1}^{z}\frac{1}{\chi_l}$. This gives:

$$F_f=\frac{\sum_{l=i^*}^{z}\prod_{1}^{l}\left(\frac{1}{\chi}\right)_l}{1+\sum_{l=1}^{z}\prod_{1}^{l}\left(\frac{1}{\chi}\right)_l} \qquad \text{(IX-68)}$$

According to the dependence (IX-63) the total flow of the monofraction in an arbitrary element of the upper branch of the cascade is:

$$r_i=\frac{F_f}{(1-k_i)}\cdot\sum_{l=1}^{i}\prod_{l}^{i}\chi_l;(1\le i\le i^*) \qquad \text{(IX-69)}$$

Similarly, from equation (IX-66) for the lower branch:

$$r_i=\frac{1}{(1-k_i)}\left(F_f\sum_{l=1}^{i}\prod_{l}^{i}\chi_l-\sum_{l=i^*+1}^{i}\prod_{l}^{i}\chi^*\right);(i^*\le i\le z) \qquad \text{(IX-70)}$$

Therefore, in the general form:

$$F_f=\frac{1+\sum_{n=i^*+1}^{z}\prod_{q=n}^{z}\chi_q}{1+\sum_{n=1}^{z}\prod_{q=n}^{z}\chi_q}=\frac{\sum_{n=i^*}^{z}\prod_{q=1}^{z}\chi_q}{1+\sum_{n=1}^{z}\prod_{q=1}^{n}\chi_q};$$

$$r_i=F_f\frac{(\chi_i+1)}{\chi_i}\sum_{n=1}^{i}\prod_{q=n}^{i}\chi_q;(1\le i\le i^*)$$

$$r_i=\frac{(\chi_i+1)}{\chi_1}\cdot\left[F_f\sum_{n=1}^{i}\prod_{q=n}^{i}\chi_q-\sum_{n=i^*+1}^{i}\prod_{q=n}^{i}\chi_q\right];(i^*\le i\le z)$$

We confirm the validity of equations (IX-68), (IX-69) and (IX-70) for the regular cascade ($\chi_i = \chi$). According to (IX-68) we have:

$$F_f=\frac{1+\chi^{z-i^*}+\chi^{z-i^*-1}+...+\chi^2+\chi}{1+\chi^z+\chi^{z+1}+...+\chi^2+\chi}$$

Using the equation for the sum of the terms of geometrical progression with the denominator $\chi \neq 1$, we obtain:

$$F_f = \frac{1-\chi^{z+1-i^*}}{1-\chi^{z+1}}$$

which corresponds to previously obtained equations.

References

1. Seader J.D. and Henley E.J., Separation process principles, Wiley, New York (1998).
2. Khonry F.M., Predicting the performance of multistage separation process, CRC Press, Boca Raton, Florida (2000).
3. Govorov A.V., Cascade and combined process of fractionation of bulk materials, a dissertation for the title of the Candidate of technical sciences, Sverdlovsk (1986).
4. Barsky M.D., Fractionation of powders, Nedra, Moscow (1980).
5. Barsky M.D. and Barsky E., General Trends of Gravity Separation, Proceedings of the XXI International Mineral Processing Congress, Elsevier, Rome (2000).
6. Barsky E. and Barsky M., Master curve of separation process. Physical Separation in Science and Engineering, Taylor and Francis, Vol. 3, No. 1 (2004).

Chapter X

COMBINED CASCADE PROCESSES

1. MAIN PARAMETERS

There are three stages in the history of operation of separation systems. In the first stage, the main attention was given to increasing the efficiency of operation of individual apparatus. This resulted in the appearance of a large number of classifiers based on a single separation act. This direction is also being developed at present, but the limited efficiency of all the currently available classification systems has resulted in the appearance of multistage separation apparatus.

Experimental investigations carried out in recent decades show that the combining identical operations of separation, taking place in separate sections into a separation cascade, has made it possible to greatly increase the efficiency of fractionation of the powders. This combination realises the cascade system of separation and forms the second stage of improvement of the operation of classifiers. The separation cascade of order I consists of 'z' sections having a set number of links between them. The simplest variant of separation cascade of order I is realised when all the separating elements (sections) are identical and operate in the same regime. A regular cascade is obtained in this case. Complicated separation cascades of order I assume all possible relationships between the sections and not identical separating elements operating in different conditions. In this case, a non-regular cascade is realised. There is no mathematical description of operation of such a cascade in a general form, and this description must be found separately for each specific case. The experimental investigations of the simple variants of separation cascade of order I show that the effect of increasing the number of sections, starting with seven, rapidly decreases. Therefore, the next direction of improvement of apparatus (third stage) is the organisation of cascades of order II of the type $z_i \times n$ or combined cascades.

In a general case, the simplest combined cascade is realised in the presence of n separating columns (cascades of order I) consisting of z_i separating elements, operating in different conditions. This case is very complicated to examine because it is characterised by a very large number of parameters acting on the process. The simplest variant of a combined cascade of type $z \times n$ assumes n identical simplest separating cascade of order I consisting of z sections with a fixed area of introduction of initial material and working in the same regime.

In this book, we restrict our examination to only homogeneous combined separating cascades (CSC). Even in this case, the CSC is characterised by a very large number of working, topologically non-isomorphous structural schemes of relationships between SC-I, having different separating capacities. It should be mentioned that the majority of the CSC systems realises a higher order of organisation of the process in comparison with separation cascades of order I. They are not equivalent to a simple increase of the number of elements of the apparatus, as confirmed by the results of experimental investigations.

To avoid cumbersome repeated formulations, it is convenient to define the following concepts:

– free output – the local flow of the material leading into a combined fine or coarse product from a separate separation column;

– the link – the local non-free output at a separate separating column (associated with some other column);

An active link for the examined column – the flow arriving in the given column from some other column, for which this link is passive;

The structural scheme – the scheme of inputs, outputs and links between the individual separating elements (cascade columns) in the CSC;

F_0 – fractional extraction of a single column (into a fine product);

F – fractional extraction of the entire CSC (into fine product);

$F(F_0)$ – the function of the link, corresponding to the examined specific structural scheme;

the isomorphous scheme – the scheme having the same link function;

the inverse scheme – a scheme formed from a given scheme by

changing the positions of some columns with complete retention of the outputs and links with previous elements; in this case, the first column is fixed with respect to feed and position and is not rearranged (it is evident that the inverse schemes are isomorphous);

the reversed scheme – the scheme in which all outputs and links with respect to the fine product become identical to outputs and links with respect to the coarse product, and vice versa. The function of the link of the inverse scheme is $F^{-1}(F_0)$. It is completely evident that for the inverse scheme, the function of the link with respect to the coarse product is identical to the function $F(F_0)$ in which the argument F_0 is replaced by $(1-F)$, i.e.

$$F_c^{-1}(F_0)=F(F_0 \Rightarrow 1-F_0)$$

Since $F^{-1}(F_0) = 1 - F_c^{-1}(F_0)$, for the inverse scheme:

$$F^{-1}(F_0)=1-F(F_0 \Rightarrow 1-F_0) \tag{X-1}$$

The working scheme of the CSC – scheme capable of operation, realised in practice (in contrast to defective schemes).

There are the following varieties of the defective schemes:

neutralising scheme – scheme neutralising a number of columns and reducing separation to a process taking place in the remaining cascade of order I. In principle, these schemes decrease the number of separating columns, taking part in the process, changing the type of CSC. Neutralisation of a single separating column occurs when both passive links of the scheme are active links of some other column (Fig.X-1,a). In this scheme, the separating element *i* is neutralised, because any active link of the element *i* becomes automatically an active link of *j*. In neutralisation of several separating elements, the latter do not contain free outputs, and all links of these elements come with the exception of two, are organised between these elements, and the two remaining links are active links of some separating column (Fig.X-1,b). This scheme neutralises the separating elements I and II and the scheme becomes isomorphous with respect to the scheme with a single separating column. In the partial case of the examined variant, neutralisation of a similar group of elements takes place when the scheme retains a single link with any elements outside this group (Fig.X-1,c);

isolating scheme – the schemes in which the groups of individual separating elements do not contain any active links with other columns (Fig.X1,d), and an individual apparatus is isolated from the flow in this scheme;

transporting scheme – the scheme in which the groups of separate separating elements contain one or several free outputs strictly into a single product, and all links are organised within the limits of this group (Fig.X-1,e). In this scheme, the group of the elements II, III, IV is a transporting group. All included in this group is transferred completely into the fine product. The scheme is isomorphous with respect to the case with a single separating column.

It should be mentioned that in all the varieties of topologically non-isomorphous schemes, there may be schemes having two defects or a complete set of defects, and also a set of these defects in any combination.

The problem of listing the number of the structural schemes of CSC of the type $z \times n$ is of primary importance because it is closely linked with the determination of the most advanced schemes. In the case of a combined cascade, including n apparatuses, the total number of free outputs and links is $2n$. The minimum of the free outputs is:

$$P_{\min} = 2 \tag{X-2}$$

Consequently, the maximum possible number of links between the n separating elements is:

$$S_{\max} = 2n - 2$$

The minimum number of links between n apparatuses is:

$$S_{\min} = n - 1$$

Consequently, the maximum number of the free outputs should be:

$$P_{\max} = n + 1 \tag{X-3}$$

In the presence of free outputs P, the number of links, which must operate in the CSC, is:

$$S = 2n - P \tag{X-4}$$

In a general case for fixed P, the number of schemes is equal to the number of methods which can be used to organise S links. For any column, any link can be produced from any of $(n-1)$ columns. Since there are S links, and in each link there are $(n-1)$ directions, the number of different schemes is:

$$N_p = (n-1)^S$$

Taking into account (X-4) we have:

$$N_p = (n-1)^{2n-p} \tag{X-5}$$

Taking into account the resultant relationships, it may be shown that the number of all possible non-isomorphous systems, including the direct, reversed, inversion and all defective systems, is expressed by the dependence:

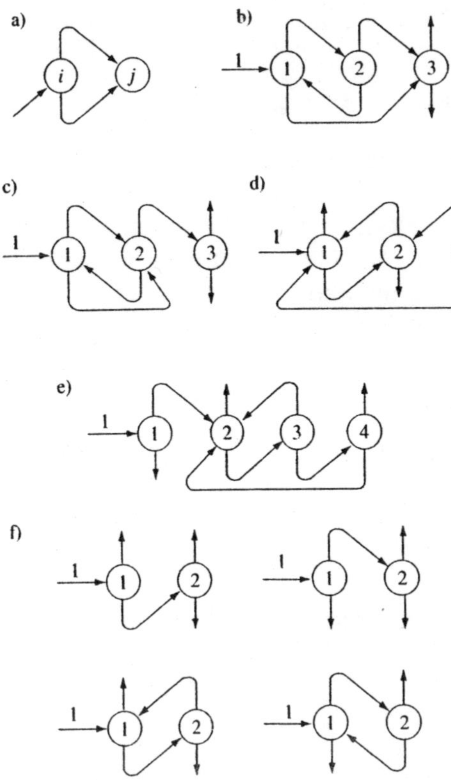

Fig, X-1 Examples of different structural diagrams of CSC: a,b,c,d) defective neutralising schemes; e) transporting schemes; f) four working variants for two-element CSC.

$$N_{\Sigma} = \sum_{P=2}^{n+1}\left[(n-1)^{2n-p}\cdot\sum_{m=1}^{P-1}(c_n^m\cdot c_n^{p-m})\right] \qquad \text{(X-6)}$$

where $c_n^m = \dfrac{n!}{m!(n-m)!}$ is the number of combinations of n elements with respect to m (according to its meaning, m is the number of free outputs into a fine product);

It should be mentioned that the number of schemes, determined from expression (X-6) is, for known reasons, considerably greater than the number of working schemes of CSC, consisting of n columns. Thus, for $n = 2$, the number of all possible schemes according to (X-6) is:

$$N_{\Sigma} = 8$$

On the other hand, the working number of schemes is $N_{pab} = 4$. For $n = 3$ we have:

$$N_{\Sigma} = 348; \qquad N_{pab} = 47$$

For $n = 4$

$$N_{\Sigma} = 30348; \qquad N_{pab} = 904$$

The structural schemes of all working variants for $n = 2$ are shown in Fig.X-1,f.

It is clear that for CSC of the type $z \times n$, the function of the link is found from a system of $2n$ combined equations: n equations are formed for fractional extractions and n equations of the material balance for n separating columns:

$$\begin{aligned}
&F_{f_1} = F_0 F_{ent_1} \\
&F_{f_2} = F_0 F_{ent_2} \\
&\dots\dots\dots\dots\dots\dots \\
&F_{f_n} = F_0 F_{ent_n} \\
&F_{f_1} + F_{c_1} = F_{ent_1} \\
&F_{f_2} + F_{c_2} = F_{ent_2} \\
&\dots\dots\dots\dots\dots\dots\dots\dots \\
&F_{f_n} + F_{c_n} = F_{ent_n}
\end{aligned} \tag{X-7}$$

where F_{f_i} is the fractional output of the fine product of the i-th column; F_{c_i} is the fractional output of the coarse product of the i-th column; F_{ent_i} is the fractional input of the i-th column (determined in accordance with a specific structural scheme).

From the system of equations (X-7), we can determine the main parameters of operation of the CSC: $F_f; F_{c_i}; F_{ent_i}; F(F_0)$

Since the following equation is valid for any column:

$$F_{f_i} = F_0 (F_{f_i} + F_{c_i})$$

For the fractional output of the coarse product we have:

$$F_{c_i} = F_{f_i} \frac{1 - F_0}{F_0} \tag{X-8}$$

We determine identical parameters of the fractionation process, based on the available parameters:

– the yield of fine and coarse products in the i-th column

$$\gamma_{f_i} = \sum_{j=1}^{m}\left[F_{f_{ij}} \cdot r_j\right];$$

$$\gamma_{c_i} = \sum_{j=1}^{m}\left[F_{c_{ij}} \cdot r_j\right] = \sum_{j=1}^{m}\left[F_{f_{ji}} \cdot \frac{(1-F_{0j})}{F_{0j}} \cdot r_j\right];$$

– total yield

$$\gamma_i = \sum_{j=1}^{m}\left[F_{ent_{ij}} \cdot r_j\right] = \sum_{j=1}^{m}\left[\frac{F_{f_{ij}}}{F_{0j}} \cdot r_j\right];$$

$$\gamma_i = \gamma_{f_i} + \gamma_{c_i};$$

–extraction of the fine class into the fine and coarse products of the i-th column:

$$\varepsilon_{Df_i} = \frac{1}{\sum_{j=1}^{j} r_j} \cdot \sum_{j=1}^{j}\left[F_{f_{ij}} \cdot r_j\right] = \frac{1}{\alpha}\sum_{j=1}^{j}\left[F_{f_{ij}} \cdot r_j\right];$$

$$\varepsilon_{Df_i} = \varepsilon_{D_i} - \varepsilon_{Df_i};$$

– extraction of the coarse classes into the fine and coarse products of the i-th column:

$$\varepsilon_{Rf_i} = \varepsilon_{R_i} - \varepsilon_{Df_i};$$

$$\varepsilon_{Rf_i} = \frac{1}{1-\alpha} \cdot \sum_{j=j_f+1}^{j}\left[F_{c_{ij}} \cdot r_j\right];$$

– the total extraction of the fine and coarse classes in the i-th column:

$$\varepsilon_{D_i} = \frac{1}{\alpha} \cdot \sum_{j=1}^{jm}\left[\frac{F_{f_{ij}}}{F_{0j}} \cdot r_j\right];$$

$$\varepsilon_{R_i} = \frac{1}{1-\alpha} \cdot \sum_{j=j_f+1}^{m}\left[\frac{F_{f_{ij}}}{F_{0j}} \cdot r_j\right];$$

– content of the fine fractions in the fine and coarse products of the i-th column:

$$D_{f_i} = \frac{1}{\gamma_{f_i}} \cdot \sum_{j=1}^{j} \left[F_{f_{ij}} \cdot r_j \right] = \frac{\alpha \cdot \varepsilon_{D_{fi}}}{\gamma_{f_i}};$$

$$D_{c_i} = \frac{\alpha \cdot \varepsilon_{D_{fi}}}{\gamma_{c_i}};$$

– the content of the coarse fractions in the fine and coarse products of the i-th column:

$$R_{f_i} = 1 - D_{f_i};$$

$$R_{c_i} = 1 - D_{c_i};$$

– the total content of the fine and coarse classes in the i-th column:

$$D_i = \frac{\alpha \cdot \varepsilon_{D_i}}{\gamma_i};$$

$$R_i = 1 - D_i;$$

– the flow of the fine product, the coarse products and total flow in the i-th column:

$$q_{f_i} = \gamma_{f_i} q;$$

$$q_{c_i} = \gamma_{c_i} q;$$

$$q_i = \gamma_i q;$$

– yield of the fine and coarse products for all CSC

$$\gamma_f = \sum_{i_f} \gamma_{f_i} = \sum_{i_f} \sum_{j=1}^{m} \left[F_{f_{ij}} \cdot r_j \right];$$

$$\gamma_c = 1 - \gamma_f;$$

– extraction of fine classes into fine and coarse products for CSC

$$\varepsilon_{D_f} = \sum_{i_j} \varepsilon_{D_i} = \frac{1}{\alpha} \sum_{i_f} \sum_{j=1}^{j_f} \left[F_{c_{ij}} \cdot r_j \right];$$

$$\varepsilon_{D_c} = 1 - \varepsilon_{D_f};$$

– extraction of the coarse fractions into fine and coarse products for the entire CSC

$$\varepsilon_{R_f} = 1 - \varepsilon_{R_c};$$

$$\varepsilon_{R_c} = \frac{1}{1-\alpha} \cdot \sum_{i_c} \sum_{j=j_f+1}^{m} \left[F_{c_{ij}} \cdot r_j \right]$$

– the content of fine classes into fine and coarse products for the CSC

$$D_f = \frac{\alpha \cdot \varepsilon_{D_f}}{\gamma_f};$$

$$D_c = \frac{\alpha \cdot \varepsilon_{D_c}}{\gamma_c};$$

– the content of coarse fractions in the fine and coarse products of the entire CSC

$$R_f = 1 - D_f;$$
$$R_c = 1 - D_c;$$

– the flow of the fine and coarse products for the entire CSC

$$q_f = \gamma_f q;$$
$$q_c = (1 - \gamma_f) q$$

where m is the number of monofractions; D is the content of fine classes in the initial product; j_f is the number of fine monofractions; q is the flow of the narrow class in the initial material kg/s; i_f is the number of columns with the yield into the fine products; i_c is the number of columns with the yield into the coarse products.

The relationships can be used to calculate any technological criteria of the classification process in the CSC and also for selecting the optimum structural scheme. The best scheme can be selected from the separation curve for the entire CSC on the basis of the function of the link:

$$F_f(x_j) = \sum_{i_f} F_{f_{ij}} = F\left[F_{0j}(x_j) \right] \tag{X-9}$$

Using the system of equations (X-7), we can determine the functions of the link for a specific structural scheme for every fixed n. However, it is interesting to determine the function of the link as the function of n: $F = F(F_0; n)$. The presence of such a dependence makes it possible to analyse the dependence in order to determine the degree of rationality of the given structural scheme and rational ranges of the values of n. We shall examine specific examples.

2. SOME VARIETIES OF CSC OF THE TYPE $z \times n$

1. The structural scheme, realising consecutive purification of the coarse product (Fig.X-2).

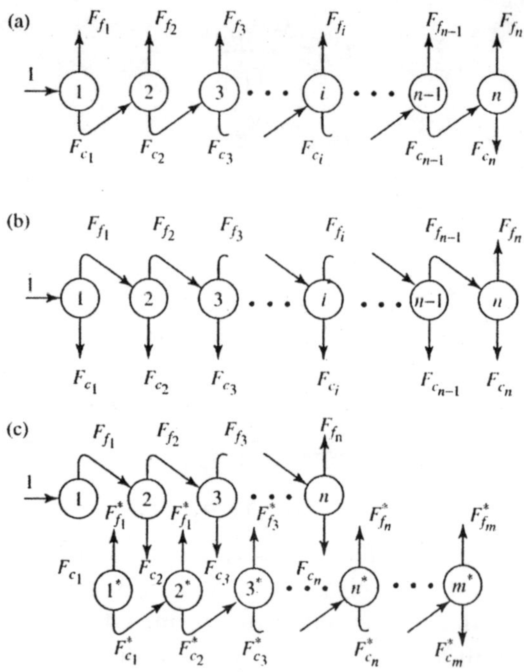

Fig.X-2 Some types of structural diagrams of CSC $z \times n$: a) scheme with gradual cleaning of the coarse product; b) scheme with gradual cleaning of the fine product; c) scheme of mixed cleaning.

From (X-7), we have the following system of equations for the given scheme:

$$
\begin{aligned}
F_{f_1} &= F_0 \\
F_{f_2} &= F_0 F_{c_1} \\
&\dots\dots\dots\dots \\
F_{f_2} &= F_0 F_{c_{i-1}}
\end{aligned}
\qquad \text{(X-10)}
$$

....

$$F_{f_n} = F_0 F_{c_{n-1}}$$

$$F_{f_2} + F_{c_2} = F_{c_1}$$

....

$$F_{f_i} + F_{c_i} = F_{c_{i-1}}$$

....

$$F_{f_n} + F_{c_n} = F_{c_{n-1}}$$

Fractional extraction into the fine product for the entire CSC according to (X-9) is determined as follows:

$$F = \sum_{i=1}^{n} F_{f_i} = 1 - F_{c_n} \tag{X-11}$$

From the system of equations (X-10), the recurrent equation for F_{f_i} has the form:

$$F_{f_i} + \frac{F_{f_{i+1}}}{F_0} = \frac{F_{f_i}}{F_0}$$

or

$$F_{f_{i+1}} - F_{f_i}(1 - F_0) = 0 \tag{X-12}$$

The recurrent equation (X-12) represents a homogeneous linear finite-differential equation of the order I with constant coefficients. Its boundary condition is $F_{f_1} = F_0$. The solution of equation (X-12) will be determined in the form:

$$F_{f_i} = c\lambda^i \tag{X-13}$$

where λ is the root of the appropriate characteristic equation

$$\lambda - (1 - F_0) = 0; \tag{X-14}$$

c is a constant determined from the boundary condition.

Taking into account (X-14), for (X-13) we obtain:

$$F_{f_i} = c(1 - F_0)^i$$

Substituting the boundary condition gives:

$$c(1 - F_0) = F_0$$

Consequently, the general solution is in the form:

$$F_{f_i} = F_0(1 - F_0)^{i-1} \tag{X-15}$$

The last equation, according to equation (X-8) gives:

$$F_{c_i} = (1 - F_0)^i$$

Substituting the above equation into (X-11), we obtain a function of the link in the form:

$$F = 1 - (1 - F_0)^n \tag{X-16}$$

The efficiency of the given structural scheme for removing the dust has been confirmed in practice (multirow apparatus).

The scheme of successive purification of the fine product

According to the above definitions, this scheme is inverse in relation to the one examined previously. Consequently, according to equation (X-1) for this scheme:

$$F = F_0^n$$

It is evident that for local fractional yields of the direct and reverse schemes we have:

$$F_{f_i} = F_{c_i}^{-1}(F_0 \Rightarrow 1 - F_0);$$

$$F_{c_i} = F_{f_i}^{-1}(F_0 \Rightarrow 1 - F_0)$$

According to the above equations for the examined scheme we obtain:

$$F_{f_i} = F_0^i;$$

$$F_{c_i} = (1 - F_0) \cdot F_0^{i-1}$$

3. THE MIXED PURIFICATION SCHEME

In a general case, we can construct a relatively large number of schemes of mixed purification of the products. For example, let us assume that we have a scheme realising n-fold consecutive purification of the fine product and purification of the coarse product in n apparatuses (Fig.X-2,c). Evidently, for the given structural scheme:

$$F_{f_i} = F_0^i;$$

$$F_{c_i} = F_0^{i-1}(1 - F_0);$$

$$F_{f_i}^* = F_0(1 - F_0)^i;$$

$$F_{c_i}^* = (1 - F_0)^{i+1}$$

The function of the link of the examined combined scheme is determined from the condition:

$$F = F_{f\cdot n} + \sum_{i=1}^{m} F^{*}_{f_i}$$

or

$$F = F_0^n + F_0(1-F_0) + F_0(1-F_0)^2 + ... + F_0(1-F_0)^m$$

Using the expression for the sum of the terms of the geometrical progression, we obtain:

$$F = F_0^n + \frac{F_0(1-F_0)\left[1-(1-F_0)^m\right]}{1-(1-F_0)}$$

Consequently, the final equation is:

$$F = F_0^n + (1-F_0) - (1-F_0)^{m+1} \qquad \text{(X-17)}$$

An interesting case is the one in which $m = n - 1$ and, consequently, from (X-17), the expression for the function of the link has the form:

$$F = F_0^n + (1-F_0) - (1-F_0)^n \qquad \text{(X-18)}$$

From equation (X-18) we obtain that at $n = 1$ and $n = 2$ the given combined scheme is isomorphous with respect to the single column $F = F_0$. For the particles of the boundary size, there is also an equality of the fractional extractions for any n:

$$F = 0.5, \quad \text{at} \quad F_0 = 0.5,$$

i.e. the given CSC displaces the separation boundaries for any n and it is therefore interesting to verify, using the separation curve, the efficiency of using the given scheme at $n > 2$. This estimation can be carried out on the basis of the curvature of the curve of separation of the combined scheme and the single column at the point $F = F_0 = 0.5$:

$$\frac{\left[\dfrac{dF}{dx}\right]_{F=0,5}}{\left[\dfrac{dF_0}{dx}\right]_{F_0=0,5}} = \left[\frac{dF}{dF_0}\right]_{F_0=0,5}$$

From (X-18) we have:

$$\left[\frac{dF}{dF_0}\right]_{F_0=0,5} = \frac{2n}{2^{n-1}} - 1 \qquad \text{(X-19)}$$

From the resultant equation it follows that for n = 1, n = 2

$$\left[\frac{dF}{dx}\right]_{F=0.5} = \left[\frac{dF_0}{dx}\right]_{F_0=0.5},$$

and starting with n = 3, the curvature of the separation curve of the given CSC becomes smaller than the curvature of the curve of a single column, since

$$\left[\frac{dF}{dF_0}\right]_{F_0=0,5;n=3} = 0,5 < 1$$

At n = 4, the separation curve of the CSC becomes horizontal in the range of the boundary separation size:

$$\left[\frac{dF}{dF_0}\right]_{F=0.5,n=4}$$

and with a further increase of $n(n > 4)$, the curvature of the separation curve in the region of the separation boundary becomes negative, according to (X-19) (in relation to the curvature of the curve of the single column) and increases with increasing n. In this case, the separation curve of the CSC loses its monotonic appearance and, consequently, becomes ambiguous (Fig.X-3). Thus, we do not obtain suitable results from the process in this case and the application of this combined scheme for separation is evidently ineffective. However, it should be mentioned that this scheme of the CSC is of considerable interest for solving the problems associated with homothetic transformation tasks.

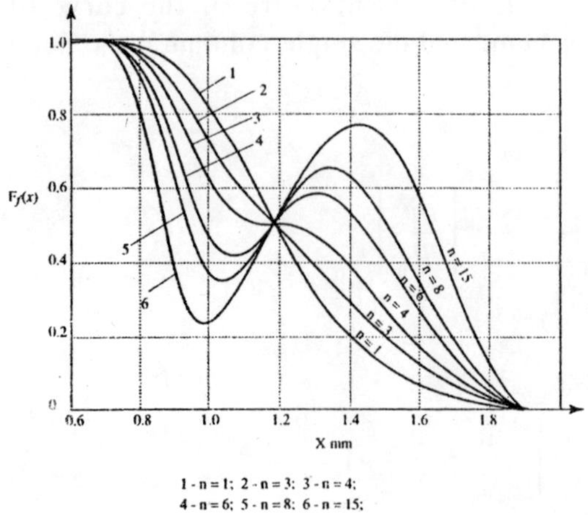

Fig.X-3 Separation curves for CSC with mixed purification.

4. COMBINED SCHEME WITH CONSECUTIVE RECIRCULATION (FIG.IX-4)

The system of equations, corresponding to the given structural scheme, has the form

$$F_{f_1} = F_{f_2} F_0$$
$$F_{f_2} = (F_{c_1} + F_{f_3})F_0$$
$$F_{f_3} = (F_{c_2} + F_{f_4})F_0$$
$$\dots\dots\dots\dots\dots\dots$$
$$F_{f_i} = (F_{c_{i^*-1}} + F_{f_{i^*-1}})F_0 + F_0 \qquad \text{(X-20)}$$
$$\dots\dots\dots\dots\dots\dots$$
$$F_{f_{n-1}} = (F_{c_{n-2}} + F_{f_n})F_0$$
$$F_{f_n} = F_{c_{n-1}} F_0$$
$$F_{f_1} + F_{c_1} = F_{f_2}$$
$$F_{f_2} + F_{c_2} = F_{c_1} + F_{f_3}$$
$$\dots\dots\dots\dots\dots\dots$$
$$F_{f_{i^*}} + F_{c_{i^*}} = F_{c_{i^*-1}} + F_{f_{i^*+1}} + 1$$
$$\dots\dots\dots\dots\dots\dots$$

It is evident that for any column of the given structural scheme the following equation is valid:

$$F_{f_i} + F_{c_i} = \frac{F_{f_i}}{F_0} = \frac{F_{c_i}}{1 - F_0} \qquad \text{(X-21)}$$

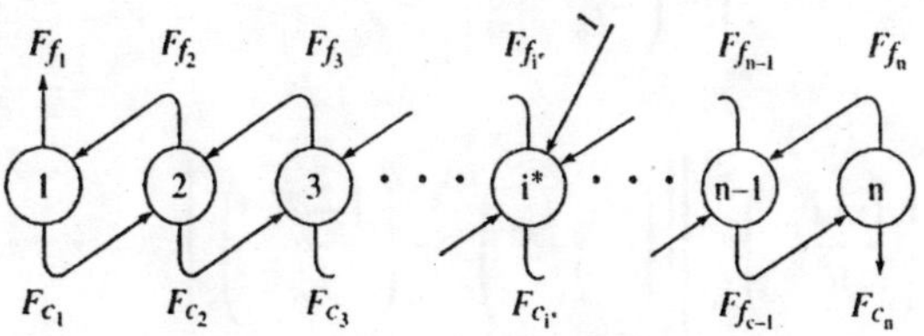

Fig. X-4 Structural diagram of CSC with consecutive recirculation of both products.

Taking into account (X-21), the recurrent equation, corresponding to the system (X-20) is written in the form:

$$\frac{F_{f_i}}{F_0}=\frac{F_{f_{i-1}}}{F_0}(1-F_0)+\frac{F_{f_{i+1}}}{F_0}\cdot F_0 \tag{X-22}$$

This is a homogeneous linear finite-difference equation of the 2nd order with constant coefficients. Its boundary conditions are:

$$\frac{F_{f_1}}{F_0}=\frac{F_{f_2}}{F_0}\cdot F_0 \tag{X-23}$$

and

$$\frac{F_{f_n}}{F_0}=\frac{F_{f_{n-1}}}{F_0}(1-F_0) \tag{X-24}$$

For a column for feeding with the initial material:

$$\frac{F_{f_{i^*}}}{F_0}=\frac{F_{f_{i^*-1}}}{F_0}(1-F_0)+\frac{F_{f_{i^*+1}}}{F_0}\cdot F_0+1 \tag{X-25}$$

A similar recurrent equation with identical boundary conditions and feed conditions has already been examined in previous chapters. It is concluded that they are identical at:

$$\frac{F_{f_i}}{F_0}=r_i;F_0=k;n=z$$

It is clear that the appropriate solutions of the examined scheme result from the previously obtained relationships ($F_0\neq 0.5$) for

$$F_{f_i}=\frac{\left[1-\left(\frac{1-F_0}{F_0}\right)^{n+1-i^*}\right]\left[1-\left(\frac{1-F_0}{F_0}\right)^{i}\right]}{\left[1-\left(\frac{1-F_0}{F_0}\right)^{n+1}\right](2F_0-1)}\cdot F_0;(1\le i\le i^*) \tag{X-26}$$

$$F_{f_i}=\frac{\left[1-\left(\frac{1-F_0}{F_0}\right)^{-i^*}\right]\left[\left(\frac{1-F_0}{F_0}\right)^{n+1}-\left(\frac{1-F_0}{F_0}\right)^{i}\right]}{\left[1-\left(\frac{1-F_0}{F_0}\right)^{n+1}\right](2F_0-1)}\cdot F_0;(i^*\le i\le n) \tag{X-27}$$

The function of the link for the given CSC scheme has the form:

$$F(F_0)=F_{f_1}(F_0)$$

For $i = 1$ from (X-26) we get

$$F(F_0)=\frac{1-\left(\frac{1-F_0}{F_0}\right)^{n+1-i^*}}{1-\left(\frac{1-F_0}{F_0}\right)^{n+1}} \qquad \text{(X-28)}$$

Correspondingly, for $F_0 = 0.5$, we have:

$$F=\frac{n+1-i^*}{n+1} \qquad \text{(X-29)}$$

Equation (X-29) shows that the absence of displacement of the boundary separation size for the combined scheme takes place only at $i^*=\frac{n+1}{2}$.

In this case, from equation (X-28) for the function of the link we obtain:

$$F(F_0)=\frac{1}{1+\left(\frac{1-F_0}{F_0}\right)^{\frac{n+1}{2}}}$$

Since there is no displacement of the boundary, the curvature of the separation curve combined scheme can be evaluated on the basis of the derivative:

$$\left[\frac{dF}{dF_0}\right]_{F_0=0,5}=\left[\frac{\frac{n+1}{2}\left(\frac{1-F_0}{F_0}\right)^{\frac{n-1}{2}}\cdot\left(\frac{1}{F_0}\right)^2}{\left\{1+\left(\frac{1-F_0}{F_0}\right)^{\frac{n+1}{2}}\right\}^2}\right]_{F_0=0,5}=\frac{n+1}{2}$$

This shows that with increasing n the efficiency of separation of the combined scheme continuously increases, exceeding the efficiency of separation of a single column, starting at $n = 3$. Thus, the combined scheme with consecutive recirculation is far more efficient in separation than the single classification column,

5. COMBINED CASCADE, REALISING THE BYPASS OF BOTH SEPARATION PRODUCTS

The diagram of such a cascade is shown in Fig.X-5.
The main system of equations for the flows, corresponding to the given structural schema, may be represented in the form:

$$\begin{aligned}
&F_1 = 1\\
&F_1^* = F_1(1-F_0)\\
&F_2 = (F_1 + F_1^*)F_0\\
&F_2^* = (F_2 + F_1^*)(1-F_0)\\
&\dots\dots\dots\dots\dots\dots\dots\dots\\
&F_i = (F_{i-1} + F_{i-1}^*)F_0\\
&F_i^* = (F_i + F_{i-1}^*)(1-F_0) \qquad\qquad (X\text{-}30)\\
&\dots\dots\dots\dots\dots\dots\dots\dots\\
&F_n = (F_{n-1} + F_{n-1}^*)F_0\\
&F_n^* = (F_n + F_{n-1}^*)(1-F_0)
\end{aligned}$$

where F_i is the total fractional flow in the i-th column; F_i^* is the flow of the particles of the same fixed narrow size class to the column i^*.

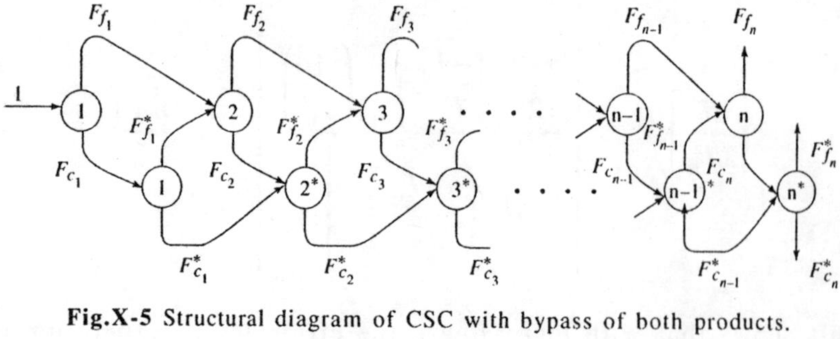

Fig.X-5 Structural diagram of CSC with bypass of both products.

Fractional extraction into the fine product for the entire CSC is:

$$F_{f_\Sigma}(n) = F_n F_0 \qquad \text{(X-31)}$$

– in the case of an odd number of columns equal to $N = 2n-1$.

$$F^*_{f_\Sigma}(n) = (F_n + F^*_n) F_0 \qquad \text{(X-32)}$$

– in the case of an even number of columns equal to $N = 2n$.

Equation (X-30) shows that

$$F_{f_n} = (F_{n-1} + F^*_{n-1}) F_0^2$$

Taking into account (X-32) gives:

$$F_{f_\Sigma}(n) = F_0 \cdot F^*_{f_\Sigma}(n-1) \qquad \text{(X-33)}$$

Thus, the problem of determination of the fractional extraction into the fine product of the entire CSC is reduced to the determination of the flows F_n and F^*_n.

The system of equations (X-30) may be transformed to the following form:

$$F_1 = 1;$$

$$F_1^* = 1 - F_1 F_0$$

This makes it possible to express $F_1 F_0$ and substitute it into the third equation of the system (X-30). Consequently, we obtain:

$$F_2 = 1 - F_1^* + F_1^* F_0 = 1 - F_1^*(1 - F_0)$$

Therefore, the value of $F^*_1(1 - F_0)$ is substituted into the fourth equation of the system (X-30):

$$F_2^* = F_2(1 - F_0) + (1 - F_2) = 1 - F_2 F_0$$

Continuing in the same manner, we obtain:

$$F_3 = 1 - F_2^*(1 - F_0);$$

$$F_3^* = 1 - F_3 F_0 \text{ and so on}$$

It is easy to confirm by the method of complete mathematical induction that in a general case:

$$F_i = 1 - F^*_{i-1}(1 - F_0); \qquad \text{(X-34)}$$

$$F_i^* = 1 - F_i F_0 \qquad \text{(X-35)}$$

The equations (X-34) and (X-35) already make it possible to form two mutually independent recurrent equations:

$$F_i = 1-(1-F_{i-1}F_0)(1-F_0)$$

After simplification, we obtain:

$$F_i - F_{i-1}F_0(1-F_0) = F_0 \tag{X-36}$$

Finally, from (X-35) taking (X-34) into account we obtain:

$$F_i^* - F_{i-1}^*F_0(1-F_0) = 1-F_0 \tag{X-37}$$

The equations (X-36) and (X-37) are inhomogeneous linear finite-difference equations of order I with the boundary conditions:

$$F_1 = 1 \text{ and } F^*_1 = 1 - F_0$$

The solution of these equations may be found by standard methods. Thus, using gradually equation (X-36) gives:

$$F(1) = 1;$$

$$F(2) = F_0 + F_0(1-F_0);$$

$$F(3) = F_0 + F_0^2(1-F_0) + F_0^2(1-F_0);$$

$$F(4) = F_0 + F_0^2(1-F_0) + F_0^3(1-F_0)^2 + F_0^3(1-F_0)^3;$$

....

$$F(i) = F_0 + F_0^2(1-F_0) + F_0^3(1-F_0)^2 + \cdots + F_0^{i-1}(1-F_0)^{i-2} + F_0^{i-1}(1-F_0)^{i-1};$$

....

$$F(n) = F_0 + F_0^2(1-F_0) + F_0^3(1-F_0)^2 + \cdots + F_0^{n-1}(1-F_0)^{n-2} + F_0^{n-1}(1-F_0)^{n-1}$$

It may easily be shown that here we are concerned with a geometrical progression.

We denote the denominator of the progression:

$$q = F_0(1-F_0)$$

and therefore

$$F(n) = F_0 + F_0q + F_0q^2 + \cdots + F_0q^{n-2} + q^{n-1}$$

or

$$F_0q^{n-1} - q^{n-1} + F(n) = F_0 + F_0q + F_0q^2 + \cdots + F_0q^{n-1}$$

Multiplying both parts of the equality by $q \neq 0$ gives

$$F_0q^n - q^n + qF(n) = F_0q + F_0q^2 + \cdots + F_0q^n$$

Transforming the right hand part, we obtain

$$F_0q^n - q^n + qF(n) = \frac{F_0q - F_0q^{n+1}}{1-q}$$

and therefore

$$F(n) = q^{n+1} + \frac{1-q^{n-1}}{1-q} \cdot F_0$$

Passing to fractional extraction of n-th column we obtain:

$$F_f(n) = F_0 q^{n-1} + F_0^2 \frac{1-q^n}{1-q} \tag{X-38}$$

Similarly, examining recurrent equation (X-37), we obtain:

$$F_f^*(n) = q\frac{1-q^n}{1-q} \tag{X-39}$$

Verification shows that equations (X-38) and (X-39), taking equations (X-31) and (X-32) into account, satisfy (X-33).

In analysis of the given CSC we restrict ourselves to a less efficient scheme of the system (incomplete), containing an odd number of columns. In this case, the fractional extraction of the entire CSC is written in the simplest form:

$$F_{f_\Sigma}(n) = F_0 q^{n-1} + F_0^2 \frac{1-q^{n-1}}{1-q} \tag{X-40}$$

Simplifying (X-40) gives

$$F_{f_\Sigma}(n) = \frac{1-F_0}{1-q} \cdot q^n + \frac{F_0^2}{1-q} \tag{X-41}$$

It should be mentioned that the fractional extraction of the single column is always greater (or equal to) than the fractional extraction of the entire CSC. Equality is observed only at $F_0 = 0$ and $F_0 = 1$. For example, this follows from (X-40). In fact:

$$F_0 - F_{f_\Sigma}(n) = F_0(1-q^{n-1})\left(1 - \frac{F_0}{1-q}\right)$$

but

$$\frac{F_0}{1-q} = \frac{F_0}{1-F_0+F_0^2} < 1 \quad \text{always at } F_0 \neq 1$$

Equation (X-41) may be used to construct the graph of the dependence $F_{f_\Sigma}(n)$ on F_0 and the number of separating columns (Fig.X-6). The graph shows that:

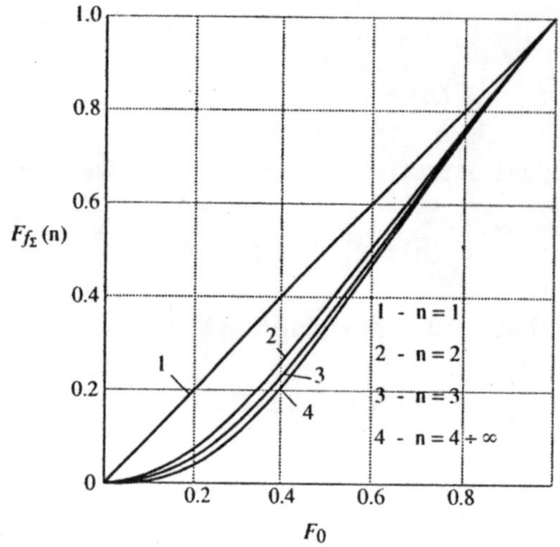

Fig.X-6 Dependence of the degree of fractional extraction of CSC on F_0 and the number of separation columns.

1. Large displacement of the regime of fixed extraction;

2. For $n > 3$ all separation curves of the CSC merge almost completely, and the separation boundary is determined by the regime in the single column $\overline{F_0} \cong 0.6$.

The accurate value of $\overline{F_0}$ may be found from the equation:

$$F_{f_\Sigma}(\overline{F_0}) = 0,5$$

For $n > 3$, from (X-41) we have:

$$\frac{\overline{F_0^2}}{1-\overline{F_0}+\overline{F_0^2}} = 0.5$$

Consequently $\overline{F_0} = 0.618$;

3. The curvature of the separation curves of the CSC at the point of the separation boundaries greatly exceed the curvature of the single column;

4. It is not efficient to use the given CSC with the number of columns greater than seven.

The quantitative evaluation of the curvature of the separation curve of the CSC with respect to the curvature of the single column may be estimated from the equation:

$$\bar{k} = 4 \cdot \left[\frac{dF_{f_\Sigma}}{dF_0} \right]_{F_0 = \overline{F_0}} \cdot \overline{F_0}(1 - \overline{F_0})$$

It should be noted (Fig.X-4,b) that with $n > 3$, the value of $\bar{k}$ is virtually independent of n, i.e. it holds that $\bar{k}(n \geq 4) \cong \bar{k}(n = \infty)$. Thus, the derivative of the first term of equation (X-41), containing q^n converts to zero, since $q \leq 0.25$. Consequently

$$\bar{k} = 4 \left\{ \frac{d}{dF_0} \left[\frac{F_0^2}{1 - q} \right] \right\}_{F_0 = \overline{F_0}} \cdot \overline{F_0}(1 - \overline{F_0})$$

After differentiation, we obtain:

$$\bar{k} = \frac{4\overline{F_0^2}\left(2 - \overline{F_0}\right)\left(1 - \overline{F_0}\right)}{\left(1 - \bar{q}\right)^2}$$

Substituting $\overline{F_0} = 0.618$, into the result, gives $\bar{k}(n \geq 4) = 1.382$..

To find $\bar{k}(n < 4)$, it is necessary to differentiate the entire equation (X-41). Omitting cumbersome derivations, we present the final result:

$$\frac{dF_{f_\Sigma}}{dF_0} = \frac{F_0 + q}{(1 - q)^2} - \frac{q^{n-1}}{(1 - q)} \left[q + (2F_0 - 1)(1 - F_0)\left(n + \frac{q}{1 - q} \right) \right] \quad \text{(X-42)}$$

In this case, the regime $\overline{F_0}$, determining the separation boundary, will depend on n.

It may easily be verified that the following holds:

$$F_{f_\Sigma}(n = 2) = 0.5 \quad \text{at} \quad \overline{F_0} = 0.597$$

$$F_{f_\Sigma}(n = 3) = 0.5 \ \text{at} \ \overline{F_0} = 0.613$$

Using these values, taking (X-42) into account gives:

$$\bar{k}(n = 2) = 1.269$$

$$\bar{k}(n = 3) = 1.350$$

Thus, it may be asserted that the examined CSC scheme is progressive.

This analysis shows that it is possible to construct a large number of interesting systems for the CSC but this cannot be investigated in the present work.

Therefore, we shall examine the scheme of a multirow apparatus which is of considerable interest for practice.

6. Multirow classifier

This apparatus is a suitable example of the CSC used in industrial practice. It is used for fractioning fine-grained potassium chloride in Uralkalli Company (Russia) and also in Machteshim Company (Israel) for separating organic powders. The first apparatus is of considerable interest because its productivity is 30 t/h, whereas that of the latter is up to 2 t/h. Therefore, we shall examine the Russian system.

The need to remove the dust from the fine-grained flotation concentrate was the result of orders to supply the material to the USA. A commercial product for export was a flotation concentrate with the amount of dust reduced to boundaries of 0.1 and 0.2 mm with a content of dust fractions not exceeding 4%, and in the initial material the content of the dust fraction was up to 20%.

The complicated nature of the solution of the task was that the existing continuous and circulation separators did not satisfy the requirements to the quality of product. These classifiers are designed for separating cleaner fine products. Investigations into the removal of dust from potassium chloride in a centrifugal separator with a productivity of up to 7 t/h showed that equipment makes it possible to reduce the content of dust fractions in the coarse product only to 8%. It was assumed that it would be possible to solve the problem for potassium fertilizers in a multirow CSC realising the process of consecutive purification of the coarse product. Therefore, to solve the problem, a multirow classifier with a total size of 1.0×2.5×4 m and with a productivity of 30 t/h was designed and assembled.

The circuit diagram of the entire equipment for the removal of dust in the processing line is shown in Fig.X-7. Equipment consists of seven identical parallel separating columns (the tray cascade of order I). Each column (row) includes six sections. In the lower part of apparatus, there is a gas-distribution grid (a flat sheet with perforated holes) inclined at a certain angle in relation to the horizon in the direction of unloading the coarse product.

The initial material is supplied from the drying pipe 1 through the distribution device 5 using screw feeder 20. The distribution device makes it possible to regulate the supply rate of material into the classifier from 5 to 45 t/h. Sealing of the apparatus on the side of loading the material is carried out with the screw feeder. The coarse product is discharged directly from the grid of the apparatus through the double gate of the distribution plate 22, ensuring the sealing of the separating chamber on the side of unloading the completed product. The fine product is trapped in seven cyclones 8. The material from the cyclone is unloaded onto the transporter 17 through double gates of the distributing plate 10 sealing the apparatus on the side of unloading the dust product. Air is supplied through the window 7 because of the rarefaction formed in the apparatus. A fan is an air blowing machine. Equipment operates under a rarefaction thus preventing the release of dust into the environment. Prior to injection into the atmosphere, the air is passed through two cleaning stages: the first stage is the separation of the solid phase in the cyclones, the second is the trapping of particles by foam apparatus, positioned behind the fan. To regulate the flow rate of air in separation channels of the apparatus, gate valves 12 are placed in gas lines between the cyclones and the general collector 9, and segment diaphragms 24 are used for inspection. In nozzles for supplying the initial material, unloading the coarse product and the products of cyclones there are nozzles for taking samples of materials.

The method of carrying out industrial testing includes the following operations:

1. A constant flow of air through the apparatus is set up.
2. The supply of initial material into the classifier and the operating regime of the apparatus are stabilised.
3. Samples for chemical and grain size analysis are taken from the flow of the initial material of the classification products using the standard procedure.
4. The material flows of the yield of products of classification are measured by means of cutting of the dust fractions and the coarse product during a specific period of time.
5. During a single experiment, inspection of the total flow rate of air in apparatus and in individual separating columns is carried out.
6. The grain size composition of the initial material and classification products are determined by screening a charge of the material in a set of sieves with mesh sizes of 0.8; 0.65; 0.4; 0.315; 0.2; 0.16; 0.1 and 0.063 mm.
7. Screening was carried out by a mechanical method according to the standard procedure.

For more detailed examination of the possibilities of separation of the material in the apparatus, conditions were determined for removing the dust from the initial material with respect to four size boundaries: 0.063; 0.1; 0.16 and 0.2 mm. In the tests, the conditions of fractionation in the given size range, the optimum productivity of the classifier was established, its separating capacity evaluated, and the grain size and chemical compositions of the separation products were determined.

Since the region of self-similarity with respect to the concentration of the solid phases in the gas flow was not determined in industrial conditions for a multirow apparatus, similar investigations were carried out in the equipment. Figure X-8 shows the dependences of the degree of fractional extraction on the consumed concentration of the material at an air-flow rate of Q = 15100 m^3/h. The graphs show that in the industrial conditions, as in laboratory experiments, the range of self-similarity with respect to the consumption concentration is 0–2.3 kg/m^3.

On the basis of the experimental data, Fig.X-9 shows the dependences of the degree of fractional extraction into the fine product in individual single columns of apparatus for different monofractions. The graphical dependences in Fig.X-9 make it possible to determine the operation of the multirow apparatus on the basis of the principle of single-row CSC, examined previously.

The modelling representations of the operation of the CSC are based on the concept of homogeneity, i.e. constancy of the coefficients of distribution of the monofractions in identical sections of the apparatus and the constancy of fraction separation of the individual columns. This position has been confirmed in our investigations. Therefore, the CSC can be calculated using the proposed models: duplex cascade, structural and combined. However, only these models are insufficient because in contrast to the previously examined apparatuses, the multirow apparatus has a special feature. It contains two types of separating elements: the distribution grid and pour-over trays, organising the separating columns.

The pour-over elements determine the operation of each section in two regimes: the aerodynamic regime in the continuous section is characterised by the coefficient of distribution of particles of the narrow size class k_1, and in the total cross section k_2. The mechanism of formation of the distribution coefficients was examined previously in the context of the structure of the moving flow. Consequently, the following expressions were obtained for a tray column with a square or rectangular cross section:

$$k_1 = 0.8678(1 - 0.5\sqrt{0.4 \cdot B})$$

$$k_2 = 0.8678(1 - \sqrt{0.4 \cdot B})$$

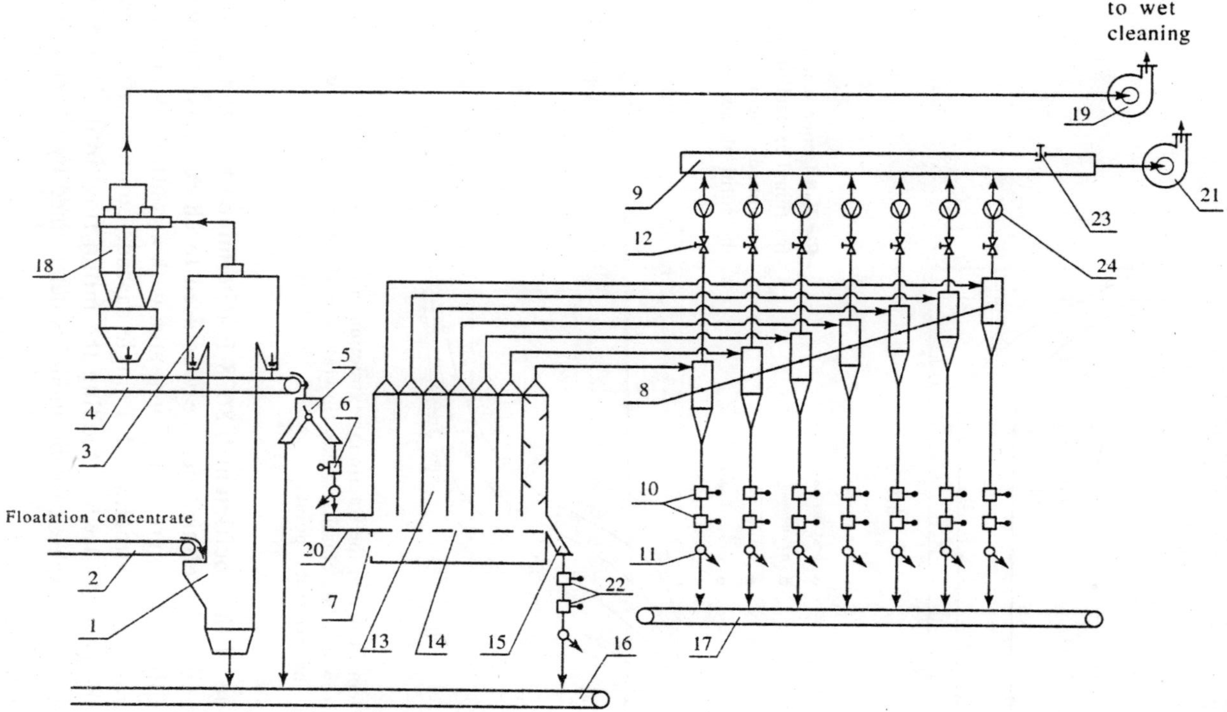

Fig.X-7 Diagram of a multirow classifier.

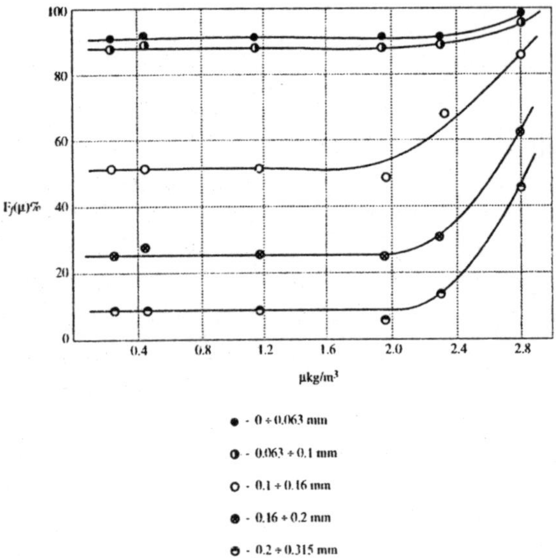

Fig X-8 Dependence of the fractional extraction on the consumed concentration of the solid phase.

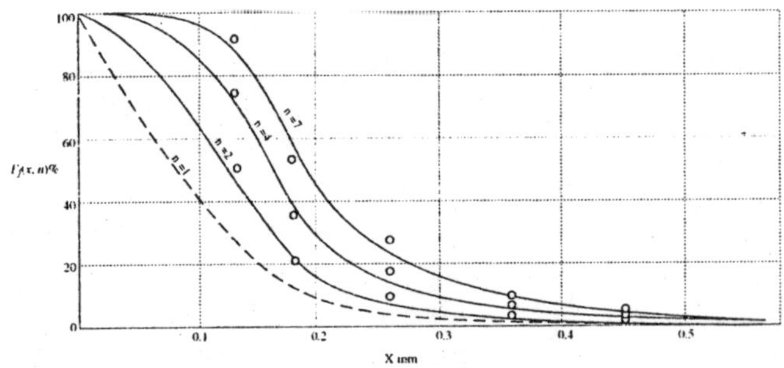

Fig.X-9 Dependence of the fractional extraction in individual separated columns of a multirow cascade classifier; gas flowrate V = 16200 m³/h. —— - calculated curves; o – experimental points.

The value of the coefficient 0.8678 is determined by the fact that the equation of the third degree was used to characterise the degree of filling of the cross sectional area with a continuous medium.

The fractional extraction of a single cascade column without the grid of the type A–B when feeding the initial flow into element B according to the duplex cascade model is determined by the equation:

$$F_{0_2} = \frac{1 - Q^{z-i^*} \cdot \dfrac{1-k_2}{k_1}}{1 - Q^z \cdot \dfrac{1-k_2}{k_1}}$$

For design of the multirow apparatus $i^* = z$ and consequently

$$F_{0_2} = \frac{1 - \dfrac{1-k_2}{k_1}}{1 - Q^6 \cdot \dfrac{1-k_2}{k_1}}$$

The separating grid having the form of a separating element operating in combination with a cascade column is characterised by the coefficient of fixed monofraction transfer from the grid to the lower section of the column λ.

For transfer coefficient λ a semi-empirical dependence, linking the coefficient with the generalised parameter of classification, was derived:

$$B = 3 \cdot \frac{\sqrt{1-\sqrt{\lambda}}}{1+2\sqrt{\lambda}}$$

The combination of the separating grid with the tray column realises the consecutive operation with partial recirculation of the monofraction from the column into the grid. The fraction of active return represents approximately 50% of the flow of the particles of the fixed narrow size class of the coarse product owing to the fact that return applies to 50% of the area of the grid below the column. The second half is simply transported to the section of the grid below the next column (Fig.X-7). In accordance with this structural scheme of the link of the grid with the column, fractional extraction of their combined operation is expressed by the dependence:

$$F_0 = \frac{\lambda F_{0_2}}{1 - \dfrac{1}{2}\lambda(1 - F_{0_2})}$$

The general expression for the fractional extraction of particles into the fine product for the entire CSC has the function of the link:

$$F = 1 - (1 - F_0)^n$$

Thus, using the previously obtained dependences, we can carry out a complete analytical calculation of the CSC.

References

1. Seader J.D. and Henley E.J., Separation process principles, Wiley Inc, New York (1998).
2. Khonry F.M., Predicting the performance of multistage separation process, CRC Press, Boca Raton, Florida (2000).
3. Govorov A.V., Cascade and combined processes of fractionation of bulk materials, a dissertation for the title of the Candidate of technical sciences, Sverdlovsk (1986).
4. Barsky M.D., Fractionation of powders, Nedra, Moscow (1980).
5. Barsky M.D. and Barsky E., General Trends of Gravity Separation, Proceedings of the XXI International Mineral Processing Congress, Elsevier, Rome (2000).
6. Barsky E. and Barsky M., Master curve of separation process. Physical Separation in Science and Engineering, Taylor and Francis, Vol. 3, No. 1 (2004).

Chapter XI

Separation Curves for Cascade Processes

PART 1. THE MAIN PROPERTIES OF SEPARATION CURVES

The separation curve is the most important characteristic of the fractionation process. We shall examine the main properties of the separation curves in fractionation of polydispersed bulk materials.

1. The degree of fractional separation into the fine product has the values in the range from 1 to 0.

In this case

$$F_f(x=0)=1;\ F_f(x_{50})=0.5;\ F_f(x\to\infty)=0 \qquad \text{(XI-1)}$$

2. The separation curve is continuous and differentiable in the entire range of values of x.

3. The separation curve is monotonically decreasing (this may be strictly confirmed on the basis of the structural model of the regular cascade), i.e.

$$\frac{dF_f(x)}{dx}\le 0,$$

and

$$\left[\frac{dF_f(x)}{dx}\right]_{x=0}=\left[\frac{dF_f(x)}{dx}\right]_{x=\infty}=0 \qquad \text{(XI-2)}$$

4. In most cases, the separation curve has two characteristic sections:

$$\frac{d^2F_f(x)}{dx^2}<0 \qquad \text{for the upper branch} \quad (x<x_{50})$$

$$\frac{d^2F_f(x)}{dx^2}>0 \qquad \text{for the lower branch} \quad (x>x_{50}).$$

5. The middle section of the curve is often approximated with sufficient accuracy by a linear dependence.

6. The length of the lower part of the curve in the size range is usually greater than the length of the upper curve.

7. Separation curves in different regimes in a relatively wide range of the separation boundary in the self-similar region of the consumption concentration ($\mu \leq 2 \div 3$ kg/m^3) are affine similar. In this case, the scale coefficients on the co-ordinate axis

$$c_y = 1; \qquad 0 < c_x \neq 1$$

This property is very important because, taking it into account, it is possible to pre-determine the results of the process in transition from one regime to another (from one separation boundary to another).

Equations (XI-1) and (XI-2) may be interpreted as the boundary conditions which must be satisfied by the typical separation curves.

8. The separation curve in the affine transformation is invariant in relation to the grain size composition of the separated material – the most important of the properties for optimising the separation process from the design viewpoint.

9. The affine properties of the separation curve predetermine the constancy, in different conditions (within the framework of a specific structure), of a number of parameters of the efficiency of the process determined on the basis of the form of the curve. The main of these are:

a. $E = -\left(\dfrac{dF_f(x)}{dx}\right)_{x_{50}} \cdot x_{50}$ – the Eder–Bokshtein criterion – a point indicator varying from 0 to ∞ (ideal process);

b. $\dfrac{F_{TP}}{x_{50}}$ – the integral Tromp criterion changing from 0 (ideal process) to ∞;

c. $\chi_{75/25} = \dfrac{x_{75}}{x_{25}}$ – the Eder–Mayer point criterion changing from 0 to 1 (ideal process), where x_{25}, x_{50}, x_{75} is the size of the particles according to the separation curves, extracted into the fine product to respectively, 25, 50 and 75%;

d. $F_{TP} = x_{50} - \int\limits_0^{x_{50}} F_f(x)dx + \int\limits_{x_{50}}^{\infty} F_f(x)dx -$ – the Tromp area.

It should be mentioned that the integral parameters include a large amount of information on the nature of the curve than the point parameters but they are more difficult to handle.

These considerations show that the affine properties of the separation curves are the most important properties and determine all other properties. We shall discuss them in greater detail. The necessary and sufficient condition for the affine similarity is the possibility of transferring from equation $F_f(x,w)$ to the equation identical for all regimes $F_f\left(\frac{x}{x_{50}}\right)$. In principle, this corresponds to the affine transformation of any initial separation curve $F_f(x,w_i)$, with a scale coefficient on the abscissa axis $\frac{1}{x_{50}}$. Previously it was shown that the transition to the unitary curve is quite distinctive in the combined examination of the structural model and the model of the regular cascade.

We shall examine in the general form the formation of the unitary curve.

It is assumed that we have some two-parameter function for the degree of fractional extraction which is written in the form:

$$F_f(x;w) = F_f\left[\left(\frac{x}{x_{50}}\cdot x_{50}\right);w\right] \tag{XI-3}$$

It is also assumed that the right-hand part of this equation may be transferred into the form:

$$F_f\left[\left(\frac{x}{x_{50}}\cdot x_{50}\right);w\right] = F_f\left[\frac{x}{x_{50}};f(x_{50};w)\right] \tag{XI-4}$$

The realisation of the condition (XI-4) will be referred to as the unitary transformation. It should be mentioned that an arbitrarily large number of two-parameter functions corresponds to the unitary transformation. We shall present examples of some of these functions, used in special literature:

$$\begin{aligned} F_f &= \varphi_1\left[x^a\cdot\varphi_2(w)\right]; \\ F_f &= \varphi_1\left\{\left[\varphi_2(w)\right]^{x^a}\right\}; \\ F_f &= \varphi_1\left[\varphi_2(w)+\lg x^a\right]; \end{aligned} \tag{XI-5}$$

$$F_f = \varphi_1\left[\varphi_2(w)\cdot\varphi_3(x)\right];$$
$$F_f = \varphi_1\left\{\left[\varphi_2(w)\right]^{\varphi_3(x)}\right\};$$
$$F_f = \varphi_1\left[\varphi_2(w)+\lg\varphi_3(x)\right]$$

where φ_1, φ_2 are the arbitrary functions of any complexity; φ_3 is any homogeneous function of the degree k, i.e. function satisfying the expression $\varphi_3(\lambda x)=\lambda^k\cdot\varphi_3(x)$.

Substituting the boundary condition (XI-1) into (XI-4), we obtain:

$$F_f\left[1; f(x_{50};w)\right]=0.5 \tag{XI-6}$$

Solving (XI-7) in relation to the function $f(x_{50};w) = c$, we obtain

$$f(x_{50};w)=c \tag{XI-7}$$

Substituting (XI-7) into (XI-4) we obtain the final form of the unitary curve

$$F_f(x;w)=F_f\left(\frac{x}{x_{50}};c\right) \tag{XI-8}$$

Thus, the affine properties of the separation curves have been ensured. These considerations show that the realisation of the unitary transformation is a sufficient condition for the affine similarity of any curves (including the separation curves), irrespective of the nature of the process. Thus, an infinite number of functions $F_f(x;w)$ ensure the automatic fulfilment of the affine properties of the separation curves. It should noted that in a partticular case (XI-5) when

$$F_f = \varphi_1\left[x^a\cdot w^b\right] \tag{XI-9}$$

the unitary transformation may be carried with respect to both the parameter x and in relation to w. In this case, the unitarisation of the separation curves will also take place in the co-ordinates of the relative velocity:

$$F_f(x;w)=F_f\left(\frac{w}{w_{50}};c'\right)$$

Thus, the affine properties of the separation curves will be observed with respect to both parameters.

In an even more particular case, when equation (XI-9) $b = -2a$, the fractional extraction will be represented by a unitary curve of the Froude parameter:

$$F_f = \varphi(Fr) \qquad \text{(XI-10)}$$

where $Fr = \frac{gx}{w^2}$ is the Froude criterion.

Equation (XI-10) corresponds to the structural model of the classification process and to a large number of experimental data. This in turn confirms the functional dependence of the degree of fractional separation of the type (XI-9).

The unitary form of the separation curve (XI-8) shows clearly the main previously mentioned properties of the separation curves (in particular, the property XI-9). Consequently, the Eder–Bokshtein criterion assumes the form:

$$E = -\left[\frac{dF_f(x)}{dx}\right]_{x=x_{50}} x_{50} = -\left[\frac{dF_f\left(\frac{x}{x_{50}};c\right)}{d\left(\frac{x}{x_{50}}\right)}\right]_{\frac{x}{x_{50}}=1} \qquad \text{(XI-11)}$$

It is clearly evident that here E = *idem*. For a specific function $F_f\left(\frac{x}{x_{50}}\right)$ all parameters are linked together. In fact, equation (XI-11) shows that

$$E = f(1;c) = const \qquad \text{(XI-12)}$$

Solving (XI-12) in relation to c gives:

$$c = c(E)$$

Substituting the result into (XI-8) gives:

$$F_f\left(\frac{x}{x_{50}};c\right) = F_f\left(\frac{x}{x_{50}};E\right) \qquad \text{(XI-13)}$$

Using (XI-13), we determine the Eder–Mayer criterion:

$$0.75 = F_f\left(\frac{x_{75}}{x_{50}};E\right), \quad \text{therefore} \quad \frac{x_{75}}{x_{50}} = f_1(E)$$

$$0,75 = F_f\left(\frac{x_{75}}{x_{50}}; E\right), \quad \text{therefore} \quad \frac{x_{25}}{x_{50}} = f_2(E)$$

$$\chi_{75/25} = \frac{x_{75}/x_{50}}{x_{25}/x_{50}} = \frac{f_1(E)}{f_2(E)} = f^*(E) = idem(w)$$

For the Tromp criterion

$$\frac{F_{TP}}{x_{50}} = \frac{x_{50} - \int_0^{x_{50}} F_f(x)dx + \int_{x_{50}}^{\infty} F_f(x)dx}{x_{50}} =$$

$$=1 - \int_0^1 F_f\left(\frac{x}{x_{50}}; E\right) \cdot d\left(\frac{x}{x_{50}}\right) + \int_1^{\infty} F_f\left(\frac{x}{x_{50}}; E\right) \cdot d\left(\frac{x}{x_{50}}\right) = f^{**}(E) = idem(w)$$

This is also valid for a number of other parameters based on the separation curves. For example, in the case of the model of the regular cascade, the quality of the classification process may be evaluated using a criterion identical with the Eder–Bokshtein parameter (Fig.XI-1)

$$Q = \left(\frac{dF}{dk}\right) \cdot (1 - k_{50}) \qquad \text{(XI-14)}$$

It will be evident that $Q = idem(w)$.

According to the structural model, the distribution coefficient has the form:

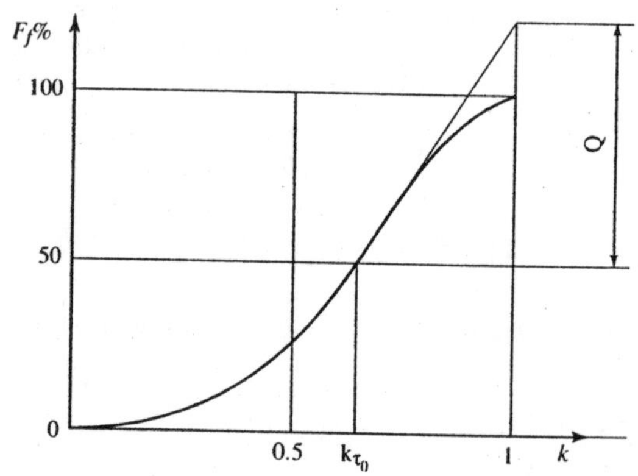

Fig. XI-I Evaluation of the model of the regular cascade.

$$k = 1 - \sqrt{A \cdot x}$$

and then

$$k_{50} = 1 - \sqrt{A \cdot x_{50}}$$

Taking the last equation into account, the equation (XI-14) will be transformed to the form:

$$Q = \left(\frac{dF}{dk}\right)_{k_{50}} \cdot \sqrt{A \cdot x_{50}} = -\left(\frac{dF}{dk}\right)_{k_{50}} \cdot \left(-\frac{A}{\sqrt{A \cdot x_{50}}}\right) \cdot x_{50}$$

or

$$Q = -\left(\frac{dF}{dk}\right)_{k_{50}} \cdot 2\left(\frac{dk}{dx}\right)_{x_{50}} \cdot x_{50} = -2\left(\frac{dF}{dx}\right)_{x_{50}} \cdot x_{50}$$

Finally, we obtain:

$$Q = 2E$$

2. APPROXIMATIONS OF SEPARATION CURVES

In this context, we shall analyse the specific approximations of the separation curves obtained on the basis of unitary transformations.

1. The exponential approximation of the type

$$F_f A + B(x^a \cdot w^b)$$

For this approximation it is not possible to fulfil simultaneously all boundary conditions and, consequently, we shall retain these conditions only for the lower part which is the longest. Using unitary transformations, we obtain:

$$F_f = A + B\left(\frac{x}{x_{50}}\right)^a \cdot (x_{50}^a \cdot w^b) \qquad \text{(XI-15)}$$

From the boundary conditions

$$0{,}5 = A + B(x_{50}^a \cdot w^b)$$

and

$$x_{50}^a \cdot w^b = \frac{0{,}5 - A}{B}$$

The resultant equation is substituted into (XI-15):

$$F_f = A + B\left(\frac{x}{x_{50}}\right)^{\frac{a(0,5-A)}{B}} = A + (0.5 - A)\cdot\left(\frac{x}{x_{50}}\right)^a$$

From the boundary condition $F_f\left(\frac{x}{x_{50}} \to \infty\right) = 0$, we have:

$$A = 0; \qquad a < 0$$

Consequently, $F_f = 0.5\left(\frac{x}{x_{50}}\right)^{-a}$ where $a > 0$

The condition with respect to the derivative is fulfilled

$$\frac{dF_f}{d\left(\frac{x}{x_{50}}\right)} = -0.5a\left(\frac{x}{x_{50}}\right)^{-(a+1)} \leq 0$$

Finally, the expression for the exponential approximation has the form

$$F_f = 0.5\left(\frac{x}{x_{50}}\right)^{-a} \qquad \text{(XI-16)}$$

and that $\left(\frac{x}{x_{50}}\right) \geq 2^{-\frac{1}{a}}$, so that $F_f \leq 1$.

The resultant expression (XI-16) may be reduced to the form (XI-13):

$$E = -\left[\frac{dF_f\left(\frac{x}{x_{50}}\right)}{d\left(\frac{x}{x_{50}}\right)}\right]_{\frac{x}{x_{50}}=1} = 0.5\cdot a$$

Consequently

$$a = 2E; \qquad F_f = 0.5\left(\frac{x}{x_{50}}\right)^{-2E} \qquad \text{(XI-17)}$$

This means that:

$$\frac{x_{75}}{x_{50}} = 1.5^{-\frac{1}{2E}}; \qquad \frac{x_{25}}{x_{50}} = 0.5^{-\frac{1}{2E}}$$

Consequently, the Eder–Mayer criterion is

$$\chi_{75/25} = 3^{-\frac{1}{2E}}$$

The Tromp criterion is calculated from

$$\frac{F_{TP}}{x_{50}} = 1 - 2^{-\frac{1}{2E}} - \int_{2^{-\frac{1}{2E}}}^{1} 0.5\left(\frac{x}{x_{50}}\right) \cdot d\left(\frac{x}{x_{50}}\right) + \int_{1}^{\infty} 0.5\left(\frac{x}{x_{50}}\right)^{-2E} \cdot d\left(\frac{x}{x_{50}}\right) =$$

$$\frac{2E}{(2E-1)}\left(1 - 2^{-\frac{1}{2E}}\right)$$

2. The exponential approximation of the type

$$F_f = \varphi_1\left[x_a \cdot \varphi_2(w)\right] = c^{\left[x^a \cdot \varphi_2(w)\right]}$$

In the unitary transformation we have

$$F_f = c^{\left(\frac{x}{x_{50}}\right)\left[x_{50}^a \cdot \varphi_2(w)\right]} \tag{XI-18}$$

We use the boundary condition:

$$0.5 = c^{\left[x_{50}^a \cdot \varphi_2(w)\right]},$$

and therefore

$$x_{50}^a \cdot \varphi_2(w) = \frac{\ln 0.5}{\ln c}$$

Substituting the last equation into (XI-18), gives:

$$F_f = \left(c^{\frac{\ln 0.5}{\ln c}}\right)^{\left(\frac{x}{x_{50}}\right)^a} = 0.5^{\left(\frac{x}{x_{50}}\right)^a}$$

Finally, we obtain

$$F_f = 2^{-\left(\frac{x}{x_{50}}\right)^a} \tag{XI-19}$$

Transforming (XI-19) to the form (XI-18) gives:

$$E = -\left[\frac{dF_f\left(\frac{x}{x_{50}}\right)}{d\left(\frac{x}{x_{50}}\right)}\right]_{\frac{x}{x_{50}}=1} = a\left(\frac{x}{x_{50}}\right)^{a-1}_{\frac{x}{x_{50}}=1} \cdot 2^{-\left(\frac{x}{x_{50}}\right)^a} \cdot \ln 2 = \frac{a \ln 2}{2}$$

or

$$a = \frac{2E}{\ln 2}$$

Taking into account (XI-19) we obtain

$$F_f = 2^{-\left(\frac{x}{x_{50}}\right)^{\frac{2E}{\ln 2}}} \qquad \text{(XI-20)}$$

This approximation satisfies all the boundary conditions (XI-1) and (XI-2).

From (XI-20) we get:

$$\frac{x_{75}}{x_{50}} = \left(\frac{\ln 0.75}{\ln 0.5}\right)^{\frac{\ln 2}{2E}}; \qquad \frac{x_{25}}{x_{50}} = \left(\frac{\ln 0.25}{\ln 0.5}\right)^{\frac{\ln 2}{2E}}$$

Consequently, the Eder–Mayer criterion is:

$$\chi_{75/25} = \left(\frac{\ln 0.75}{\ln 0.25}\right)^{\frac{\ln 2}{2E}} = 2.974^{\frac{1}{2E}}$$

The Tromp criterion is:

$$\frac{F_{TP}}{x_{50}} = 1 - \int_0^1 2^{-y^{\frac{2E}{\ln 2}}} dy + \int_1^\infty 2^{-y^{\frac{2E}{\ln 2}}} dy = f(E)$$

3. The exponential approximation of the type

$$\varphi_1\left[x^a \cdot \varphi_2(w)\right] = c \cdot e^{-[ax \cdot \varphi_2(w)]};$$

$$\varphi_2(w) = w^b;$$

$$F_f = c \cdot e^{-a\left(\frac{x}{x_{50}}\right)\left(x_{50} \cdot w^b\right)} \qquad \text{(XI-21)}$$

We use the boundary conditions

$$\frac{1}{2}=c\cdot e^{-a(x_{50}w^{b})};\qquad c=\frac{1}{2}e^{a(x_{50}w^{b})}.$$

$$E=-\left[\frac{dF_f\left(\frac{x}{x_{50}}\right)}{d\left(\frac{x}{x_{50}}\right)}\right]_{\frac{x}{x_{50}}=1}=ac(x_{50}w^{b})\cdot e^{-a(x_{50}w^{b})} \qquad \text{(XI-22)}$$

Substituting into the result:

$$E=\frac{1}{2}a(x_{50}w^{b})$$

gives

$$x_{50}w^{b}=\frac{2E}{a} \qquad \text{(XI-23)}$$

Taking into account (XI-22) we have:

$$c=\frac{1}{2}e^{2E}$$

Substituting the last equation and (XI-23) into (XI-21) we finally obtain:

$$F_f=\frac{1}{2}e^{2E\left(1-\frac{x}{x_{50}}\right)} \qquad \text{(XI-24)}$$

The approximation (XI-24) satisfies the boundary conditions

$$F_f\left(\frac{x}{x_{50}}\to\infty\right)=0;\qquad F_f'\left(\frac{x}{x_{50}}\to\infty\right)=0;\qquad F_f'\left(\frac{x}{x_{50}}\right)<0$$

This equation describes the discontinuous unitary curve. In order to ensure that the degree of fractional extraction does not exceed unity, it is necessary to fulfil the following condition:

$$\frac{x}{x_{50}}\geq 1-\frac{\ln 2}{2E}$$

In accordance with (XI-24) we obtain:

$$\frac{x_{75}}{x_{50}}=1-\frac{\ln 1.5}{2E};\qquad \frac{x_{25}}{x_{50}}=1-\frac{\ln 0.5}{2E}$$

Consequently, the Eder-Mayer criterion is returned to the form:

$$\chi_{75/25} = \frac{2E - \ln 1.5}{2E + \ln 2}$$

The Tromp criterion is:

$$\frac{F_{TP}}{x_{50}} = 1 - \left(1 - \frac{\ln 2}{2E}\right) - \int_{1-\frac{\ln 2}{2E}}^{1} \frac{1}{2} e^{2E\left(1-\frac{x}{x_{50}}\right)} d\left(\frac{x}{x_{50}}\right) + \int_{1}^{\infty} \frac{1}{2} e^{2E\left(1-\frac{x}{x_{50}}\right)} d\left(\frac{x}{x_{50}}\right) = \frac{\lg 2}{2E}$$

We have introduced a new parameter linearly linked with E: $\varphi = \dfrac{2E}{\ln 10}$.

Consequently, equation (XI-24) transforms to the form:

$$F_f = \frac{1}{2} e^{\ln 10 \cdot \varphi\left(1-\frac{x}{x_{50}}\right)} = \frac{1}{2} \cdot 10^{\varphi\left(1-\frac{x}{x_{50}}\right)}$$

Transferring in the final equation to the degree of fractional extraction, expressed in per cent, gives:

$$F_f \% = 50 \cdot 10^{\varphi\left(1-\frac{x}{x_{50}}\right)} \qquad \text{(XI-25)}$$

Taking the logarithm of (XI-25) with a base 10 gives:

$$\lg F_f = 1.7 + \varphi - \varphi \frac{x}{x_{50}} \qquad \text{(XI-26)}$$

The dimensionless size is represented in the form:

$$\frac{x}{x_{50}} = \frac{hxw^b}{hx_{50}w^b} \qquad \text{(XI-27)}$$

where h is some constant. We denote $Fr^* = hxw^b$.

Consequently, according to (XI-27):

$$\frac{x}{x_{50}} = \frac{Fr^*}{Fr^*_{50}}$$

Substituting this equation into (XI-26) gives:

$$\lg F_f = 1.7 + \varphi - \frac{\varphi}{Fr^*_{50}} \cdot Fr^* \qquad \text{(XI-29)}$$

(XI-23) shows that $Fr^*_{50} = \dfrac{h2E}{a} = \text{const}$.

Consequently, $\dfrac{\varphi}{Fr^*_{50}}=\dfrac{a\varphi}{2ER}=\dfrac{a}{h\ln 10}=k(\text{const})$.

In this case, the relationship (XI-29) has the form:

$$\lg F_f = 1.7+\varphi-kFr^*_{50} \qquad \text{(XI-30)}$$

The graphic interpretation of approximation (XI-30) is shown in Fig.XI-2. In a particular case when 'b' in equation (XI-21) b = 2, we obtain a semi logarithmic dependence of the degree of fractional extraction on the Froude criterion. Here $\varphi = kFr_{50} = \dfrac{2E}{\ln 10}$ is the parameter of the efficiency of separation proportional to the Eder–Bokshtein criterion in the case of the exponential approximation of the separation curve. These parameters were previously found by experiments.

Hyperbolic approximation of the type

$$\varphi_1\left[x^a\cdot\varphi_2(w)\right]=\frac{b}{c+\left[x^a\cdot\varphi_2(w)\right]};$$

$$F_f=\frac{b}{c+\left[x^a\cdot\varphi_2(w)\right]} \qquad \text{(XI-31)}$$

We use the boundary condition $F_f(x = 0) = 1$, and therefore $c = b$.

(XI-31) will be written in the transformation

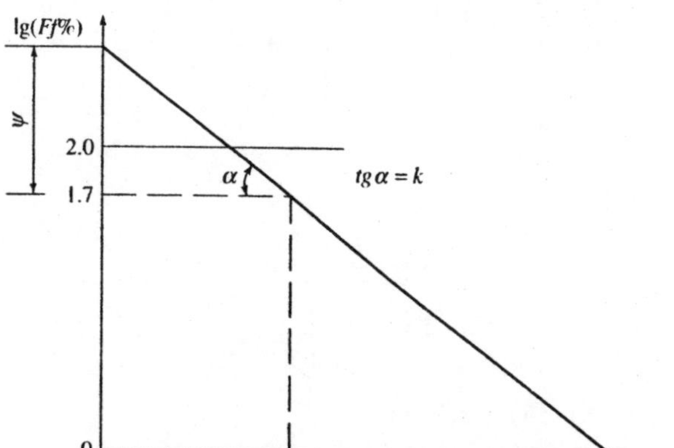

Fig.XI-2 Interpretation of the exponential approximation in semi-logarithmic coordinates.

$$F_f = \frac{b}{b + \left(\frac{x}{x_{50}}\right)^a \left[x_{50}^a \cdot \varphi_2(w)\right]} \qquad \text{(XI-32)}$$

We use the condition

$$F_f\left(\frac{x}{x_{50}} = 1\right) = \frac{1}{2};$$

and taking this into account

$$\frac{1}{2} = \frac{b}{b + \left[x_{50}^a \cdot \varphi_2(w)\right]},$$

and consequently $x_{50}^a \cdot \varphi_2(w) = b$.

Taking this into account, equation (XI-32) has the form:

$$F_f = \frac{1}{1 + \left(\frac{x}{x_{50}}\right)^a} \qquad \text{(XI-33)}$$

We obtain the well-known Plitt approximation satisfying all boundary conditions.

Equation (XI-33) will be presented in the form (XI-13):

$$E = -\left[\frac{dF_f\left(\frac{x}{x_{50}}\right)}{d\left(\frac{x}{x_{50}}\right)}\right]_{\frac{x}{x_{50}}=1} = \left\{\frac{a\left(\frac{x}{x_{50}}\right)^{a-1}}{\left[1 + \left(\frac{x}{x_{50}}\right)^a\right]}\right\}_{\frac{x}{x_{50}}=1} = \frac{a}{4}$$

Consequently:

$$F_f = \frac{1}{1 + \left(\frac{x}{x_{50}}\right)^{4E}} \qquad \text{(XI-34)}$$

From equation (XI-34) we have:

$$\frac{x_{75}}{x_{50}} = 3^{-\frac{1}{4E}}; \qquad \frac{x_{25}}{x_{50}} = 3^{\frac{1}{4E}}$$

Consequently, the Eder–Mayer criterion is determined as follows:

$$\chi_{75/25} = 3^{-\frac{1}{8E}} \qquad \text{(XI-35)}$$

The general form of the Tromp criterion is:

$$\frac{F_{TP}}{x_{50}} = 1 - \int_0^1 \frac{1}{1+y^{4E}}\,dy + \int_1^{\infty} \frac{1}{1+y^{4E}}\,dy = f(E)$$

The comparison of the approximating unitary functions (XI-20), (XI-24) and (XI-34) with the actual curves is shown in Fig.XI-3.

The power-exponential approximation of the type

$$F_f = \left[b \cdot x^a \cdot \varphi_2(w)\right]^{-cx^a \cdot \varphi_2(w)}$$

or in the unitary transformation

$$F_f = \left[b\left(\frac{x}{x_{50}}\right)^a \cdot x_{50}^a \cdot \varphi_2(w)\right]^{-c\left(\frac{x}{x_{50}}\right)^a \cdot x_{50}^a \cdot \varphi_2(w)}$$

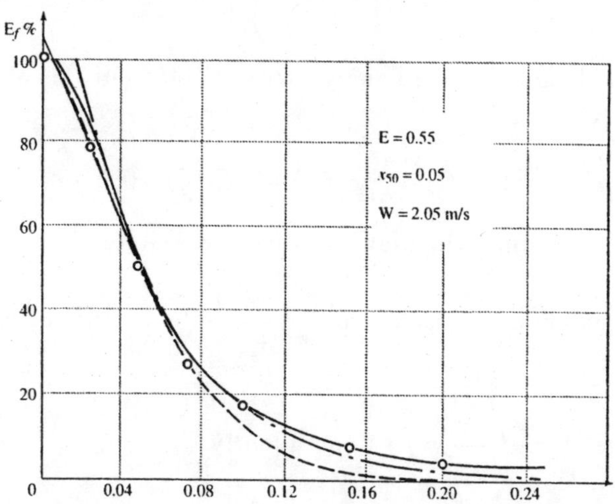

X mm experimental values;
--- - approximation (XI-20);
–·– - approximation (XI-24);
____ - Plitt approximation (XI-34).

Fig.XI-3 Separation curve of a single column of the multirow classifier with a productivity of 35 t/h (n = 7; z = 6; i^* = 6; μ = 2 kg/m³; ρ = 2000 kg/m³; $a + b$ = 0.88 × 0.35 m²).

From the boundary condition we get:

$$\frac{1}{2}=\left[b\cdot x_{50}^{a}\cdot\varphi_2(w)\right]^{-cx_{50}^{a}\cdot\varphi_2(w)}$$

and consequently

$$\ln 2=cx_{50}^{a}\cdot\varphi_2(w)\cdot\ln\left[bx_{50}^{a}\varphi_2(w)\right]$$

or

$$bx_{50}^{a}\cdot\varphi_2(w)=e^{\frac{\ln 2}{cx_{50}^{a}\varphi_2(w)}}$$

We denote $cx^{a}_{50}\cdot\varphi_2(w)=\rho$, and consequently $bx_{50}^{a}\cdot\varphi_2(w)=e^{\frac{\ln 2}{\rho}}=2^{\frac{1}{\rho}}$.

The resultant equation will be substituted into $F_f\left(\frac{x}{x_{50}}\right)$:

$$F_f=\left[2^{\frac{1}{\rho}}\cdot\left(\frac{x}{x_{50}}\right)^{a}\right]^{-\rho\left(\frac{x}{x_{50}}\right)} \qquad \text{(XI-36)}$$

The approximation of the unitary curve (XI-36) satisfies all boundary conditions:

$$F_f(0)=1;F_f(1)=\frac{1}{2};F_f(\infty)=0;F_f'(0)=0;F_f'(\infty)=0$$

However, this approximation has a unique feature. In order to show this, we determine the derivative $F_f'\left(\frac{x}{x_{50}}\right)$. To simplify considerations, we denote $2^{\frac{1}{\rho}}=d;\frac{x}{x_{50}}=y$. Consequently:

$$F_f'(y)=\frac{d}{dy}\left[dy^{a}\right]^{-\rho y^{a}}=-\rho a y^{a-1}\left[dy^{a}\right]^{-\rho y^{a}}\cdot\left[1+\ln(dy^{a})\right] \qquad \text{(XI-37)}$$

(XI-37) shows that $F_f(y)=0$ at two values of y_0:

1. $y_0=0$;

2. $1+1\ln(dy^{a}_{0})=0$, and consequently $dy_0^{a}=\frac{1}{e}$, or

$$y_0\left(\frac{1}{de}\right)^{\frac{1}{a}}=\left(e\cdot 2^{\frac{1}{\rho}}\right)^{-\frac{1}{a}} \tag{XI-38}$$

For $y_0 = 0$, the value $F_f(y_0) = 1$.

According to (XI-38) for y_0

$$F_f(y_0)=e^{\frac{\rho}{e\cdot 2^{\frac{1}{\rho}}}}>1$$

Thus, the maximum value of the approximating function is obtained at $y_0 > 0$ and not at $y = 0$ and is

$$F_{f\max}=\exp\left(\frac{\rho}{e\cdot 2^{\frac{1}{\rho}}}\right) \tag{XI-39}$$

Consequently, at $\rho \to 0$; $F_{f\max} \to 1$, and at $\rho = 0$ this approximation is transformed into a power approximation (XI-19).

Using these methods, it is possible to determine the general characteristic also for other approximations of the unitary separation curves.

3. EFFICIENCY OF SEPARATION IN THE CASCADE

We shall estimate the separating capacity of the cascade systems based on the Eder–Bokshtein criterion and the cascade-structural model. We introduce the concept of the reversed cascade:

$$F_{rev}(k)=1-F_{dir}(1-k) \tag{XI-40}$$

It is completely evident that $k_{rev}=1-k_{dir}$; $\chi_{rev}=\dfrac{1}{\chi_{dir}}$.

In accordance with the model of the regular cascade, for (XI-40) we obtain:

$$F_{rev}(k)=1-\frac{1-\left(\dfrac{1}{\chi}\right)^{z+1-i}}{1-\left(\dfrac{1}{\chi}\right)^{z+1}}=\frac{1-\chi^{i}}{1-\chi^{z+1}} \tag{XI-41}$$

Relationship (XI-41) may be written in the form:

$$F_{rev}(k) = \frac{1-\chi^{z+1-i_{rev}}}{1-\chi^{z+1}} \tag{XI-42}$$

where $i_{rev} = z + 1 - i$ where i is the section of feed for the reverse cascade.

Thus, in accordance with (XI-42) the reversed cascade is a direct cascade with the feed section replaced by symmetric section (Fig.XI-4). According to (XI-40), the graphic interpretation of the curves $F_{dir}(k)$ and $F_{rev}(k)$ is shown in Fig.XI-5.

We determine the curvature of the curve $F_f(k)$ for the direct and reversed cascade at the points of the distribution k and k_{rev} = $1 - k$.

$$\left[\frac{dF_{rev}(k)}{dk}\right]_{k_{rev}} = \left\{\frac{d\left[1-F_{dir}(1-k)\right]}{d(1-k)}\right\}_{1-k_{rev}} \cdot \frac{d(1-k)}{dk} = \left[\frac{dF_{dir}(k)}{dk}\right]_{k_{dir}}$$

Thus, for the direct and reversed cascades we have the equality of the tangents of the angles of the inclination of the curves $F_{dir}(k)$ and $F_{rev}(k)$ in appropriate points of the values of the distribution coefficients (Fig.XI-5).

This also applies at the points of the boundary coefficient of distribution at the points k' and $1 - k'$, and also at points $k_{50}dir$ and $k_{50rev} = 1 - k_{50dir}$, since $F_{dir}\left(k_{50dir}\right) = \frac{1}{2}$, and in accordance with (XI-40).

$$F_{rev}(k_{50rev}) = F_{rev}(1-k_{50dir}) = 1 - F_{dir}(k_{50dir}) = \frac{1}{2}$$

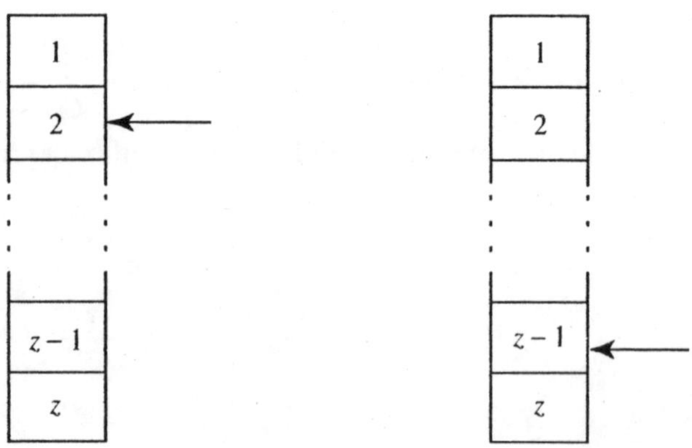

Fig.XI-4 Explanation of the direct and reverse cascades.

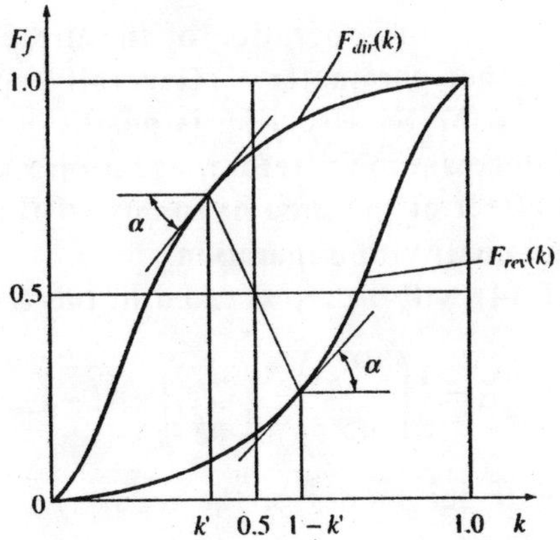

Fig.XI-5 Dependence of the fractional extraction on the separation factors for direct and reverse cascades.

We determine the Eder–Bokshtein criterion for the direct and reversed cascades:

$$E_{dir}=-\left(\frac{dF_f}{dx}\right)_{x_{50}}\cdot x_{50}=-\left(\frac{dF_f}{dk}\right)_{k_{50dir}}\cdot\left(\frac{dk}{dx}\right)_{x_{50}}\cdot x_{50} \qquad \text{(XI-43)}$$

According to the structural model

$$k=1-\sqrt{A\cdot x};$$

$$\left(\frac{dk}{dx}\right)_{x_{50}}\cdot x_{50}=-\frac{A\cdot x_{50}}{2\sqrt{A\cdot x_{50}}}=-\frac{1}{2}\sqrt{A\cdot x_{50}}=-\frac{1}{2}(1-k_{50})$$

The results are substituted into (XI-43):

$$E_{dir}=\frac{1}{2}\cdot\left(\frac{dF_f}{dk}\right)_{k_{50dir}}\cdot(1-k_{50dir}) \qquad \text{(XI-44)}$$

Correspondingly, for the reversed cascades we have:

$$E_{rev}=\frac{1}{2}\left(\frac{dF_f}{dk}\right)_{k_{50dir}}\cdot k_{50dir}$$

Thus, the evaluation of the efficiency of the operation of direct and reversed cascades on the basis of the Eder–Bokshtein criterion in the context of the cascade-structural model is non-equivalent. If for the direct cascade $i<\frac{z+1}{2}$, then $k_{50dir}<\frac{1}{2}$, which means that the efficiency of the operation of the direct cascade is characterised by a higher parameter E. Generally speaking for an arbitrary cascade, if the feed section is positioned lower, the value of parameter E decreases. Therefore, equation (XI-44) can be used to evaluate the effect of the area of supply of the material into the apparatus on the quality of separation.

Equation (XI-44) will be represented in the form:

$$E=\frac{1}{2}\left(\frac{dF_f}{d\chi}\right)_{x_{50}}\cdot\left(\frac{d\chi}{dk}\right)_{k_{50}}\cdot(1-k_{50}) \qquad \text{(XI-45)}$$

In accordance with the model of the regular cascade

$$\left(\frac{dF_f}{d\chi}\right)_{\chi_{50}}=\left[\frac{d}{d\chi}\left(\frac{1-\chi^{z+1-i}}{1-\chi^{z+1}}\right)\right]=$$

$$=-\frac{1}{2\left(1-\chi_{50}^{z+1}\right)\chi_{50}}\left[1(z+1)\chi_{50}^{z+1-i}-(z+1)\chi_{50}^{z+1}-2i\chi_{50}^{z+1-i}\right]$$

The above equation is transformed taking into account that:

$$2\chi_{50}^{z+1-i}-\chi_{50}^{z+1}=1$$

Consequently:

$$\left(\frac{dF_f}{d\chi}\right)_{\chi_{50}}=-\frac{z+1}{2}\cdot\frac{1}{\left(1-\chi_{50}^{z+1}\right)\chi_{50}}\left[1-\frac{2i}{z+1}\chi_{50}^{z+1-i}\right]; \qquad \text{(XI-46)}$$

$$\left(\frac{d\chi}{dk}\right)_{k_{50}}=\left(\frac{d\left(\frac{1}{k}\right)}{dk}\right)_{k_{50}}=-\frac{1}{k_{50}^2}=-(1+\chi_{50})^2; \qquad \text{(XI-47)}$$

$$1-k_{50}=\frac{\chi_{50}}{1+\chi_{50}} \qquad \text{(XI-48)}$$

In order to determine the boundary parameter of distribution, equations (XI-46), (XI-47) and (XI-48) and substituted into (XI-45):

$$E=\frac{z+1}{4}\cdot\frac{1+\chi_{50}}{(1-\chi_{50}^{z+1})}\cdot\left[1-\frac{2i}{z+1}\chi_{50}^{z+1-i}\right];$$

Table XI-1. Results of calculations of criterion E

i^*	1	2	3	4	5	6	7
χ_{50}	1.9920	1.3562	1.1394	1.0	0.8777	0.7374	0.5020
E	0.7298	0.9525	1.0235	2.0	0.8980	0.7023	0.3665

$$\frac{1-\chi_{50}^{z+1-i}}{\chi_{50}^{z+1}}=\frac{1}{2} \qquad \text{(XI-49)}$$

The resultant system of equations can be used to carry out the quantitative evaluation of operation of an arbitrary regular cascade. Table XI-1 gives the results of calculations of criterion E for an apparatus consisting of seven sections with the material supplied into each section.

We note a characteristic unique feature of the curve $E(i^*)$. At the point $i^* = 4$, as a result of the symmetry the curve shows a discontinuity (a finite jump from $E = 1.0$ to $E = 2.0$). At all remaining points, the curve $E(i)$ smoothly changes and initially increases to the maximum value $E = 1.0235$ (within the framework of discrete i) and then progressively decreases. It will be shown that at an arbitrary infinitely small difference of χ_{50} from unity criterion E turns to unity.

We set $\chi_{50} = 1 + \alpha$, where α is an infinitely small value; $\chi^{Z+1}_{50} = (2 + \alpha)^8 = 1 + 8\alpha + 28\alpha^2$ with the accuracy to the infinitely small values of the second order.

Equation (XI-49) shows that:

$$z+1-i=\frac{\ln\left[\frac{1}{2}\left(1+\chi_{50}^{z+1}\right)\right]}{\ln\chi_{50}} \qquad \text{(XI-50)}$$

$\ln\left[\frac{1}{2}\left(1+\chi_{50}^{z+1}\right)\right]=\ln(1+4\alpha+14\alpha^2)$ – is expanded into a series with the accuracy to the infinitely small values of the second order:

$$\ln(1+4\alpha+14\alpha^2)=(4\alpha+14\alpha^2)-\frac{(4\alpha+14\alpha^2)^2}{2}=4\alpha+6\alpha^2 \qquad \text{(XI-51)}$$

Consequently

$$\ln\chi_{50} = \ln(1+\alpha) = \alpha - \frac{\alpha^2}{2} \qquad \text{(XI-52)}$$

Taking into account (XI-51) and (XI-52), expression (XI-53) is written in the form:

$$z+1-i = \frac{4\alpha+6\alpha^2}{\alpha - \frac{\alpha^2}{2}} = 4+8\alpha+4\alpha^2+\cdots \qquad \text{(XI-53)}$$

The co-factor χ_{50}^{z+1-i} is expanded into a series:

$$(1+\alpha)^{4+8\alpha+4\alpha^2} = 1+(4+8\alpha+4\alpha^2)\ln(1+\alpha)+$$

$$+\frac{1}{2}(4+8\alpha+4\alpha^2)^2\left[\ln(1+\alpha)\right]^2 = 1+\left(4+8\alpha+4\alpha^2\right)\left(\alpha-\frac{\alpha^2}{2}\right)+$$

$$\frac{1}{2}(4+8\alpha+4\alpha^2)^2\left(\alpha-\frac{\alpha^2}{2}\right)^2 = 1+4\alpha+14\alpha^2$$

and we restrict ourselves to the infinitely small values of the second order.

Equation (XI-53) gives:

$$i = 4-8\alpha-4\alpha^2$$

Consequently:

$$\frac{2i}{z+1}\chi_{50}^{z+1-i} = (1-2\alpha-\alpha^2)(1+4\alpha+14\alpha^2) = 1+2\alpha+5\alpha^2$$

Therefore, the relationship (XI-49) has the form:

$$E = \frac{8}{4}\cdot\frac{(2+\alpha)}{\left(\frac{8\alpha+28\alpha^2}{2\alpha+5\alpha^2}\right)} = \frac{4+2\alpha}{4+4\alpha} \qquad \text{(XI-54)}$$

Equation (XI-54) tends to unity, at $\alpha \to 0$, i.e. at $\chi_{50} \to 1$. Since the distribution parameter χ_{50} in a real apparatus cannot be determined with infinite accuracy, equal to unity, it is then clear that the jump on the curve $E(i)$ is of no interest to us. Thus, on the basis of the cascade structural model, the optimum area of the supply of material into the apparatus is enclosed between the upper and middle section (in the examined example it is section 3). For $z = 3$, $i_{opt} = 1$, for $z = 5$, $i_{opt} = 2$. This conclusion, obtained from modelling representations, has been efficiently confirmed in practice. Identical results are obtained if we carry out evaluation on the basis of the Eder-Mayer criterion:

$$E_{75/25} = \frac{x_{75}/x_{50}}{x_{25}/x_{50}}$$

According to the structural model, we have:

$$k = 1 - (1 - k_{50})\sqrt{\frac{x}{x_{50}}}$$

Consequently

$$E_{75/25} = \left(\frac{1-k_{75}}{1-k_{25}}\right)^2 = \left[\frac{\chi_{75}}{\chi_{25}} \cdot \frac{(1+\chi_{25})}{(1+\chi_{75})}\right]^2 \qquad \text{(XI-55)}$$

The results of calculations using equation (XI-55) for the previous example are shown in Table XI-2.

The optimum area of supply of the material, as in the previous case, is section 3, the results of calculations using the two criteria differ by the normalising factor.

4. EVALUATION OF THE EFFICIENCY OF COMBINED CASCADES

We shall evaluate the efficiency of operation of combined separation cascades (CSC). Let us assume that F_0 is the degree of fractional extraction into the fine product in a single column, and $F(F_0)$ is the

Table XI-2. The Eder–Mayer criterion $E_{75/25}$ ($z = 7$, $i = 1 - 7$)

i^*	1	2	3	4	5	6	7
χ_{75}	1.2712	1.0	0.8662	0.7598	0.6494	0.5030	0.2500
χ_{25}	3.9998	1.9880	1.5400	1.3161	1.1545	1.0	0.7868
$E_{75/25}$	0.4895	0.5650	0.586	0.577	0.540	0.448	0.206

link function for the entire CSC. By analogy with the reversed cascade, the function of the link of the reversed CSC $F_{rev} = 1 - F_{dir}(1 - F_0)$. It may easily be seen that this expression describes the operation of a structural scheme in which all types of links with respect to coarse and fine products have changed their places (have been reversed) in the constant operating regime of all columns.

In fact, the degree of fractional extraction into the fine product with a reversed scheme is equal to the degree of fractional extraction into the coarse product in the direct scheme in which F_0 and $(1 - F_0)$ change places in all columns.

We shall examine several properties of the direct and reversed link functions.

1. The property of orthogonal symmetry (Fig.XI-6a).

From the conversion condition we obtain

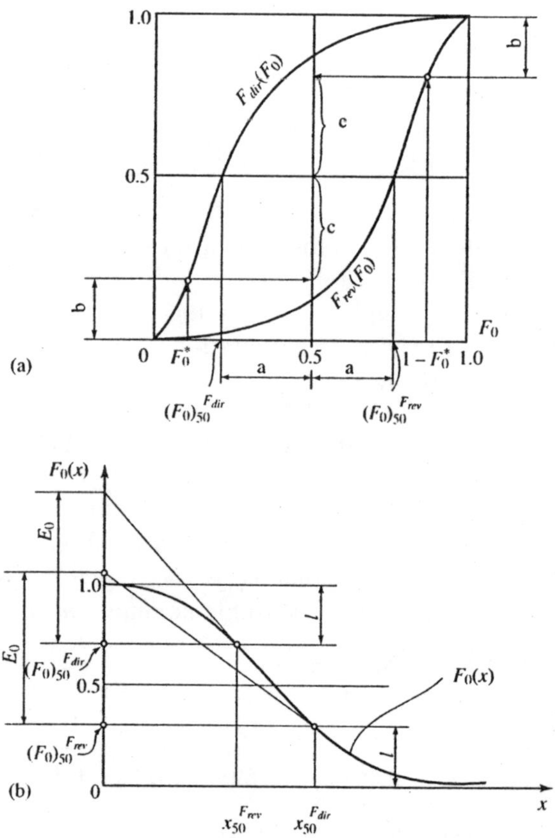

Fig.XI-6 Evaluation of direct and reversed CSC schemes. a) direct and reversed connecting functions; b) curve of separation of a single column of CSC.

$$F_{rev}(F_0 = 1 - F_0^*) = \left[1 - F_{dir}(1 - F_0)\ \right]_{F_0 = 1 - F_0^*} = 1 - F_{dir}(F_0^*)$$

Thus

$$F_{rev}(1 - F_0) = 1 - F_{dir}(F_0)$$

or

$$F_{dir}(F_0) = 1 - F_{rev}(1 - F_0)$$

This also holds for argument F_0

$$\underbrace{F_{dir}\left[F_0^{dir}(F)\right]}_{F} = 1 - F_{rev}\left[1 - F_0^{dir}(F)\right]$$

Consequently

$$F_{rev}\underbrace{\left[1 - F_0^{dir}(F)\right]}_{F_0^{rev}(1-F)} = 1 - F$$

Therefore, it may be assumed that

$$F_0^{rev}(1 - F) = 1 - F_0^{dir}(F);$$

$$F_0^{rev}(F) = 1 - F_0^{dir}(F)$$

Therefore, the attribution to any curve of the symbol of the direct or reversed function is completely conditional. In other words: it is difficult to determine which of the two curves (link functions) is direct and which is reversed.

2. Differentiation of the direct and reversed link functions. Let

$$\frac{dF(F_0)}{dF_0} = f(F_0);$$

$$\left(\frac{dF}{dF_0}\right)_{F_0^*} = f(F_0^*)$$

where f is some function of argument F_0; F_0^* is an arbitrary value of the argument.

For the reversed scheme of the CSC it holds that

$$\frac{dF_{rev}}{dF_0} = \frac{d}{dF_0}\left[1 - F(1 - F_0)\right] = \frac{d\left[F(1 - F_0)\right]}{d(1 - F_0)} = f(1 - F_0),$$

and then

$$\left(\frac{dF_{rev}}{dF_0}\right)_{1-F_0^*} = f(F_0^*) = \left(\frac{dF}{dF_0^*}\right)_{F_0^*} \qquad \text{(XI-56)}$$

is the property identical to the property of the reversed cascade.

3. Integration of the direct and reversed link functions. We set

$$\int F(F_0)dF_0 = F(F_0); \qquad \int_{F_0'}^{F_0''} F(F_0)dF_0 = F(F_0'') - F(F_0')$$

For the reversed function:

$$\int F_{rev}(F_0)dF_0 = \int [1 - F(1-F_0)]dF_0 = \int dF_0 + \int F(1-F_0)d(1-F_0) =$$
$$= \int dF_0 + F(1-F_0)$$

Consequently:

$$\int_{F_0'}^{F_0''} F_{rev}(F_0)dF_0 = F_0'' - F_0' + F(1-F_0'') - F(1-F_0')$$

Selecting this or other criterion $I(F_0)$ (point or integral) for evaluating the operation of the CSC, one can obtain identical or different parameters for the direct and reversed schema. It should be mentioned that they appear to be mirror reflected (symmetric in relation to each other) schemes of organisation of the process. In formulating partial problems of the type of restrictions of technological parameters (contamination, extraction, yield, etc.), the efficiency of these schemes differs. However, in formulating the task of improving the separation curve, characterising the process as a whole, without any reference to specific technological parameters, invariantly in relation to the initial grain size composition, it is efficient to utilise the fact that the direct and reversed schemes of CSC cannot be separated from the viewpoint of evaluation of the efficiency of their operation. Evaluating the efficiency of the combined cascade by the Eder–Bokshtein criterion gives:

$$E = -\left(\frac{dF}{dF_0}\right)_{(F_0)_{50}} \cdot \left(\frac{dF_0}{dx}\right)_{x_{50}^F} \cdot x_{50}^F \qquad \text{(XI-57)}$$

In this equation, the first co-factor does not distinguish between the direct and reversed schema.

The condition (XI-56) assumes that:

$$(F_0)_{50}^{F_{rev}} = 1 - (F_0)_{50}^{F_{dirr}}$$

In this case, the derivative in expression (XI-57) is taken at a point which is such that (Fig.XI-6,a):

$$F_{dir}\left[F_0=(F_0)_{50}^{F_{dir}}\right]=\frac{1}{2}$$

Consequently

$$F_{rev}\left[F_0=(F_0)_{50}^{F_{rev}}\right]=1-F_{dir}\left[F_0=(F_0)_{50}^{F_{dir}}\right]=\frac{1}{2}$$

Thus, for the separation curve of a single column it holds that (Fig.XI-6,b):

$$F_0(x_{50}^{F_{dir}})+F_0(x_{50}^{F_{rev}})=1 \qquad \text{(XI-58)}$$

For the second co-factor of the expression (XI-57), to ensure that the direct and reversed schemes can be separated, it is necessary to fulfil the condition (Fig.XI-6,b):

$$E_0=-\left(\frac{dF_0}{dx}\right)_{x_{50}^{F_{dir}}}\cdot x_{50}^{F_{dir}}=-\left(\frac{dF_0}{dx}\right)_{x_{50}^{F_{rev}}}\cdot x_{50}^{F_{rev}} \qquad \text{(XI-59)}$$

It should be mentioned that the realisation of different combined schemes determines different link functions $F_i(F_0)$, characterised by different values of $(F_0)_{50i}$. In turn, different values of $(F_0)_{50i}$ correspond to different values of $(x_{50})_i$ with respect to the separation curve of a single column. Thus, for fixed $F_0(x)$ relationships (XI-58) and (XI-59) should be fulfilled for an arbitrary "x":

$$F_0(x_i^{dir})+F_0(x_i^{rev})=1;$$

$$\left(\frac{dF_0}{dx}\right)_{x_i^{dir}}\cdot x_i^{dir}=\left(\frac{dF_0}{dx}\right)_{x_i^{rev}}\cdot x_i^{rev} \qquad \text{(XI-60)}$$

We introduce notations:

$$x_i^{dir}=x;\quad x_i^{rev}=z;\quad \frac{dF_0}{dx}=f(x) \qquad \text{(XI-61)}$$

Consequently, it may be written that:

$$\int_0^{F_0(x)} dF_0=F_0(x)=\int_\infty^x f(x)dx;$$

$$\int_0^{F_0(z)} dF_0=F_0(z)=\int_\infty^z f(x)dx$$

Therefore, the system of equations (XI-60) may be represented in the form:

$$f(x)\cdot x = f(z)\cdot z\,; \tag{XI-62}$$

$$\int_{\infty}^{x} f(x)dx + \int_{\infty}^{z} f(x)df = 1 \tag{XI-63}$$

The last equation will be differentiated with respect to "x":

$$\frac{d}{dx}\int_{\infty}^{x} f(x)dx + \frac{d}{dz}\left[\int_{\infty}^{z} f(x)dx\right]\cdot\frac{dz}{dx} = 0$$

or

$$f(x) + f(z)\frac{dz}{dx} = 0$$

Substituting into (XI-62):

$$f(z)\frac{z}{x} + f(z)\frac{dz}{dx} = 0$$

This gives:

$$\frac{dx}{x} = -\frac{dz}{z}$$

Integrating the last equation, we obtain:

$$x = \frac{c}{z} \tag{XI-64}$$

where c is some constant.

Constant c is determined on the basis of the following considerations: there are some combined schemes which do not displace the separation boundaries (the direct and reversed link functions are identical). For these $F(F_0)$ it can be written that:

$$(F_0)_{50}^{dir} = (F_0)_{50}^{rev} = \frac{1}{2}$$

since

$$(F_0)_{50}^{dir} + (F_0)_{50}^{rev} = 1$$

However, $(F_0)_{50} = \frac{1}{2}$ on the separation curve of a single column corresponds to:

$$x = z = x_{dir} = x_{rev} = x_{50}$$

In this case, (XI-64) shows that $c = x^2_{50}$.
Thus

$$x_{rev} = \frac{x_{50}^2}{x_{dir}}$$

For the unitary separation curve we get:

$$F_0\left(\frac{x_{rev}}{x_{50}}\right) = F_0\left(\frac{x_{50}}{x_{dir}}\right) \qquad \text{(XI-65)}$$

Denoting

$$\frac{x}{x_{50}} = y$$

Consequently, taking into account (XI-60) and (XI-65) finally shows that for the unitary separation curve of a single column it is necessary to fulfill the following condition in order to ensure that the direct and reversed combined schemes do not differ in the quantitative evaluation of the separation capacity:

$$F_0\left(\frac{x}{x_{50}} = y\right) + F_0\left(\frac{x}{x_{50}} = \frac{1}{y}\right) = 1 \qquad \text{(XI-66)}$$

Since the true functional dependence, describing the unitary curve of the classification process is not available, it is necessary to use approximations. Analysis shows that the relationship (XI-66) is not satisfied by any of the available approximations, with the exception of the Plitt approximation. For the Plitt approximation, we obtain:

$$F_0(y) + F_0\left(\frac{1}{y}\right) = \frac{1}{1+y^a} + \frac{1}{1+\left(\frac{1}{y}\right)^a} = \frac{2+y^a+\left(\frac{1}{y}\right)^a}{2+y^a+\left(\frac{1}{y}\right)^a} = 1$$

In this case, the Plitt approximation will also automatically satisfy the relationship (XI-62). Using the invariant Plitt's approximation, we obtain

$$\left(\frac{dF_0}{dx}\right)_{x_{50}^F}\cdot x_{50}^F=\left[\frac{dF_0}{d\left(\frac{x}{x_{50}^{F_0}}\right)}\right]_{\frac{x}{x_{50}^{F_0}}=\frac{x_{50}^F}{x_{50}^{F_0}}}\cdot\frac{x_{50}^F}{x_{50}^{F_0}}=-4E_0\frac{\left(\frac{x_{50}^F}{x_{50}^{F_0}}\right)}{\left[1+\left(\frac{x_{50}^F}{x_{50}^{F_0}}\right)^{4E_0}\right]^2}\qquad\text{(XI-67)}$$

It is well known that:

$$\left(\frac{x_{50}^F}{x_{50}^{F_0}}\right)^{4E_0}=\frac{1}{(F_0)_{50}}-1$$

Substituting the results into (XI-67) gives:

$$\left(\frac{dF_0}{dx}\right)_{x_{50}^F}\cdot x_{50}^F=-4E_0\cdot(F_0)_{50}\left[1-(F_0)_{50}\right]$$

Consequently, for equation (XI-57) we obtain a final estimate of the separating capacity of CSC on the basis of the Eder–Bokshtein criterion using Plitt's invariant approximation:

$$E=4E_0\cdot\left(\frac{dF}{dF_0}\right)_{(F_0)_{50}}\cdot(F_0)_{50}\left[1-(F_0)_{50}\right]\qquad\text{(XI-68)}$$

In a partial case in which $F = F_0$, $(F_0)_{50} = 1/2$, expression (XI-68) is reduced to the separating capacity of a single column E_0. Equation (XI-68) is in good agreement with the indiscernible estimate of the efficiency of the direct and reversed combined cascades. On the basis of invariant Plitt's approximation we have an indiscernible estimate of the efficiency of direct and reversed combined schemes according to the Eder–Mayer criterion.

For the direct scheme:

$$E_{75/25}=\frac{x_{75}^F}{x_{25}^F}=\frac{y_{75}^F}{y_{25}^F}$$

For Plitt's approximation we have:

$$(F_0)_{75}=\frac{1}{1+\left(y_{75}^F\right)^{4E_0}}$$

Consequently

$$y_{75}^F = \left[\frac{1}{(F_0)_{75}} - 1\right]^{1/4E_0}$$

In the same manner

$$y_{25}^F = \left[\frac{1}{(F_0)_{25}} - 1\right]^{1/4E_0}$$

Consequently

$$E_{75/25} = \left[\frac{\dfrac{1}{(F_0)_{75}} - 1}{\dfrac{1}{(F_0)_{25}} - 1}\right]^{\frac{1}{4E_0}} \qquad \text{(XI-69)}$$

For the reversed schema:

$$E_{75/25}^{rev} = \left[\frac{\dfrac{1}{(F_0)_{75}^{rev}} - 1}{\dfrac{1}{(F_0)_{25}^{rev}} - 1}\right]^{\frac{1}{4E_0}}$$

According to the property of orthogonal symmetry

$$(F_0)_{75}^{rev} = 1 - (F_0)_{25}; \qquad (F_0)_{25}^{rev} = 1 - (F_0)_{75}$$

Consequently

$$E_{75/25}^{rev} = \left[\frac{\dfrac{1}{1-(F_0)_{25}} - 1}{\dfrac{1}{1-(F_0)_{75}} - 1}\right]^{\frac{1}{4E_0}} = \left[\frac{\dfrac{1-(F_0)_{75}}{(F_0)_{75}}}{\dfrac{1-(F_0)_{25}}{(F_0)_{25}}}\right]^{\frac{1}{4E_0}}$$

This gives an equation identical with (XI-69).

The Eder–Mayer criterion for a single column according to the previous considerations is written in the form

$$\left(E_{75/25}\right)_0 = 9^{-\frac{1}{4E_0}}$$

Consequently

$$\frac{1}{4E_0} = -\frac{\ln\left(E_{75/25}\right)_0}{\ln 9}$$

Therefore, the relationship (XI-69) has the form:

$$E_{75/25} = \left[\frac{\dfrac{1}{(F_0)_{75}} - 1}{\dfrac{1}{(F_0)_{25}} - 1}\right]^{-\frac{\ln\left(E_{75/25}\right)_0}{\ln 9}} \qquad \text{(XI-70)}$$

The dependence (XI-70) for a single column satisfies the trivial result $\left(E_{75/25}\right)_0$.

The equations (XI-69) and (XI-70) are not suitable for evaluating the combined schemes with the single column because the unknown efficiency of the single column is included in the estimate as an exponent. A new parameter is introduced in order to linearise the criterion with respect to E_0:

$$I = -\frac{\frac{1}{4}\ln 9}{\ln\left(E_{75/25}\right)}$$

Consequently

$$I = \frac{E_0 \ln 9}{\ln\left[\frac{\dfrac{1}{(F_0)_{25}} - 1}{\dfrac{1}{(F_0)_{75}} - 1}\right]}$$

The resultant parameter, invariant for the direct and reversed combined schemes, on the basis of the Eder–Mayer parameter, the Eder–Bokshtein parameter for the single column and the invariant Plitt approximation changes from 0 to ∞ (ideal process). For the single column we obtain a trivial result E_0.

Since the complexes

$$\left(\frac{dF}{dF_0}\right)_{(F_0)_{50}};\ \left(\frac{dF}{dF_0}\right)_{0.5};\ \frac{\dfrac{1}{(F_0)_{75}}-1}{\dfrac{1}{(F_0)_{25}}-1};\ \frac{\dfrac{1}{F(0.75)}-1}{\dfrac{1}{F(0.25)}-1};$$

$$(F_0)_{50}\left[1-(F_0)_{50}\right];\ \ F(0.5)\cdot\left[1-F(0.5)\right]$$

do not distinguish between the direct and reversed schemes, one can propose different invariant criteria from their combinations. For example, the following criterion may be a modification of (XI-69):

$$I=9\left[\frac{\dfrac{1}{(F_0)_{75}}-1}{\dfrac{1}{(F_0)_{25}}-1}\right] \qquad \text{(XI-71)}$$

Coefficient "9" normalises the results with respect to unity for the single column. It should be mentioned that expression (XI-68) indicates that the estimate of the anomalous link function is associated with the absence of any separation process, since for all schemes $1-(F_0)_{50}=0$. There is no argument $(F_0)_{75}$ for expression (XI-71). Taking this into account, it is possible to attempt several other invariant criteria normalised with respect to unity for the single column, in particular:

$$Q=\left(\frac{dF}{dF_0}\right)_{(F_0)_{50}}\cdot\left(\frac{dF}{dF_0}\right)_{0.5}; \qquad \text{(XI-72)}$$

$$Q^*=\frac{1}{2}\left[\left(\frac{dF}{dF_0}\right)_{(F_0)_{50}}+\left(\frac{dF}{dF_0}\right)_{0.5}\right] \qquad \text{(XI-73)}$$

We shall estimate the limiting values of the efficiency for criteria E and I. For the combined schemes $F=F_0^n$. According to (XI-68) we have:

$$E=2n\left[1-\left(\frac{1}{2}\right)^{\frac{1}{n}}\right] \qquad \text{(XI-74)}$$

$$E_{n\to\infty} = \lim_{n\to\infty} \frac{1-\left(\frac{1}{2}\right)^{\frac{1}{n}}}{\frac{1}{2n}} = \lim_{n\to\infty} \frac{\left(\frac{1}{2}\right)^{\frac{1}{n}} \ln 2}{\frac{1}{2}} = 2\ln 2 = 1.386 \quad \text{(XI-74)}$$

Thus, using the CSC of the type F_0^n the efficiency of fractioning can be increased by a maximum value of 38.6%, with 95% of the limiting value of the separating capacity obtained at:

$$n = 7; \ (E_{n=7} = 1.32)$$

From (XI-71) for $F - F_0^n$ we have:

$$I = 9\left[\frac{\left(\frac{4}{3}\right)^{\frac{1}{n}} - 1}{4^{\frac{1}{n}} - 1}\right] \quad \text{(XI-75)}$$

$$I_{n\to\infty} = 9\lim_{n\to\infty} \frac{\left(\frac{4}{3}\right)^{\frac{1}{n}} \ln\frac{4}{3}}{4^{\frac{1}{n}} \ln 4} = 9\frac{\ln\frac{4}{3}}{\ln 4} = 1.868$$

In this case, 95% of the maximum separating capacity is achieved at the number of columns (n = 11; $I_{n=11}$ = 1.776). It is not rational to use more than 7÷11 columns in a multirow apparatus.

Returning to the evaluation of the separating capacity of anomalous criteria E and I, it is still possible to use some transformation for normalisation of the anomalous curve. In this case, the transformation should not contradict the relationship (XI-40), in order to ensure that the estimate of the direct and reversed schemes is identical. We show that this requirement is satisfied by the affine transformation on the ordinate with the scale coefficient $1/a$ of the curve, having $F(1) = a$, $(a < 1)$, and the affine transformation of the reversed curve (with the same coefficient), with the equidistant transfer on the ordinate by $\left(\frac{1}{a} - 1\right)$.

Let us assume that $F(F_0)$, for which $F(0) = 0$ at $F(1) = a$, has a direct CSC schema. Consequently, affine normalised curve is written in the form:

$$F^*_{dir}(F_0) = \frac{1}{a} \cdot F_{dir})(F_0) \qquad \text{(XI-76)}$$

Taking (XI-40) into account gives:

$$F^*_{rev}(F_0) = 1 - \frac{1}{a} \cdot F_{dir}(1 - F_0) \qquad \text{(XI-77)}$$

From (XI-40) we also get:

$$F_{dir}(1 - F_0) = 1 - F_{rev}(F_0)$$

The last equation is substituted in (XI-76):

$$F^*_{rev}(F_0) = 1 - \frac{1}{a}[1 - F_{rev}(F_0)]$$

or

$$F^*_{rev}(F_0) = \frac{1}{a} F_{rev}(F_0) - \left(\frac{1}{a} - 1\right) \qquad \text{(XI-78)}$$

Since $F_{rev}(0) = 1 - a$, from (XI-78) we obtain:

$$F^*_{rev}(0) = \frac{1}{a}(1 - a) - \left(\frac{1}{a} - 1\right) = 0$$

$$F^*_{rev}(1) = \frac{1}{a} - \left(\frac{1}{a} - 1\right) = 1$$

Thus, the invariant estimate of the direct and reversed schemes is realised.

References

1. Seader J.D. and Henley E.J., Separationg Process Principles, Wiley, New York (1998).
2. Khonry F.M., Predicting the Performance of Multistage Separation Process, CRC Press, Boca Raton, Florida (2000).
3. Govorov A.V., Cascade and Combined Processes of Fractioning of Bulk Materials, Dissertation for the title of Candidate of Technical Sciences, Sverdlovsk (1986).
4. Barsky M.D., Fractionation of Powders, Nedra. Moscow (1980).
5. Barsky E. and M. Barsky M.D., Master Curve of Separation Process. Physical Separation in Science and Engineering, Taylor and Francis, Vol. 13, No. 1 (2004).

Chapter XII

SPECIAL PROCESSES OF FRACTIONATION OF POWDERS

1. MULTIPRODUCT SEPARATION

Separation with the yield of the product in different stages

In many technological processes it is desirable to separate simultaneously powders into narrow size ranges. This is essential, for example, for the development of the potential of powder metallurgy and production of high-density refractory materials.

Knowledge of the main relationships of the cascade separation process makes it possible to expand greatly the possibilities of classification in this direction.

Usually, a classifier is designed only for separating the initial powder into two products. In this case, the process is regulated and set only in relation to a single boundary size. Therefore, the multifractional separation of powders is usually carried out using classification devices operating in sequence, with each device set for a specific boundary size. Because of the absence of high-efficiency classifiers (as already mentioned, cascade apparatus is not yet used widely in industrial practice) these attempts have usually been unsuccessful. In practice, the separation of powders into narrow classes is at present possible only in the ranges which enable screening to be carried out, i.e. for particles larger than 1–2 mm. For particles smaller than 1 mm it is difficult to organise accurate separation using sieves, in particular in the conditions of industrial production. It should be remembered that screening usually ensures the accurate separation of only the undersieve fine product, whereas the coarse products remain contaminated with the fine one. The degree of contamination increases with a decrease in the mesh size of the separating surface.

The cascade principle of the organisation of the process also enables this problem to be solved most efficiently. Firstly, separation may be organised in such a manner that at the outlet of apparatus we obtain not two final products but a large number of these products. Secondly, it is possible to ensure any (previously set) efficiency of the process.

Figure (XII-1) shows the schematic diagram of a cascade apparatus with the yield of part of the product in each stage.

The principle of operation is that in each stage (or in each ap-

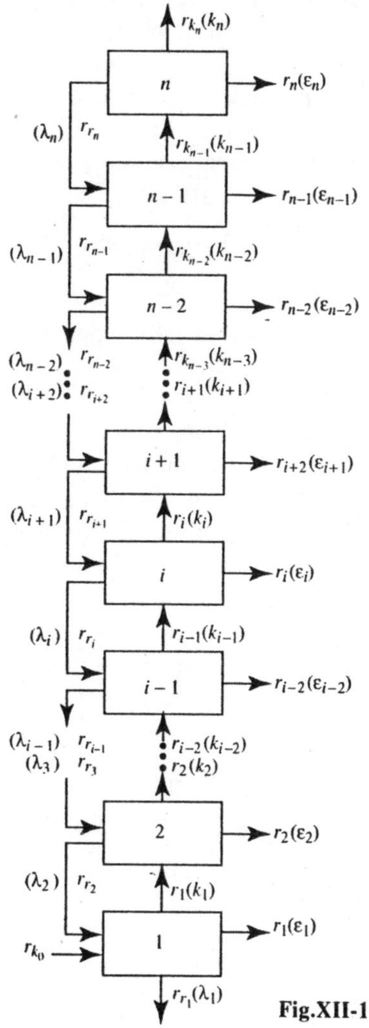

Fig.XII-1 Calculation diagram of multiproduct separation.

paratus) we create different technological regimes, and the yield of the product is organised only in stages where it will correspond to the given conditions.

We shall now clarify the scheme. Each element of the scheme will be referred to as a block, taking into account that its design should correspond to the effective classification device enabling sufficiently accurate separation. The meaning of the notations in the diagram is as follows:

i – the number of the blocks in a column; r_{k_0} – the content of the narrow size class in the initial material, supplied to the first block; r_i – the amount of this narrow size class extracted in the i-th block; r_{k_i} – the amount of the narrow size class transferred from the i-th to $(i + 1)$ –th block; r_{r_i} – the amount of the narrow size class transferred from the i-th to $(i - 1)$ block; r_{s_i} – the amount of the narrow size class in the initial feed of the i-th block; w_i – the technical regime in the i-th block (for example, the flow rate of air).

The appropriate coefficients, determining the fractional separation of the material in each block will be denoted as follows:

$k_i = r_{k_i} / r_{s_i}$ – the coefficient of fractional extraction to a subsequent block;

$\varepsilon_i = r_i / r_{s_i}$ – the coefficient of fractional extraction from apparatus in the i-th block;

$\lambda_i = r_{r_i} / r_{s_i}$ – the coefficient of fractional return.

The initial feed, supplied for separation into the multistage system represents the polyfraction material. Calculations will be carried out for some j-th fixed narrow size class. Subsequently, the data obtained as a result of this calculation must be generalised for the entire size range. In other words, if the problem can be solved for one narrow class, it will also be solved in the same manner for the entire starting material. According to the accepted notations, the following relationship holds for every block:

$$\varepsilon_j + k_j + \lambda_j = 1$$

We shall examine the balance of the last block. This block received the product only from the penultimate block and it can therefore be written that:

$$r_{s_n} = r_{s_{n-1}} k_{n-1}$$

The return to the previous block is:

$$r_{r_n} = r_{s_n} \lambda_n \qquad \text{(XII-1)}$$

hence

$$r_{r_n} = r_{s_{n-1}} k_{n-1} \lambda r$$

The initial product in the previous block consists of two flows:

$$r_{s_{n-1}} = r_{r_n} + r_{k_{n-2}} = r_{s_n} \lambda_n + r_{s-2} k_{n-2} \qquad \text{(XII-2)}$$

Equation (XII-1) is substituted into (XII-2):

$$r_{r_n} = (r_{s_{n-2}} k_{n-2} + r_{r_n}) k_{n-2} \lambda_n$$

This relationship is solved with respect to r_{r_n} :

$$r_{r_n} = r_{s_{n-2}} \frac{k_{n-2} k_{n-1} \lambda_n}{1 - k_{n-1} \lambda_n} \qquad \text{(XIII-3)}$$

The balance of $(n-1)$-th block is expressed by the dependence (XII-2). This dependence will be analysed taking relationship (XII-3) into account:

$$r_{s_{n-1}} = r_{s_{n-2}} k_{n-2} + r_{s_{n-2}} \frac{k_{n-2} k_{n-1} \lambda_n}{1 - k_{n-1} \lambda_n} \qquad \text{(XII-4)}$$

Consequently

$$r_{s_{n-1}} = r_{s_{n-2}} \frac{k_{n-2}}{1 - k_{n-1} \lambda_n} \qquad \text{(XII-5)}$$

We examine the balance of the $(n-2)$-th block:

$$r_{s_{n-2}} = r_{s_{n-3}} k_{n-3} + r_{s_{n-1}} \lambda_{n-1}$$

In accordance with (XII-4) we obtain:

$$r_{s_{n-2}} = r_{s_{n-3}} k_{n-3} + r_{s_{n-2}} \frac{k_{n-2} \lambda_{n-1}}{1 - k_{n-1} \lambda_n}$$

Solving this relationship with respect to $r_{s_{n-2}}$ gives:

$$r_{s_{n-2}} = r_{s_{n-3}} \frac{k_{n-3}}{1 - \dfrac{k_{n-2} \lambda_{n-1}}{1 - k_{n-1} \lambda_n}} \qquad \text{(XII-6)}$$

We examine the following $(n-3)$ block:

$$r_{s_{n-3}} = r_{s_{n-2}} k_{n-4} + r_{s_{n-2}} \lambda_{n-2}$$

Substituting equation (XII-6) into this dependence and solving with respect to $r_{s_{n-3}}$, gives:

$$r_{s_{n-3}} = r_{s_{n-4}} \frac{k_{n-4}}{1-\cfrac{k_{n-3}\lambda_{n-2}}{1-\cfrac{k_{n-2}\lambda_{n-1}}{1-k_{n-1}\lambda_n}}} \tag{XII-7}$$

It is clear that the denominator of the appropriate expression is a continued fraction, and the initial composition in the i-th block is expressed by the dependence

$$r_{s_i} = r_{s_{i-1}} \frac{k_{i-1}}{\underbrace{1-\cfrac{k_i\lambda_{i+1}}{1-\cfrac{k_{i+1}\lambda_{i+2}}{1-k_{i+2}\lambda_{i+2} \ddots 1-\cfrac{k_{n-2}\lambda_{n-1}}{1-k_{n-1}\lambda_n}}}}_{(n-i)\ steps}} \tag{XII-8}$$

In the second block, the initial composition would be expressed by the dependence

$$r_{s_2} = \frac{r_{s_3}k_1}{\underbrace{1-\cfrac{k_2\lambda_3}{1-\cfrac{k_3\lambda_4}{1-k_4\lambda_5 \ddots 1-\cfrac{k_{n-2}\lambda_{n-1}}{k_{n-1}\lambda_n}}}}_{(n-2)\ steps}} \tag{XII-9}$$

The balance for the first block is:

$$r_{s_2} = r_{k_0} + r_{s_2}\lambda_2$$

Using equation (XII-9), it may be written that:

$$r_{s_2} - \cfrac{r_{s_1} k_1 \lambda_2}{1 - \cfrac{k_2 \lambda_3}{1 - k_3 \lambda_4}} = r_{k_0}$$

$$1 - k_{n-1}\lambda_n$$

and consequently

$$r_{s_1} = \cfrac{r_{k_0}}{\underbrace{1 - \cfrac{k_1\lambda_2}{1 - \cfrac{k_2\lambda_3}{1 - k_3\lambda_4 \ddots 1 - \cfrac{k_{n-2}\lambda_{n-1}}{1 - k_{n-1}\lambda_n}}}}_{(n-1)\ steps}} \qquad \text{(XII-10)}$$

In order to simplify equations, we introduce the conventional notation for continued fractions in the form of a symbol with two indices D^p_n, where n is the total number of blocks; p is the number of the first index at k standing in the first line of the denominator.

Consequently, the fraction for the dependence (XII-6) will be written as D^{n-3}_n, for (XII-8) as D^{i-1}_n, and for (XII-9) as D^1_n. The value of the upper and lower indices in the notation of the continued fraction also shows the range of selection of the appropriate values of coefficients k and λ, and the number of lines in the continued fraction is $n-p$.

The resultant dependences determine the composition of the initial product of each block only through the composition of the previous block. However, it is necessary to express the content of the material in each block in relation to the initial feed.

The initial composition of the product of the second block is expressed by the dependence:

$$r_{s_2} = r_{s_1} k_1 + r_{s_3} \lambda_3$$

and taking into account relationship (XII-10) for r_{s_2} this expression is written in the form

$$r_{s_2} = \frac{r_{k_0} k_1}{D_n^1} + \frac{r_{s_2} k_2 \lambda_3}{D_n^3},$$

and consequently

$$\left.\begin{aligned} r_{s_2}\left(1 - \frac{k_2 \lambda_3}{D_n^3}\right) &= \frac{r_{k_0} k_1}{D_n^1}; \\ r_{s_2} &= \frac{r_{k_0} k_1}{D_n^1 D_n^2} \end{aligned}\right\} \qquad \text{(XII-11)}$$

For the initial composition in the third block

$$r_{s_3} = r_{s_2} k_2 + r_{s_4} \lambda_4 = r_{s_2} k_2 + \frac{r_{s_3} k_3 \lambda_4}{D_n^4},$$

from which

$$r_{s_3} = \frac{r_{k_0} k_{1_2}}{D_n^1 D_n^2 D_n^3} \qquad \text{(XII-12)}$$

It is evident that for any i-th block it may be written that:

$$r_{s_i} = \frac{r_{k_0} k_1 k_2 k_3 \ldots k_{i-1}}{D_n^1 D_n^2 D_n^3 \ldots D_n^i} \qquad \text{(XII-13)}$$

For the penultimate block

$$r_{s_{n-1}} = \frac{r_{k_0} k_1 k_2 \ldots k_{n-2}}{D_n^1 D_n^2 \ldots D_n^{n-1}} \qquad \text{(XII-14)}$$

Since the initial composition of the product in the last block is expressed by the dependence

$$r_{s_n} = r_{s_{n-1}} k_{n-1}$$

For the n-stage we obtain:

$$r_{s_n} = \frac{r_{k_0} k_1 k_2 \ldots k_i \ldots k_{n-2} k_{n-1}}{D_n^1 D_n^2 \ldots D_n^i \ldots D_n^{n-1}} \qquad \text{(XII-15)}$$

In calculations, it is necessary to know the ratio of the initial composition in adjacent blocks. We determine the values of these ratios:

$$\frac{r_{s_1}}{r_{s_2}} = \frac{r_{k_0} D_n^1 D_n^2}{D_n^1 r_{k_0} k_1} = \frac{D_n^2}{k_1};$$

$$\frac{r_{s_2}}{r_{s_3}} = \frac{D_n^3}{k_2};$$

$$\frac{r_{s_i}}{r_{s_{i+1}}} = \frac{D_n^{i+1}}{k_i};$$

$$\frac{r_{s_{n-2}}}{r_{s_{n-1}}} = \frac{D_n^{n-1}}{k_{n-2}};$$

$$\frac{r_{s_n}}{r_{s_{n-1}}} = k_{n-1}$$

It is then possible to determine the ratios for calculating parameters k_1 in each block and, consequently, the value of r_{k_i} .

We determine the separation parameters. For this purpose, we examine initially the expression for determining r_{s_1} :

$$\frac{r_{k_0}}{r_{s_1}} = 1 - \frac{k_1 \lambda_2}{D_n^2},$$

and consequently

$$\frac{k_1 \lambda_2}{D_n^2} = 1 - \frac{r_{k_0}}{r_{s_1}} = \frac{r_{s_1} - r_{k_0}}{r_{s_1}} = \frac{r_{r_2}}{r_{s_1}}$$

Therefore

$$r_{r_2} = r_{s_1} \frac{k_1 \lambda_2}{D_n^2}$$

The dependence for r_{s_2} may be presented in the form

$$r_{s_2} = \frac{r_{s_1} k_1}{1 - \frac{k_2 \lambda_3}{D_n^3}},$$

which gives

$$\frac{r_{s_1}}{r_{s_2}} k_1 = 1 - \frac{k_2 \lambda_3}{D_n^3}$$

and

$$\frac{k_2 \lambda_3}{D_n^3} = 1 - \frac{r_{s_1} k_1}{r_{s_2}} = \frac{r_{s_2} - r_{s_1} k_1}{r_{s_2}} = \frac{r_{r_3}}{r_{s_2}}$$

This means that

$$r_{r_3} = r_{s_2} \frac{k_2 \lambda_3}{D_n^3}$$

We examine the dependence

$$r_{s_3} = \frac{r_{s_2} k_2}{1 - \frac{k_3 \lambda_2}{D_n^4}}$$

Similarly,

$$r_{r_4} = r_{s_3} \frac{k_3 \lambda_4}{D_n^4}$$

and it may also be written that

$$r_{ri} = r_{s_{i-1}} \frac{k_{i-1} \lambda_i}{D_n^i};$$

$$r_{r_{n-1}} = r_{s_{n-2}} \frac{k_{n-2} n_{n-1}}{D_n^{n-1}};$$

$$r_{r_n} = r_{s_{n-1}} k_{n-1} \lambda_n$$

The examined scheme of multifraction separation represents the most general model.

2. MULTIPRODUCT SEPARATION IN APPARATUS ASSEMBLED FROM IDENTICAL BLOCKS

Multiproduct apparatuses with identical blocks, operating in general technological regimes, may produce powders of different grain size composition. This has been confirmed with sufficient reliability by experiments.

In this case, for the main parameters characterising the operation of apparatus, we may write the relationship

$$k_{n-1}\lambda_n = a$$

Here k_j and λ_j can change from block to block, and it is only important that their product remains constant. Taking this into account, the continued fraction, consisting of n elements may be written in the form:

$$D^{I}_{n} = \cfrac{a}{n\left\{ 1 - \cfrac{a}{1 - \cfrac{a}{1 - a \ddots 1 - \cfrac{a}{1}}} \right.}$$

This expression may be simplified taking into account that the dependence $D^{I}_{n+1} = 1 - \dfrac{a}{D^{I}_{n}}$ is a recurrent function.

We introduce the following notations:

$$D^{I}_{n} = \frac{x_n}{y_n}; \qquad D^{I}_{n+1} = 1 - \frac{a}{\dfrac{x_n}{y_n}},$$

and consequently

$$D^{I}_{n+1} = \frac{x_n - ay_n}{x_n} = \frac{x_{n+1}}{y_{n+1}}$$

Therefore,

$$x_n = y_{n+1}; \quad x_{n-1} = y_n$$

and

$$x_n - ax_{n-1} = x_{n+1}$$

From this we obtain a recurrent equation.

$$x_{n+2} - x_{n+1} - ax_n = 0$$

This equation may be solved introducing the so-called characteristic equation

$$z^2 - z - a = 0$$

The roots of this equation are the dependences

$$z_1 = \frac{1}{2} + \sqrt{\frac{1}{4} - a} \quad \text{and} \quad z_2 = \frac{1}{2} - \sqrt{\frac{1}{4} - a}$$

and consequently

$$x_n = C_1 z_1^{n-1} + C_2 z_2^{n-1},$$

i.e.

$$x_n = C_1 \left[\frac{1}{2} + \sqrt{\frac{1}{4} - a} \right]^{n-1} + C_2 \left[\frac{1}{2} - \sqrt{\frac{1}{4} - a} \right]^{n-1} \qquad \text{(XI-16)}$$

The unknown parameters C_1 and C_2 are determined from the following considerations. At $x_1 = 1$, $C_1 + C_2 = 1$, i.e. $C_1 = 1 - C_2$.

For

$$x_2 = 1 - a = (1 - C_2) \left[\frac{1}{2} + \sqrt{\frac{1}{4} - a} + C_2 \frac{1}{2} - \sqrt{\frac{1}{4} - a} \right],$$

and consequently

$$\left. \begin{aligned} C_2 &= \frac{a - \frac{1}{2} + \sqrt{\frac{1}{4} - a}}{2\sqrt{\frac{1}{4} - a}}; \\ C_1 &= \frac{\frac{1}{2} - a + \sqrt{\frac{1}{4} - a}}{2\sqrt{\frac{1}{4} - a}} \end{aligned} \right\} \qquad \text{(XI-17)}$$

Substituting (XII-16) into (XII-17) gives:

$$x_n = \frac{\frac{1}{2} - a + \sqrt{\frac{1}{4} - a}}{2\sqrt{\frac{1}{4} - a}} \left[\frac{1}{2} + \sqrt{\frac{1}{4} - a} \right]^{n-1} +$$

$$+ \frac{a - \frac{1}{2} + \sqrt{\frac{1}{4} - a}}{2\sqrt{\frac{1-a}{4}}} \left[\frac{1}{2} - \sqrt{\frac{1}{4} - a} \right]^{n-1}$$

Carrying out appropriate transformations taking into account that

$$a = \left[\frac{1}{2} + \sqrt{\frac{1}{4} - a} \right]^{n+1} - \left[\frac{1}{2} - \sqrt{\frac{1}{4} - a} \right]^{n+1},$$

we obtain

$$D_n^1 = \frac{\left[\frac{1}{2} + \sqrt{\frac{1}{4} - a} \right]^{n+1} - \left[\frac{1}{2} - \sqrt{\frac{1}{4} - a} \right]^{n+1}}{\left[\frac{1}{2} + \sqrt{\frac{1}{4} - a} \right]^{n} - \left[\frac{1}{2} - \sqrt{\frac{1}{4} - a} \right]^{n}} \qquad \text{(XII-18)}$$

Expression (XII-18) shows that the following relationship is always valid:

$$k\lambda < 0.25 \qquad \text{(XII-19)}$$

This is also clear on the intuitive level because if it is assumed that $k = 0.9$ then $\lambda < 0.1$ and then the product corresponds to the condition (XII-19).

We examine a case in which $a = k\,(1-k)$, i.e. two-product separation. For this case

$$\sqrt{\frac{1}{4} - a} = \sqrt{\frac{1}{4} - k(1-k)} = \sqrt{\frac{1 - 4k - 4k^2}{4}} = \frac{1-2k}{2},$$

and this means that

$$k = \frac{1}{2} - \frac{1-2k}{2};$$

$$1-k=\frac{1}{2}+\frac{1-2k}{2}$$

Taking this into account, the dependence (XII-18) is transformed to the form

$$D_n^1=\frac{(1-k)^{n+1}-k^{n+1}}{(1-k)^n-k^n},$$

which corresponds to the previously derived expression for the cascade two-product separation process.

Thus, the examined model of multiproduct separation is the most general model of multistage fractionation whose particular case is cascade two-product separation.

In accordance with expression (XII-18) it may be shown that

$$D_n^2=\frac{\left[\frac{1}{2}+a\right]^n-\left[\frac{1}{2}-\sqrt{\frac{1}{4}-a}\right]^n}{\left[\frac{1}{2}+\sqrt{\frac{1}{4}-a}\right]^{n-1}-\left[\frac{1}{2}-\sqrt{\frac{1}{4}-a}\right]^{n-1}},$$

and in the general case

$$D_n^i=\frac{\left[\frac{1}{2}+\sqrt{\frac{1}{4}-a}\right]^{n+2-i}-\left[\frac{1}{2}-\sqrt{\frac{1}{4}-a}\right]^{n+2-i}}{\left[\frac{1}{2}+\sqrt{\frac{1}{4}-a}\right]^{n+1-i}-\left[\frac{1}{2}-\sqrt{\frac{1}{4}-a}\right]^{n+1-i}}$$

Taking these relationships into account, the previously derived dependences can be greatly simplified:

$$r_{s_1}=r_{k_0}\frac{\left[\frac{1}{2}+\sqrt{\frac{1}{4}-a}\right]^{n+1}-\left[\frac{1}{2}-\sqrt{\frac{1}{4}-a}\right]^{n+1}}{\left[\frac{1}{2}+\sqrt{\frac{1}{4}-a}\right]^{n}-\left[\frac{1}{2}-\sqrt{\frac{1}{4}-a}\right]^{n}}$$

$$r_{s_2}=\frac{r_{k_0}k_1}{D_n^2 D_n^1}=r_{k_0}k_1\frac{\left[\frac{1}{2}+\sqrt{\frac{1}{4}-a}\right]^{n+1}-\left[\frac{1}{2}-\sqrt{\frac{1}{4}-a}\right]^{n+1}}{\left[\frac{1}{2}+\sqrt{\frac{1}{4}-a}\right]^{n-1}-\left[\frac{1}{2}-\sqrt{\frac{1}{4}-a}\right]^{n-1}}$$

Then similarly

$$r_{s_3}=r_{k_0}k_1k_2\frac{\left[\frac{1}{2}+\sqrt{\frac{1}{4}-a}\right]^{n+1}-\left[\frac{1}{2}-\sqrt{\frac{1}{4}-a}\right]^{n+1}}{\left[\frac{1}{2}+\sqrt{\frac{1}{4}-a}\right]^{n-2}-\left[\frac{1}{2}-\sqrt{\frac{1}{4}-a}\right]^{n-2}}$$

In the general case

$$r_{s_i}=r_{k_0}k_1k_2\ldots k_{i-1}\times\frac{\left[\frac{1}{2}+\sqrt{\frac{1}{4}-a}\right]^{n+1}-\left[\frac{1}{2}-\sqrt{\frac{1}{4}-a}\right]^{n+1}}{\left[\frac{1}{2}+\sqrt{\frac{1}{4}-a}\right]^{n+1-i}-\left[\frac{1}{2}-\sqrt{\frac{1}{4}-a}\right]^{n+1-i}}$$

For $(n-1)$-th block, the initial composition may be written in the form:

$$r_{s_{n-1}}=r_{k_0}k_1k_2\ldots k_{n-2}\times\frac{\left[\frac{1}{2}+\sqrt{\frac{1}{4}-a}\right]^{n+1}-\left[\frac{1}{2}-\sqrt{\frac{1}{4}-a}\right]^{n+1}}{\left[\frac{1}{2}+\sqrt{\frac{1}{4}-a}\right]^{2}-\left[\frac{1}{2}-\sqrt{\frac{1}{4}-a}\right]^{2}},$$

and consequently

$$r_{s_{n-1}}=r_{k_0}k_1k_2\ldots k_{n-2}\times\frac{\left[\frac{1}{2}+\sqrt{\frac{1}{4}-a}\right]^{n+1}-\left[\frac{1}{2}-\sqrt{\frac{1}{4}-a}\right]^{n+1}}{\sqrt{1-4a}}$$

For the n-th block

$$r_{s_n} = r_{s_{n-1}} k_{n-1}$$

It should be mentioned that the resultant model of multiproduct separation has a number of partial cases interesting for practical realisation. It has already been shown that if it is assumed for all stages that $\varepsilon_i = 0$ (with the exception of outer stages), and k = const, we obtain a general solution of the two-product cascade. If for all stages it is accepted that $\lambda_i = 0$, then the resultant model will describe a number of separation systems operating in sequence without recirculation. Of greatest interest is the possibility of realisation the concept of the model of multiproduct separation and its partial cases in a single system.

3. EQUIPMENT FOR MULTIPRODUCT SEPARATION OF POWDERS

We shall examine several concepts for the realisation of a model of multiproduct separation. It can be realised most completely on a facility whose schematic diagram is presented in in equipment whose schematic diagram is presented in Fig. XII-2.

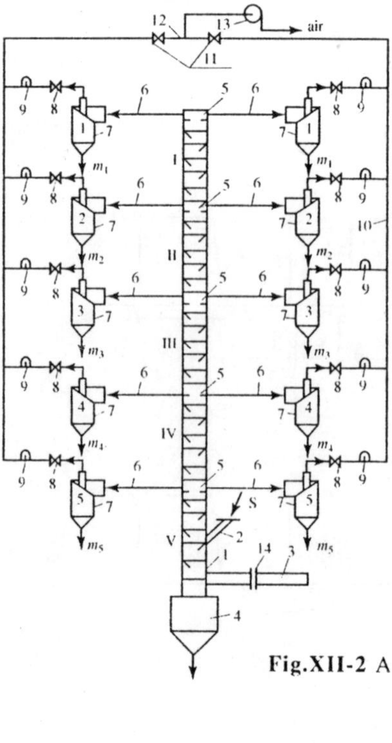

Fig.XII-2 A multiproduct classifier with recirculation.

Equipment consists of five identical units 1, represented by a shelf cascade classifier. Each unit consists of five sections. The initial product for separation is supplied along the hopper 2 into the lower fifth unit to third stage. Equipment operations under refraction generated by the fan 13. The input of air into equipment is ensured through the pipe 3 with the double normal diaphragm 14 used for measuring the total flow rate of air through the equipment. The individual units are connected together by a means of a special transition piece 5 containing two outlets. Each outlet is connected with cyclones 7 in which the material taken out of the equipment settles. Behind each cyclone there is the regulating valve 8 and the flow rate diaphragm 9 so that specific air flow rates may be set in each circuit 6. As indicated by the diagram, there are five pairs of cyclones. The air behind the cyclones is collected in the general collectors 10 connected into the box 12. The gate valves 11 placed on the collectors are used for simplifying the regulation of the flow rate of air through the equipment. The coarse material is collected in the bunker 4.

During operation of equipment the air flow rate is changed in transition from unit to unit as a result of consecutive removal of a specific amount of air in each group of cyclones. Equipment is assembled in such a manner that the air flow rate decreases in movement from bottom to top. Investigations were carried out using quartzite powder. The results of two experiments are presented in Table XII-1.

The task in this paragraph is not to present a comprehensive solution of the problems of multifraction separation. No attention has been given to the problems of optimisation of the design of equipment, the optimum relationship of its units, the required relationship of the speeds, etc. This will be discussed later.

Therefore, the data presented in Table XII-1 are far from optimum and are restricted to only two experiments. Examination of the Table shows that multiproduct separation is possible. Thus, in the first experiment, fractions smaller than 0.3 mm concentrate mainly in the first cyclone, smaller than 0.6 mm in the second cyclone, smaller than 1.5 mm in the fourth and fifth cyclones, the largest in the bunker. In the second experiment, the distribution pattern is the same. The quality of the produced powders can be greatly improved by organising the input of the material on the level of the third or fourth block. This results in a certain departure from the examined mathematical model where the input is in the lower unit, but it has a beneficial effect on the separation results.

A partial case of the general model at $\lambda = 0$, i.e. equipment operating

Table XII-1. The results of separation of quartzite powder in vertical multiproduct equipment

Experiment No.	w, m/s	Product exit area	γ,%	Partial residue (%) on screens with the mesh, mm						
				2.5	1.5	1.06	0.6	0.3	0.088	-0.088
1	2	cyclone 1	2.6	0	0	0	0	5.4	64.2	30.04
	4	cyclone 2	14.7	0	0	2.3	17.9	44.2	30.8	5.8
	6	cyclone 3	28.7	0	16.2	35.8	25.0	10.8	9.5	2.3
	8	cyclone 4	17.7	4.1	46.2	25.7	11.5	6.6	4.6	0
	10	cyclone 5	20.0	4.0	35.9	24.3	14.3	10.4	8.54	0
	-	hopper	16.3	19.4	63.2	15.1	2.3	0	0	0
2	4	cyclone 1	8.2	0	0	0	7.4	27.0	44.4	20.5
	8	cyclone 2	41.5	0.9	28.1	29.8	19.1	12.7	6.7	2.7
	12	cyclone 3	22.4	11.0	40.1	18.9	11.9	8.9	7.4	1.8
	16	cyclone 4	10.9	2.5	34.4	24.2	13.7	11.6	9.2	4.4
	20	cyclone 5	12.8	5.3	30.5	20.7	14.8	12.7	12.3	3.7
	–	hopper	4.2	3.1	51.3	32.0	9.9	3.2	0	0

Table XII-2. Results of separation of potassium chloride in a multirow cascade classifier with a grating

Experiment No.	w, m/s	Product exit area	γ,%	Partial residues in (%) on screens with the mesh, mm							
				0.63	0.4	0.25	0.2	0.16	0.1	0.063	-0.063
1	4.0	hopper	82.3	4.5	22	38.5	13.5	10.0	10.5	1.0	0
		cyclone 5	1.7	0	0	0	1.6	8.1	80.4	9.7	0
		cyclone 4	0.9	0	0	0	0	0	70.6	29.1	0
		cyclone 3	1.8	0	0	0	0	1.5	60.3	38.2	0
		cyclone 2	3.4	0	0	0	0	0	38.1	57.1	4.8
		cyclone 1	9.9	0	0	0	0	0	7	57.5	35.5
2	4	hopper	75.6	6.5	26.0	41.0	13.5	8.0	5.0	0 .	0
	8	cyclone 5	3.1	0	0	1.0	7.6	29.0	60.2	2.2	0
	12	cyclone 4	0.8	0	0	0	0	11.5	8.8	7.7	0
	16	cyclone 3	2.1	0	0	0	1.5	6.3	79.7	12.5	0
	20	cyclone 2	4.4	0	0	0	0.8	3.3	63.9	30.8	1.5
	–	cyclone 1	14.0	0	0	0	0	1.0	19.5	52.0	27.5

without a circulation (Fig.XII-3), is of considerable interest for practice.

It is equipment 1 where each channel has a separate output to cyclone 2. The air flow in each channel is regulated by the gate valve 3 and recorded with a flow rate meter connected to the diaphragm 4. The output of all cyclones are joined by a common chest which is connected to the fan 5. In the diagram, item 6 shows the stage of sanitary dust purification prior to discharge of air into the atmosphere.

The possibilities of this equipment are extensive when specifying different speeds in each of its channels. However, because of the above reasons, this will not be discussed here. Multiproduct separation can be realised even at the same flow speed in all channels of equipment.

Table XII-2 presents the results of fractionation of potassium chloride at air flow speed rates in all sections of equipment of 4 and 4.5m/s.

The Table shows that in these conditions the coarse classes are collected in the bunker. Their separation is highly accurate. The finest

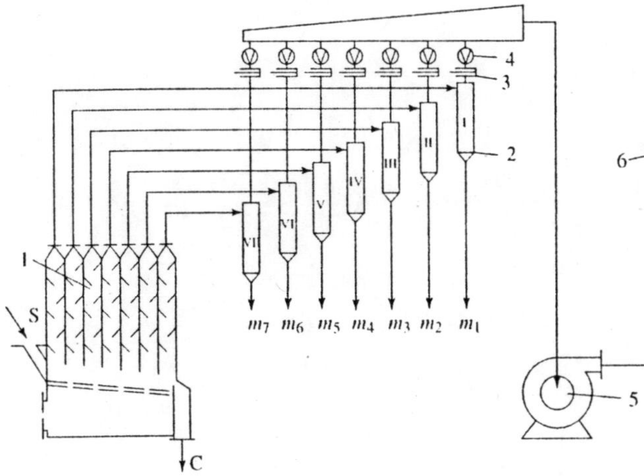

Fig. XII-3. Schematic of a multiproduct combined classifier with a grating without recirculation.

classes concentrate in the first product combining the products of the first two cyclones. Intermediate products are separated quite accurately in the remaining cyclones. This Table also shows that in even in these conditions it is possible to separate quite efficiently the initial powder into 3–4 products. It should be stressed that the possibilities of this equipment are considerable when specifying different flow speeds of the medium in adjacent cleaning channels.

A different schema of the organisation of the multiproduct separation process without recirculation is also possible (Fig.XII-4). Equipment has no grid. Initial feed is introduced into the central part of the first block, the fine fraction, separated here, falls into the first cyclone, and coarse fraction is transferred into the central part of the second and then third block.

Different air flow speeds are set in every block. The results of several experiments, obtained in experimental examination of this equipment, presented in Table XII-3.

This series of multiproduct equipment could be extended. However, our task is slightly different. It is necessary to assemble multiproduct equipment in such a manner and specify technological operating conditions in order to ensure the most efficient separation. The solution of this task is linked unambiguously with the problem of criterial evaluation of the quality of this type of separation. Only after determining the method of optimisation it is possible to consider formulation of the problem of optimisation of multiproduct separation.

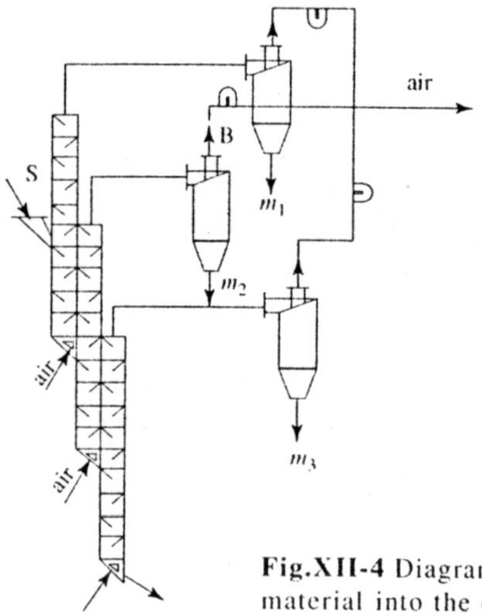

Fig.XII-4 Diagram of a multiproduct classifier with input of material into the central part of each column.

Table XII-3. Multiproduct separation of quartzite powder in a multirow cascade classifier without a grid

Experiment No.	Product exit area	ω, m/s	γ, %	Partial residues (%) on screens with mesh, mm						
				2.0	1.2	0.5	0.25	0.102	0.075	−0.075
1	cyclone 1	5.5	10.1	0	0	0.5	22.3	22.7	29.5	25.0
	cyclone 2	6.6	0.7	0	0	1.5	50	36.8	11.7	0
	cyclone 3	10.4	27.6	0	2.3	44.6	48.3	3.8	0	0
	Hopper		61.6	39.6	34.7	24.8	0.8	0	0	0
2	cyclone 1	6.1	11.6	0	0	1.3	41.4	22.8	20.0	14.5
	cyclone 2	7.9	1.21	0	0	7.0	78.4	13.1	1.5	0
	cyclone 3	12.2	33.9	1	6.8	57.4	33.8	0.8	0	0
	Hoppe		53.4	32.8	38.9	27.9	0.35	0	0	0
3	cyclone 1	6.1	9.0	0	0	0.8	29.8	23.9	26.7	18.8
	cyclone 2	11.9	28.9	0.2	1.7	44.6	47.5	5.0	0.85	0
	cyclone 3	13.0	15.8	4.3	24.9	68.15	2.6	0	0	0
	Hoppe		46.2	41.6	42.1	16.3	0	0	0	0
4	cyclone 1	6.1	10.5	0	0	1.1	35.9	23.8	23.9	15.3
	cyclone 2	12.7	31.2	0	2.8	50.8	45.1	0	0	0
	cyclone 3	13.5	16.0	6.0	29.8	62.8	1.35	0	0	0
	Hoppe		42.3	48.6	41.12	10.28	0	0	0	0
5	cyclone 1	5.1	6.3	0	0	0	19.2	25.4	31.7	24.5
	cyclone 2	10.8	8.9	0	0	14.2	63.9	14.9	7.3	0
	cyclone 3	14.0	41.9	3.53	11.79	56.95	26.09	1.6	0	0
	Hoppe		42.6	54.0	36.6	9.4	0	0	0	0
6	cyclone 1	6.3	11.8	0	0	0.21	40.6	22.5	22.0	13.5
	cyclone 2	12.6	28.2	0	1.8	52.2	42.8	2.6	0	0
	cyclone 3	14.4	22.4	10.9	34.8	53.2	1.1	0	0	0
	Hoppe	37.6	55.5	36.4	8.1	0	0	0	0	0

4. CRITERION OF THE QUALITY OF SEPARATION INTO *N* COMPONENTS

We shall attempt to formulate a criterion for evaluating the quality of separation into any number of components. If the material is divided into n components in such a manner that each component contains only particles of the required class size without impurities, this separation is ideal. This means as the produced components become more homogeneous, the separation becomes closer to ideal.

In practice, there is no equipment ensuring ideal separation and, consequently, it is necessary to develop a criterion estimating the proximity of the actual separation to ideal separation. This criterion should reflect the extent by which the homogeneity of the produced components increased after separation, in comparison with the initial material.

Let us assume that it is required to separate material into n fractions (with the particle size within each fraction equal to the mean size of the particles constituting this fraction). Let us arrange all the particles of the material in a row and count the number of transpositions with returns for these particles. The number of transpositions with returns will give objective information on the degree of inhomogeneity of the system (it may be assumed that each size of the particle is some letter, and we obtain reports consisting of the same letters in different orders from, as in this case, the number of arrangements with returns is the number of reports which can be formed from these letters).

Let us assume that G is the total number of particles in the initial material. The number of particles of each fraction will be denoted by N_1, N_2, ...N_n. Consequently, the number of transpositions with returns for the same material is $m = \frac{G!}{\prod_{i=1}^{n} N_i!}$. This means that m is the number of possible states of the system. It is well known that the logarithm of the number of states of a system corresponds to the amount of information about it.

We denote $I = \ln m$.

According to the Stirling formula $\lg A! \approx A(\ln A - 1)$ for sufficiently high values of A. Consequently

$$I = \ln m = G(\ln G - 1) - \sum_{i=1}^{n} N_i(\ln N_i - 1) =$$

$$= G\ln G - G - \sum_{i=1}^{n} N_i \ln N_i + \sum_{i=1}^{n} N_i = G\ln G - \sum_{i=1}^{n} N_i \ln N_i$$

The possibility of choosing at random a particle of size class j from the initial material is $P_j = \frac{N_j}{G}$. Consequently $I = -G\sum_{i=1}^{n} P_i \ln P_i$. It is well known that the amount of information for a single element of the system, i.e.

$$H = \frac{I}{G}$$

Consequently

$$H = -\sum_{i=1}^{n} P_i \ln P_i$$

This function also objectively reflects the degree of heterogeneity (indeterminacy) of the system.

The proposed criterion for the quality of separation should satisfy the following two boundary conditions:

1. In the case of ideal separation this criterion should have the maximum value;

2. In the case of separation without any change of the fraction composition the criterion should be equal to zero.

Let us assume that H_s is the entropy of the initial material, and H_1, H_2, ..., H_n are the entropies of each component after separation, respectively. We shall verify whether the following function is suitable as a criterion of the quality of separation, $E = H_s - \sum_{i=1}^{n} \mu_i H_i$, where μ_i is the relative amount of each component after separation, i.e.

$$\sum_{i=1}^{n} \mu_i = 1.$$

We verify whether the initial conditions are fulfilled:

1. For ideal separation we shall examine some component under number i. It consists of particles of the size class i, and the probability of extraction of the particle of size class i from this component is

1. Consequently $H_i = -\frac{N_i}{N_i}\ln\frac{N_i}{N_i} - 1\ln 1 = 0$. This gives $E = H_s$, and it is clear that this is the maximum efficiency which can be obtained for a specific composition of initial material.

2. In separation by an absolutely random manner without any change of the fractional composition of the material we obtain (it is assumed that the distribution of particles of each size class, included in the component i in relation to the amount of this size class in the initial material, will be in the same proportion as the yield of the entire component i in relation to the initial material):

$$H_i = -\sum_{k=1}^{n}\frac{\mu_i N_k}{\mu_i G}\ln\frac{\mu_i N_k}{\mu_i G} = -\sum_{k=1}^{n}\frac{N_k}{G}\ln\frac{N_k}{G} = H_s$$

Consequently $E = H_s - \sum_{i=1}^{n}\mu_i H_s = 0.$

This means that function E is suitable as a criterion for evaluating the quality of separation. However, the following problem may arise here: how to compare the efficiencies of separation of different materials (maximum efficiency of separation of each material is its initial entropy). We define a new function

$$E_1 = \frac{E}{H_s}$$

and verify whether it is suitable for evaluating the quality (efficiency) of separation.

E_1 is determined when $H_s \neq 0$ ($H_s = 0$ when either the initial material is homogeneous, or is not available at all, and in both cases there is no need to carry out separation). For the initial condition 1 (ideal separation) $E_1 = 1$. For condition 2 (in the absence of any change in the composition in separation) $E_1 = 0$. We denote $E = E_1$. Consequently

$$E = \left| \begin{array}{l} 1 - \dfrac{\sum_{i=1}^{n}\mu_i H_i}{H_s} \quad H_s \neq 0 \\ H_s = 0 \end{array} \right. \tag{XII-20}$$

is suitable as a criterion for evaluating the quality (efficiency) of separation.

We examine the application of this criterion in two cases.

a. Binary separation: the results of a real experiment will be processed using the proposed criterion.

The characteristic of the initial composition of the investigated material is:

R, %	r, %	d, mm
2.44	2.44	1.35
21.96	19.52	0.8
70.23	48.27	0.45
92.47	22.24	0.25
95.7	3.23	0.165
98.88	3.28	0.125
100	1.02	0.05

In the initial stage, we calculate the entropy of the starting material. The material can be treated as two size classes in relation to each boundary:

$$H_s = -(P_1 \ln P_1 + P_2 \ln P_2), \text{ where } P_1 = \frac{R_s}{G}, P_2 = 1 - \frac{R_s}{G}, G = 1.$$

The following Table gives the results of calculation of H_s for the starting material (the entropy for the 0.05mm boundary is not determined; it will not be considered because it is the finest size class and there are no finer particles, i.e. it is not included in the range of distribution of the product between the classification yields):

d	H_s
1.35 mm	0.114701
0.8 mm	0.526401
0.45 mm	0.608903
0.25 mm	0.267138
0.165 mm	0.177364
0.125 mm	0.056919

We determine total residues in the material transferred into the fine product for all experimental velocities. The final line of this Table gives the amount of material in μ_f, % transferred into the fine product

d	3.5m/sec	3m/sec	2.5m/sec	2m/sec	1.5m/sec	1m/sec	0.75m/sec
0.135mm	0.002497	2.92E-08	0	0	0	0	0
0.8mm	1.498068	0.185896	0.001514	0	0	0	0
0.45mm	30.4928	12.78878	2.092821	0.037337	0	0	0
0.25mm	51.48011	31.292	13.35112	2.191417	0.020404	0	0
0.165mm	54.844	34.39825	16.12573	3.712483	0.145103	0	0
0.125mm	57.93196	37.63118	19.23394	6.178821	0.703821	0.000428	0
0.05mm	58.95167	38.65045	20.25163	7.188648	1.645776	0.266742	0.009714
μ_f, %	58.95167	38.65045	20.25163	7.188648	1.645776	0.266742	0.009714

We shall now calculate the entropy (H_f) of the material transferred into the fine product, for all separation boundaries and velocities using the equation $H = \sum_{i=1}^{n} P_i \ln P_i$, where P_i is the probability of taking the particle of the narrow size class i from the material (the ratio of the amount of this size class to the initial amount of the material).

d	3.5m/sec	3m/sec	2.5m/sec	2m/sec	1.5m/sec	1m/sec	0.75m/sec
1.35mm	0.000469	0.00166	0	0	0	0	0
0.8mm	0.118412	0.030468	0.000785	0	0	0	0
0.45mm	0.692552	0.634802	0.332362	0.032502	0	0	0
0.25mm	0.380143	0.48678	0.641524	0.614913	0.066749	0	0
0.165mm	0.260457	0.34655	0.505532	0.692607	0.298276	0	0
0.125mm	0.087326	0.121894	0.199257	0.405825	0.682642	0.011936	0 μ_c,%
μ_f, %	0.589517	0.386501	0.202516	0.071886	0.0004	0.002667	0.00005

μ_f is the yield of the material into the fine product.

We now determine the amount of each size class transferred into the fine product for all velocities

$$r_c = r_s - r_f$$

where μ_c,% is the amount of material in percent, transferred into the coarse product

d	3.5m/sec	3m/sec	2.5m/sec	2m/sec	1.5m/sec	1m/sec	0.75m/sec
1.35mm	2.437503	2.44	2.44	2.44	2.44	2.44	2.44
0.8mm	18.02443	19.3341	19.51849	19.52	19.52	19.52	19.52
0.45mm	19.27527	36.66711	46.17869	48.23266	48.27	48.27	48.27
0.25mm	1.252688	3.736783	10.9817	20.08592	22.2196	22.24	22.24
0.165mm	0.041673	0.123749	0.455139	1.708934	3.105301	3.23	3.23
0.125mm	0.01648	0.047072	0.171791	0.813662	2.721282	3.279572	3.28
0.05mm	0.000288	0.000733	0.00231	0.010174	0.078045	0.753687	1.01028
μ_c,%	41.04833	61.34995	79.74837	92.81135	98.35422	99.73326	99.99029

The following table gives the results of calculation of the total residues for the coarse product by analogy with the table for the fine product.

d	3.5m/sec	3m/sec	2.5m/sec	2m/sec	1.5m/sec	1m/sec	0.75m/sec
1.35mm	2.437503	2.44	2.44	2.44	2.44	2.44	2.44
0.8mm	20.46193	21.7741	21.95849	21.96	21.96	21.96	21.96
0.45mm	39.7372	57.44122	68.13718	70.19266	70.23	70.23	70.23
0.25mm	40.98989	61.178	79.11888	90.27858	92.4496	92.47	92.47
0.165mm	41.03156	61.30175	79.57427	91.98752	95.5549	95.7	95.7
0.125mm	41.04804	61.34882	79.74606	92.80118	98.27618	98.97957	98.98
0.05mm	41.04833	61.34955	79.74837	92.81135	98.35422	99.73326	99.99029
μ_c,%	41.04833	61.34955	79.74837	92.81135	98.35422	99.73326	99.99029

We calculate the entropy of the coarse product (H_c) for all velocities and separation boundaries.

d	3.5m/sec	3m/sec	2.5m/sec	2m/sec	1.5m/sec	1m/sec	0.75m/sec
1.35mm	0.225262	0.167219	0.136809	0.121599	0.116204	0.119942	0.11471
0.8mm	0.693143	0.650438	0.588505	0.54714	0.531021	0.527145	0.526428
0.45mm	0.141426	0.237045	0.414996	0.555314	0.598484	0.607282	0.608844
0.25mm	0.010755	0.019233	0.04608	0.12519	0.227062	0.260891	0.266912
0.165mm	0.003596	0.006356	0.015557	0.050772	0.129353	0.169341	0.177075
0.125mm	9.06E-05	0.000147	0.000332	0.001109	0.006458	0.044447	0.056478
μ_f, %	0.410483	0.613496	0.797484	0.928114	0.983542	0.997333	0.999903

Subsequently, we determine the efficiency of separation for all velocities and separation boundaries using the equation $E = 1 - \frac{\mu_f H_f + \mu_c H_c}{H_s}$. The table indicates the rate resulting in high efficiency for each separation boundary.

d	3.5m/sec	3m/sec	2.5m/sec	2m/sec	1.5m/sec	1m/sec	0.75m/sec
1.35mm	0.191442	0.105605	0.04881	0.016174	0.003574	0.000575	2.09E-05
0.8mm	0.326884	0.219574	0.108129	0.035322	0.007825	0.001258	4.57E-05
0.45mm	0.234157	0.358226	0.345936	0.149731	0.033287	0.005322	0.000193
0.25mm	0.144578	0.251544	0.37099	0.399582	0.164007	0.02599	0.00094
0.165mm	0.125977	0.222833	0.352829	0.4536	0 282685	0.047778	0.001723
0.125mm	0.094895	0.170709	0.286395	0.469373	0.888363	0.220642	0.00783

b. Multiproduct separation

Attention will now be given to the material consisting of 7 size classes. We shall estimate the efficiency of separation of the material into seven components in respect of 6 boundaries. The table gives the composition of the material in percent.

d	$r_{s,f}$, %
0.55mm	1.1
0.356mm	31.36
0.181mm	20.375
0.128mm	23.015
0.09mm	12.74
0.064mm	5.65
0.0265mm	5.76

The entropy of the initial composition of the material is equal to:

$$H_s = -\sum_{i=1}^{7} (r_{s.f}/100)\ln(r_{s.f}/100) = 1.66476.$$

Now follows the table of results of separation into seven components in respect of 6 boundaries by means of successive separation in respect of all boundaries (in percent)

d	7	6	5	4	3	2	1
0.55mm	1.1	0	0	0	0	0	0
0.356mm	25.757	2.08	1.72	0.42	0.47	0.34	0.14
0.181mm	8.47	2.47	3.34	1.08	1.395	1.83	0.39
0.128mm	2.5	2.04	3.7	3.85	3.13	3.1	0.735
0.09mm	0.16	0.42	1.39	1.63	3.8	3.94	1.4
0.064mm	0	0.05	0.17	0.5	1.1	6.74	1.09
0.0265mm	0	0.01	0.023	0.11	0.47	2.5	4.44
μ_{ri},%	37.987	7.07	10.343	7.59	10.365	18.45	8.195

μ_{ri}, % is the amount of material in percent in each component. The entropies of the individual components were also calculated:

7	6	5	4	3	2	1
0.800173	1.298014	1.381915	1.352805	1.517987	1.570894	1.332966

Consequently, the efficiency of separation is:

$$E = 1 - \frac{\sum_{i=1}^{7} \mu_i H_i}{H_s} = 0.463115$$

5. ALGORITHMS OF OPTIMISATION OF SEPARATION INTO n COMPONENTS

In chapter VII we described the mathematical models of separation into two components. Attention will now be given to the algorithms of separation of material into n components for ensuring the highest possible efficiency.

It is assumed that we have a material consisting of particles with the size from a_0 to a_n and it is required to separate the material into n components in respect of the given boundaries. The method of separation may be described as follows: the material is separated in respect of some boundary into two components and, subsequently, every components is separated in respect to one of the internal boundaries into new two components, and so on, until the material is separated in respect of all boundaries. In this case, it is necessary to solve the problem of determination of the order of the separation boundaries for attaining the maximum efficiency of separation. The efficiency is calculated from the equation:

$$E = 1 - \frac{\sum_{i=1}^{n} \mu_i H_i}{H_s}.$$

Let us examine the first algorithm of solving this problem.

Algorithm 1. Complete sorting

Initially, we determine the results of separation for each of the boundaries into two components for all parameters of equipment and the process (the area of introduction of the material into apparatus i^*, the number of stages in apparatus z, the velocity of the airflow w). Subsequently, for each of the separation products determined in the calculations, it is necessary to calculate the result of separation into two components in respect of all internal boundaries for all parameters of the apparatus and the process and continue in this way until we consider all possible methods of separation of the specific material into n components for all possible parameters of the equipment and the process. Subsequently, for all the selected methods we determine the efficiency of separation of all the methods and select the highest efficiency and, consequently, we determine the most efficient method of separation of the material into n components.

The algorithm described previously is characterised by the global maximum of the efficiency of separation, but its operating time is very long, $O(n!)$. The diagram of such a separation process is shown in the form of a graph (Fig. XII-5) indicating the graph (the tree) for the case in which the initial material is separated in respect of the four boundaries (for another number of the separation boundaries we can use the same approach). In the graph, every letter indicates the boundary in respect of which the separation is carried out. After separation in respect of each boundary, we obtain the fine and coarse products. If they contain the internal boundaries of separation, examination of the fine product in the graph is continued along the edge 1, and of the coarse product along the edge 2.

We shall examine the second algorithm of determination of the method of maximum separation efficiency.

Algorithm 2. 'Greedy algorithm'

In the first stage, we calculate the results of separation into two

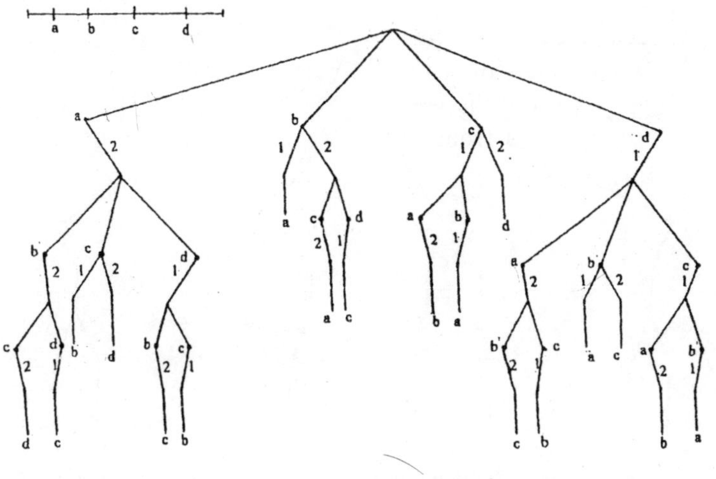

Fig.XII-5 Graph of complete sorting.

components for all boundaries and parameters of apparatus and every time we calculate the efficiency. This is followed by selecting the maximum efficiency and, consequently, determination of the first boundary in the order of the separation boundaries and the required parameters of the process and the apparatus. Subsequently, for each of the two obtained components the same operations are carried out with respect to the determined boundary until the material is divided into n components. It is assumed that for the initial material it has been determined that the maximum efficiency of separation into two components is obtained at the boundary a_j, the airflow velocity w_i, the number of stages of apparatus z_i and the number of stages of introduction of material into apparatus i^*_1. Each of the produced components receives particles from all narrow size classes of the initial material, but part of them is in the form of 'contamination' indicated for other components. In the first of the produced components, it is necessary to find the separation boundary (to maximise efficiency) between the boundaries a_0 and a_{j-1}, and in the second component between the boundaries a_{j+1} and a_n. It is thus necessary to continue until the initial material is divided in respect of all required boundaries.

This algorithm provides the local maximum of the efficiency of separation, but the operating time of this algorithm is $O(n\ln n)$ and this is considerably shorter than the operating time of the first algorithm.

In the cases in which the calculation of the order of the separation boundaries for the determination of the maximum efficiency

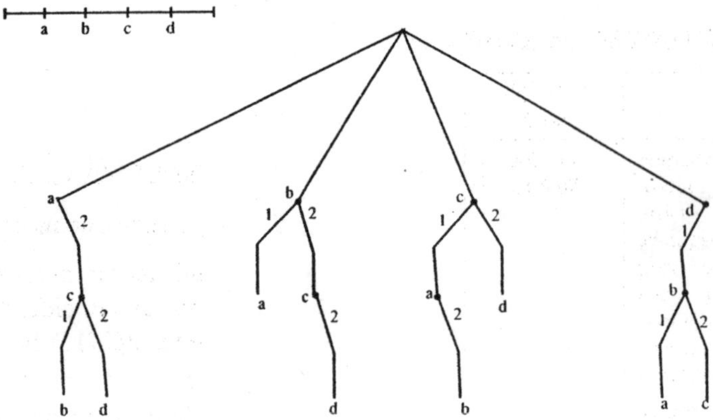

Fig.XII-6 Mixed algorithm for determining the maximum efficiency of separation into *n* components.

of separation using the first algorithm is very long in time, it is possible to transfer from some calculation stage to the determination of the local maximum of efficiency using the second algorithm.

We shall examine Fig. XII-6 showing the example of separation of the material examined in Fig. XII-5, into four components, but the first algorithm is used only in the first stage (complete sorting) and, starting with the second step, it is necessary to find the local maximum of the efficiency of separation using the second algorithm.

We shall present the results of calculations of the order of the separation boundaries for obtaining maximum efficiency on a specific example where it may be seen that the 'greedy algorithm' is not so 'bad', i.e. its results are close to the results of complete sorting. It will also be shown that the calculations of the order of the separation boundaries for obtaining maximum efficiency of separation can be started from complete sorting and this can be followed by transition to finding the local maximum.

Optimisation of separation of the given material into four components

The initial material will be represented by phosphates. The density of the material is ρ_m = 2800 kg/m^3, the density of air ρ_a = 1.2 kg/m^3. It is also assumed that the velocity of the airflow entering the apparatus from the bottom is w = 1.8 *m*/s. The number of stages of separation in the apparatus is z = 9. The number of the stage of supply of material into apparatus is i^*= 5. (Without restricting general nature of the considerations, we assume the constant velocity of the airflow, the number of separation stages and the area of introduction of material into apparatus. In the general case of optimisation of separation, these parameters must be varied).

We shall examine separation in respect of three boundaries at constant parameters of the apparatus. The grain size composition of the initial material is given in the table:

Mesh size in mm	0	0.053	0.074	0.105
Mean size of of narrow class on sieve d, mm	0.0265	0.0635	0.0875	0.1275
Partial residue of narrow classes on sieve, in %	11.4	13.4	23	52.2
Number of narrow class	4	3	2	1

We determine the entropy of the initial material, examining the material as two narrow size classes for each of the three separation boundaries. Initially, we calculate the total residue of the material for each separation boundary in%:

Number of narrow class	R_s, %	r_s,%
1	52.2	52.2
2	75.5	23
3	88.6	13.4
4	100	11.4

The initial entropy $H_s = -(P_1 \ln P_1 + P_2 \ln P_2)$, where $P_1 = \frac{R_s\%}{100\%}$, $P_2 = 1 - P_1$.

The boundaries of separation will be denoted by letters in the following manner (the numbers of the narrow classes are given at the top):

4		3		2		1
	a		b		c	

The initial entropy for each of the three separation boundaries is:

c	0.692
b	0.56
a	0.355

The coefficient of separation for all narrow size classes (k) in accordance with equation (VIII-35) is:

1	2	3	4
0.40018	0.497452	0.576695	0.40018

We determine the degree of fractional extraction for each narrow class (F_f) using the equation (VIII-17):

1	2	3	4
0.116769	0.487262	0.824351	0.992503

The amount of the material, transferred into the fine and coarse products after separation into two components for each of the three boundaries of separation is: $r_f\% = F_f r_s\ \%$; $r_c\% = r_s\% - r_f\%$.

The table of the yield of the fine product (μ_f):

Amount of materials in component, in %	1	2	3	4
39.67	6.1	11.21	11.05	11.31

The table of the yield of the coarse product (μ_c):

Amount of materials in component, in %	1	2	3	4
60.33	46.1	11.79	2.35	0.09

We determine which of the three separation boundaries results in the highest efficiency of separation into two components.

The tables of the total residues for the coarse and fine products will be presented.

The table for the fine product ($R_f\%$):

Amount of materials in component, in %	1	2	3	4
39.67	6.1	17.31	28.36	39.67

The table for the coarse product ($R_c\%$):

Amount of materials in component, in %	1	2	3	4
60.33	46.1	57.89	60.24	60.33

We determine the entropy of the large and fine products in respect of all separation boundaries.

For the fine product (H_f):

a	b	c
0.598	0.69	0.429

For the coarse product (H_c):

a	b	c
0.011	0.0776	0.389

We determine the efficiency of separation into two components in respect of all boundaries:

$$E = 1 - \frac{\mu_f H_f + \mu_c H_c}{H_s}, \mu_f = 0,3967, \mu_c = 0,6033.$$

This is followed by the results of calculation of efficiencies (E):

a	b	c
0.313	0.428	0.415

The degree of fractional extraction for every narrow size class remains unchanged, because the parameters of the process do not change.

We shall examine the separation of the initial material, as separation in respect of the boundary *a*.

The fine product ($r_f\%$):

Amount of materials in component, in %	1	2	3	4
39.67	6.1	11.21	11.05	11.31

The coarse product ($r_c\%$):

Amount of materials in component, in %	1	2	3	4
60.33	46.1	11.79	2.35	0.09

In the resultant coarse product, we find the boundary resulting in the maximum efficiency of separation into two components. It may be verified that the maximum efficiency of separation is obtained at boundary *b*. The results of separation of the coarse product into two components are found from the equations $r_f\,\% = F_f\, r_s\,\%$ and $r_c\,\% = r_s\,\% - r_f\,\%$. We obtain the following two components (separation in respect of the boundary *b*):

The fine product ($r_f\%$)

Amount of material in component, in %	1	2	3	4
13.156	5.382	5.745	1.94	0.089

The coarse product ($r_c\%$)

Amount of material in component, in %	1	2	3	4
47.174	40.718	6.045	0.41	0.001

We calculate the separation of the resultant coarse product in respect of boundary *c*.

The fine product ($r_f\%$):

Amount of materials in component, in %	1	2	3	4
8.04	4.754	2.954	0.33798	0.000993

The coarse product ($r_c\%$):

Amount of materials in component, in %	1	2	3	4
39.613	35.96	3.1	0.07202	0.000007

We determine the efficiency of separation for the examined order of the separation boundaries (*a*, *b*, *c*).

$$H_1=1.36,\ H_2=1.0272,\ H_3=0.813,\ H_4=0.\quad E=1-\frac{\sum_{i=1}^{4}\mu_i H_i}{H_s}=0.285$$

We examine the separation of the initial material for the following order of the boundaries (a, b, c). We shall use the results of separation of the coarse product initially in respect of the boundary c and subsequently, the result obtained in respect of boundary b.

The result of separation in respect of brand c

Fine product (r_f%):

Amount of material in component, in %	1	2	3	4
13.156	5.382	5.745	1.94	0.089

The coarse product (r_c%):

Amount of material in component, in %	1	2	3	4
47.174	40.718	6.045	0.41	0.001

We determine the separation of the last fine product in respect of the boundary b.

The fine product (r_f%)

Amount of material in component, in %	1	2	3	4
5.1176	0.628	2.8	1.6	0.0883

The coarse product (r_c%):

Amount of material in component, in %	1	2	3	4
8.04	4.754	2.954	0.34	0.0006

We calculate efficiency for the examined order of the boundaries (a, b, c):

$H_1 = 1.36$, $H_2 = 0.433$, $H_3 = 1.022$, $H_4 = 0.812$. $E = 1 - \frac{\sum_{i=1}^{4} \mu_i H_i}{H_s} = 0.282$.

We examine the separation of the initial material into two components as separation in respect of boundary c.

The fine product (r_f%)

Amount of material in component, in %	1	2	3	4
39.67	6.1	11.21	11.05	11.31

The coarse product (r_c%)

Amount of material in component, in %	1	2	3	4
60.33	46.1	11.79	2.35	0.09

Subsequently, we examine the separation of the fine product in respect of the boundary *b*. We obtain the following two components.

The fine product (r_f%)

Amount of material in component, in %	1	2	3	4
26.514	0.712	5.46	9.11	11.23

The coarse product (r_c%)

Amount of material in component, in %	1	2	3	4
13.16	5.39	5.75	1.94	0.08

We examine the separation of the resultant fine product in respect of boundary *a*

The fine product ($r_f\%$)

Amount of material in component, in %	1	2	3	4
21.4	0.083	2.66	7.51	11.14

The coarse product ($r_c\%$)

Amount of material in component, in %	1	2	3	4
5.113	0.629	2.8	1.6	0.084

We calculate the efficiency of separation for the resultant order of the boundaries (a, b, c):

$$H_1 = 0.66,\ H_2 = 1.04,\ H_3 = 1.1,\ H_4 = 0.984.\ E = 1 - \frac{\sum_{i=1}^{4} \mu_i H_i}{H_s} = 0.333.$$

We calculate the results of separation of the initial material for the order of the boundaries (*a*, *b*, *c*). We use the available results of separation in respect of boundary *c* and examine the separation of the fine product initially in respect of boundary *a* and then that of the remaining product in respect of boundary *b*.

The result of separation in respect of boundary *a*

The fine product ($r_f\%$)

Amount of material in component, in %	1	2	3	4
26.514	0.712	5.46	9.11	11.23

The coarse product ($r_c\%$)

Amount of material in component, in %	1	2	3	4
13.16	5.39	5.75	1.94	0.08

We calculate the separation of this coarse product in respect of boundary *b* and obtain the following two components.

The fine product (r_f%)

Amount of materials in component, in %	1	2	3	4
8.05	4.76	2.95	0.34	0.0006

We now determine the efficiency for the examined order of the boundaries (a, b, c)

$$H_1 = 0.66,\ H_2 = 1.55,\ H_3 = 1.015,\ H_4 = 0.812.\quad E = 1 - \frac{\sum_{i=1}^{4} \mu_i H_i}{H_s} = 0.315.$$

Consequently, it may be seen that if in the algorithm of complete sorting after separation of the initial material in respect of the boundary c we would have used the algorithm for finding the local maximum separation, we would have obtained that the local maximum gives the order of the separation boundaries (a, b, c).

Now follows the table including the efficiencies of separation for all possible orders of the separation boundaries:

Order of boundaries	b.a.c (obtained in search for maximum efficiency)	a.b.c (local maximum, starting with the second step)	a.c.b	c.b.a (local maximum, starting with the second step)	c.a.b
Efficiency	0.347	0.285	0.282	0.333	0.315

These and other examples show that the algorithm for finding the local maximum of the efficiency of separation is quite efficient in comparison with the algorithm of complete sorting.

6. THE MATHEMATICAL MODEL OF SEPARATION INTO *n* COMPONENTS

In chapter 7, we described a new method of calculating the results of separation of the bulk material into two products (components). By analogy with this method, we describe a method of calculating the results of separation of such a material into n components in respect of the $(n-1)$ separation boundary.

It is assumed that the initial material is given. The particle size of the material is in the range from a_1 to a_n and the material should be separated in respect of the $(n-1)$ boundary.

We assume, without restricting the general nature of the problem, that the algorithms have provided the following sequence of the separation boundaries for obtaining the maximum efficiency of separation:

$a_{i_1}, a_{i_2}, a_{i_3}, ..., a_i, a_{i+1}, ..., a_{i_{n-2}}, a_{i_{n-1}}$

(i.e. in this case, the separation in respect of the boundary a_i takes place earlier than in respect of the boundary a_{i+1}, and if the situation were reversed, the approach would be exactly the same). Also, for the separation in respect of every boundary, we obtain the parameters of apparatus (i.e. z is the number of stages, i^* is the number of the stage of supply of the material into the apparatus, w is the rate of the airflow entering the apparatus from the bottom, these parameters differ for all boundaries). It should be assumed that it is required to calculate the separation resultd in component i+1. According to the resultant sequence of the separation boundaries, the results of separation in component i+1 will be available after separation in respect of the boundary a_{i+1}. In the graph in Fig. XII-7, C is the path connecting the tips a_{i_1} and a_{i+1}. It is also assumed that $r_{s,j}$ is the amount of the material with the size j in the initial material. Consequently, after separation of the initial material in respect of the boundary a_{i_1}, the fine product contains $r_{s,i}F_{1,f,j}$ of the material with the size j, and the coarse product $r_{s,j}(1-F_{1,f,j})$, where $F_{1,f,j}$ is the degree of fractional extraction of the size class j for the first apparatus in which separation takes place. The material transferred to the fine product should be

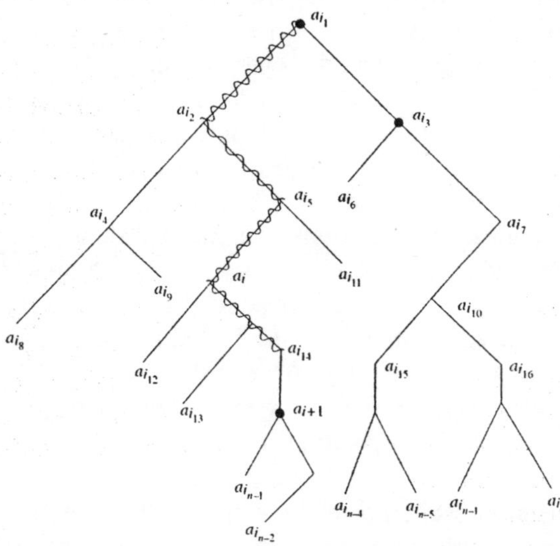

Fig.XII-7 Order of separation boundaries.

subsequently separated in respect of the internal separation boundaries, like the material, transferred into the coarse product. The determination of the amount of the material with the size class j, included in the large and fine products, after separation in respect of each boundary will be determined in the same manner as in the first separation step. Consequently:

$$r_{i+1,f,j} = r_{s,j} F_{i+1,f,j} \prod_{a_k \in C^r} (1 - F_{k,f,j}) \prod_{a_k \in C^l} F_{k,f,j}$$

where $r_{i+1,f,j}$ is the amount of the material of the size class j, transferred into the i+1 components; $F_{k,f,j}$ is the degree of fractional extraction of the material of class j into the k-th components; C is the path between a_{i_1} and a_{i+1}, $a_k \in C^r$ or $a_k \in C^l$, if the edge in the path C after the tip a_k is directed to the right or left, respectively (see Fig. XII-7).

8. CONDITIONS OF OPTIMISATION OF SEPARATION OF BINARY MIXTURES

The justification of the entropy approach to the evaluation of the separation processes creates new possibilities for explaining the objective conditions of optimisation of separation of binary mixtures. Using this approach, we shall try to solve the problem of the conditions of optimisation of mixtures of this type, which has been discussed in special literature for more than 100 years using the new approach.

We shall examine the phenomena taking place in such separation in the general form. It is assumed that a certain bulk material has the initial composition, and the grain size curve of the material in partial residues is represented by the curve ABC (XII-8). On the basis of the technological considerations, this material should be separated in respect of the boundary size of x_0 microns. The graph area, restricted by the curve ABC and the axes of the coordinates, corresponds on a specific scale to the total amount of the initial material. This amount will be regarded as unity and the curve will be denoted by $Q(x)$.

In the ideal case, separation should take place in respect of Bx_0. This line divides the initial composition into two parts: D_s– fine product, R_s – coarse product.

In the real process, separation does not take place in the ideal manner because part of the fine fractions is transferred into the coarse product, and part of the large fractions into the fine product.

It is assumed that in the real process, the fine product is described in these coordinates by the curve $g(x)$, and the coarse product by $n(x)$.

The method of construction of these curves shows that the following relationship will be valid at any point x_i:

$$Q(x_i) = g(x_i) + n(x_i) \qquad \text{(XII-21)}$$

for the degree of fractional separation

$$F_f(x_i) + F_c(x_i) = 1 \qquad \text{(XII-22)}$$

where $F_f(x_i) = \dfrac{g(x_i)}{Q(x_i)}$ – is the extraction of the narrow class into the fine product; $F_c(x_i) = \dfrac{n(x_i)}{Q(x_i)}$ – is the extraction of the narrow class into the coarse product.

There are a number of methods of direct determination of the conditions of optimality of the process without calculating its efficiency. They link the quality of the separation process with the 'boundary grain size', with a specific physical meaning given to the latter. According to this concept, for any ratio of the regime and design parameters of separation, it is always possible to select a narrow class of the size – the boundary grain size, for which the given process is optimum.

Rubinchik determined the boundary grain size using relationships characteristic of the ideal process and proposed to determine the optimum efficiency on the basis of the fine class whose content in the initial material is equal to the yield of the fine product:

$$D_s = \gamma_f$$

Bond determines the boundary grain size by the content of the fines in the fine product which is equal to the total residue of the coarse sizes in the coarse product:

$$\frac{D_f}{\gamma_f} = \frac{R_c}{\gamma_c}$$

where γ_f and γ_c are the yields of the fine and coarse products, respectively.

Povarov proposed to determine the boundary grain size as the value of the narrow size class whose relative content in the initial material and both separation products is identical:

$$Q(x) = \frac{q(x)}{\gamma_f} = \frac{n(x)}{\gamma_c}$$

Shteinmetser proposed that the value of the boundary grain size should be represented by the narrow size class which is divided into halves between the two products:

$$F_f(x) = F_c(x)$$

None of these methods has been verified and, consequently, their accuracy is still the subject of discussion.

We shall try to solve this situation, on the basis of the principle of the problems, solved by separation.

In a general case, the area of the graph in Fig. XII-8 is divided by the lines of the ideal and real processes into four sections: D_f – the fine material in the yield of the fine product; R_f – the large material in the yield of the fine product; R_c – the large material in the yield of the coarse product; D_c – the fine material in the yield of the coarse product.

It is clear that for all these parts the following relationships will be valid:

Partial residues

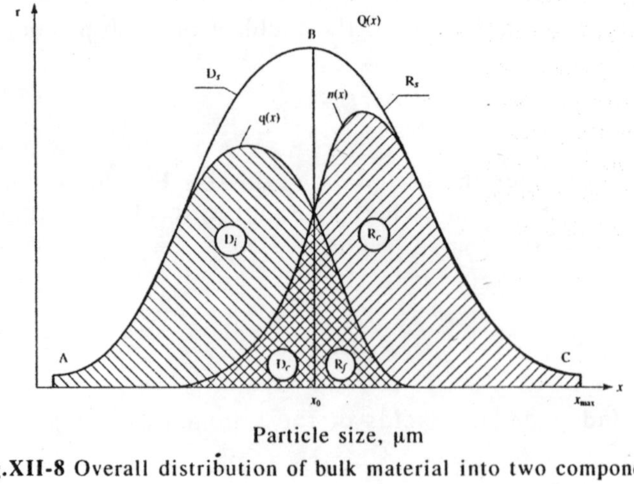

Particle size, µm

Fig.XII-8 Overall distribution of bulk material into two components.

$$D_s + R_s = 1$$

$$D_f + D_c = D_s$$

$$R_c + R_f = R_s \qquad \text{(XII-23)}$$

$$R_f + D_f = \gamma_f$$

$$R_c + D_c = \gamma_c$$

On the basis of these parameters, we can usually formulate several simple characteristics:

$\varepsilon_f = \frac{D_f}{D_s}$ – extraction of the fine product;

$\varepsilon_c = \frac{R_c}{R_s}$ – extraction of the coarse product; (XII-24)

$k_f = \frac{R_f}{R_s}$ – contamination of the fine product;

$k_c = \frac{D_c}{D_s}$ – contamination of the coarse product.

In order to determine the conditions for the maximum fractional difference in the separation products, it is necessary to optimise the relationship:

$$E_I = D_f + R_c \qquad \text{(XII-25)}$$

or, which is the same, minimise it:

$$E_{II} = R_f + D_c$$

The optimisation conditions according to these relationships are:

$$\frac{dE_I}{dx} = 0; \qquad \frac{dE_{II}}{dx} = 0$$

It may easily be seen that both relationships give the same result. We shall expand them:

$$\frac{dE_l}{dx}=\frac{dD_f}{dx}+\frac{dR_c}{dx}=\frac{d\int_0^x g(x)dx}{dx}+\frac{d\int_x^{x_{max}} n(x)dx}{dx}=0$$

It is well-known that the derivative of the definite integral with the variable upper limit and the constant lower limit is equal to the integrand expression at the upper limit point. Consequently

$$g(x)-n(x)=0, \text{ i.e. } g(x)=n(x) \quad \text{(XII-26)}$$

Taking equation (XII-21) into account, we obtain that at the optimum point, the narrow size class is divided into halves, i.e. the conditions of optimality correspond to:

$$F_f = F_s = 0{,}5 \quad \text{(XII-27)}$$

In Fig. XII-8, this condition corresponds to only one point – the intersection of the curves $g(x)$ and $n(x)$. The ordinate corresponding to this point is x_0.

Thus, in any separation process, it is always possible to find the boundary grain size corresponding to the optimum result. The validity of the dependence (XII-27) has been clearly confirmed here for the first time.

However, if this parameter is required for the determination of the optimum of the process, it is clearly evident that the parameter is insufficient for this purpose. Actually, the quantities D_c and R_f may be higher or lower but they do not determine the extent of completion of the separation process in the numerical expression. To evaluate this extent, a large number of quality criteria have been formulated. Their analysis shows that two criteria correspond more sufficiently to the separation problems:

1. The Hancock criterion:

–for the fine product

$$E_f = \frac{D_f}{D_s} - \frac{R_f}{R_s} \quad \text{(XII-28)}$$

–for the coarse product

$$E_c = \frac{R_c}{R_s} - \frac{D_c}{D_s};$$

2. The entropy criterion

$$E = -\{[D_s \ln D_s + R_s \ln R_s] - \gamma_f \left[D_f^* \ln D_f^* + R_f^* \ln R_f^*\right] -$$
$$-\gamma_c \left[R_c^* \ln R_c^* + D_c^* \ln D_c^*\right] \} \qquad \text{(XII-29)}$$

where D_f^*; R_f^*; D_c^*; R_c^* are the parameters related to the yield of the product. The following relationships were used in expression (XII-29),

$$R_f^* = \frac{R_f}{\gamma_f}; \; D_f^* = \frac{D_f}{\gamma_f}$$

$$R_c^* = \frac{R_c}{\gamma_c}; \quad D_c^* = \frac{D_c}{\gamma_c}$$

$$\text{(XII-30)}$$

$$R_f^* + D_f^* = 1; \; R_c^* + D_c^* = 1$$

It is interesting to note that the parameter (XII-28) is generally recognised and is used widely in special literature and practice. As regards criterion (XII-29), it was formulated relatively recently and is not widely known amongst the experts.

Both these criteria have a set of properties making them most objective from the viewpoint of problems solved by separation:

- they are monotonic, and in the case of the large fractional difference they provide high parameters;
- in ideal separation they provide the maximum possible values;
- in separation of the initial material into parts without variation of the fractional composition, both parameters give the zero result;
- they have the unambiguity property, i.e. give the same result regardless of the type of product for which efficiency is determined.

The latter is evident for (XII-29). For equation (XII-28) this was shown in chapter II.

Using the derived relationships (XII-26) and (XII-27), we shall try to analyse these criteria for correspondence to the optimum conditions.

The optimality criterion is written in the following form:

$$\frac{dE}{dx}=0$$

Initially, the condition will be specified for the dependence (XII-29). It is presented in the following form:

$$E=-\Big[(R_s \ln R_s + D_s \ln D_s)-(R_c \ln\frac{R}{\gamma_c}+D_c \ln\frac{D_c}{\gamma_c}+ +D_f \ln\frac{D_f}{\gamma_f}+R_f \ln\frac{R_f}{\gamma_f})\Big] \qquad \text{(XII-31)}$$

The derivative of (XII-31) after several transformations has the following form:

$$\frac{dE}{dx}=Q(x)\ln R_s - Q(x)\ln D_s - n(x)\ln\frac{R_c}{\gamma_c}+n(x)\ln\frac{D_c}{\gamma_c}+ +g(x)\ln\frac{D_f}{\gamma_f}-g(x)\ln\frac{R_f}{\gamma_f}=0$$

Consequently,

$$Q(x)\left[\ln R_s - \ln D_s\right]=n(x)\left[\ln\frac{R_c}{\gamma_c}-\ln\frac{D_c}{\gamma_c}\right]+g(x)\left[\ln\frac{R_f}{\gamma_f}-\ln\frac{D_f}{\gamma_f}\right]$$

or

$$\ln\frac{R_s}{D_s}=F_c(x)\ln\frac{R_c}{D_c}+F_f(x)\ln\frac{R_f}{D_f}$$

We shall carry out several transformations of this expression

$$\ln\frac{R_s}{D_s}-\ln\frac{R_c}{D_c}=F_f(x)\left[\ln\frac{R_f}{D_f}-\ln\frac{R_c}{D_c}\right]$$

and, consequently

$$\ln \frac{R_s \cdot D_c}{D_s \cdot R_c} = F_f(x) \ln \frac{R_f \cdot D_c}{R_c \cdot D_f}$$

According to (XII-24), we may write:

$$\ln \frac{k_c}{\varepsilon_c} = F_f(x) \ln \frac{k_f \cdot k_c}{\varepsilon_c \cdot \varepsilon_f}$$

Taking into account the fact that in the optimum regime $F_f(x) = \frac{1}{2}$, we obtain:

$$\frac{k_c}{\varepsilon_c} = \sqrt{\frac{k_f \cdot k_c}{\varepsilon_f \cdot \varepsilon_c}} \qquad \text{(XII-32)}$$

Equation (XII-32) can be valid only in one case when:

$$\varepsilon_f = \varepsilon_c \qquad \text{(XII-33)}$$

This, according to (XII-24) is automatically equalised, and

$$k_c = k_f$$

since in the optimum regime $0.5 \leq \varepsilon_f;\ \varepsilon_c \leq 1$.

Thus, by analysis of the optimality conditions in respect of the relatively complicated entropy criterion, we obtain the simple relationship (XII-33), expressing the simple condition of optimality for the separation process. Figure XII-9 shows a specific example of the application of condition (XII-33) for the determination of the optimum. In this example, the optimum size corresponds to 275 μm, the value of equal extraction is 84%.

We shall now examine the conditions of optimality for the Hancock criterion (XII-28):

$$\frac{dF_f}{dx} = 0; \qquad \frac{dE_c}{dx} = 0$$

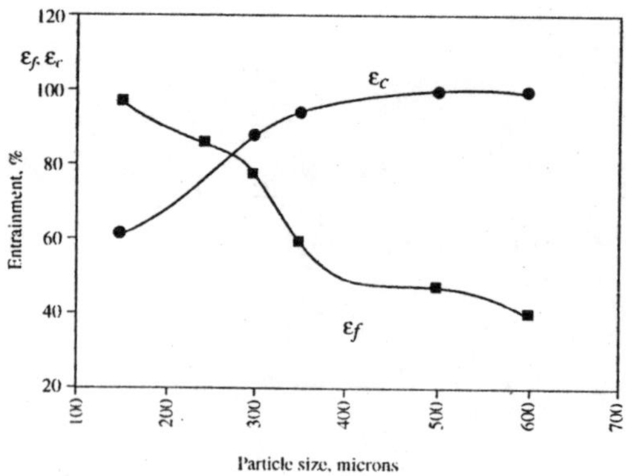

Fig.XII-9 Dependence of extraction of fine and coarse products for the cascade classifier at an airflow velocity of w = 1.5 m/s.

We expand the first of these conditions:

$$\frac{g(x)\cdot D_s - D_f Q(x)}{D_s^2} - \frac{-g(x)R_s + Q(x)R_f}{R_s^2} = 0$$

Consequently

$$\frac{F_f(x)}{D_s} - \frac{\varepsilon_f}{D_s} = \frac{k_f}{R_s} - \frac{F_f(x)}{R_s}$$

From the last condition:

$$F_f(x) = k_f D_s + \varepsilon_f R_s = R_s(\varepsilon_f - k_f) + k_f$$

It should be noted that the brackets contain the expression for the efficiency (XII-28). Consequently, we may write

$$F_f(x) = E_f \cdot R_s + k_f$$

Similarly, for E_c we obtain:

$$F_c(x) = E_c D_s + k_c$$

From these expressions, we derive relationships for the efficiency:

$$E_f = \frac{F_f(x) - k_f}{R_s}$$

$$E_c = \frac{F_c(x) - k_c}{D_s}$$

In all cases it holds that $E_f = E_c$ and, consequently:

$$\frac{F_f(x) - k_f}{R_s} = \frac{F_c(x) - k_c}{D_s} \quad \text{(XII-34)}$$

We have determined that in the optimum conditions:

$$F_f(x) = F_c(x) \text{ and } k_f = k_c$$

Taking this into account, expression (XII-34) may be valid only in the single particular case when:

$$D_s = R_s = 0.5$$

In all remaining cases the Hancock criterion cannot be used for optimising the separation processes.

Thus, as a result of the analysis results, we have formulated and justified new (relatively simple and suitable for practical application) conditions of optimality for the processes of separation of the bulk materials and we have also rejected the widely used quantitative criterion of optimisation of these processes.

8. FRACTIONATION IN A RAREFIED GAS

1. Physical fundamentals of dynamics of single particles

This problem was partially discussed in chapter III. Here, we shall examine this problem in greater detail because it is the essential condition for the separation process in a rarefied gas.

At first sight, it may appear that all the main problems, associated with the settling of the particles in a still medium (simplified case) have been examined and solved since this problem has been studied by experts for more than 100 years. In fact, the problem is very complicated and has not as yet been completely solved. There is only a single solution of the Stokes equation for the case Re$\ll$1. As regards the remaining part of the range of the values of Re, it is necessary to use empirical and semi-empirical relationships. Recently, a large number of experimental material has been obtained for the settling of single particles in different liquids. For gas media, the number of these data is considerably smaller, and for a rarefied gas (ΔP = 380÷10 mm Hg) there are almost no experimental data. However, this area is of considerable interest for the fractionation of fine-dispersion materials. It is therefore important to examine the dynamics of single particles in a rarefied gas. In addition to this, in the literature, especially applied literature, there have been many cases of insufficiently efficient theoretical analysis and even a large number of erroneous equations and conclusions. All these factors indicate that it is important to examine the given problem in a considerable detail.

In solving the problem of flow around particles, two main equations of motion of the solid medium are usually used:

1. The continuity equation

$$\frac{d\rho}{dt} + \rho div\vec{V} = 0$$

where ρ is the density of the medium; $\vec{V}$ is the vector of the motion velocity; t is time.

This equation represents the law of conservation of mass. For the case of an incompressible medium (ρ = const), the equation has the following form:

$$div\vec{V} = 0$$

or in respect of components:

$$\frac{\partial V_x}{\partial X} + \frac{\partial V_y}{\partial y} + \frac{\partial V_z}{\partial z} = 0$$

2. The Navier–Stokes equation of motion of a viscous incompressible liquid.

This equation represents the second Newton's law applied to the continuous medium on the condition of a linear relationship between the stress tensor and the strain tensor. The latter circumstance determines the rheological parameters of the medium, in particular, Newton liquid, to which all gases relate. The Newton rheological law has the following form:

$$P = a\dot{S} + b\varepsilon$$

where P is the stress tensor; $\dot{S}$ is the strain rate tensor; ε is the unit tensor; a, b are constants.

In particular,

$$P = 2\mu defV - p\varepsilon$$

where μ is the dynamic viscosity coefficient; p is pressure, or in respect of the components:

$$P_{ij} = \begin{cases} \mu\left(\dfrac{\partial V_i}{\partial X_j} + \dfrac{\partial V_j}{\partial X_i}\right) & i \neq j \\ -p + 2\mu\dfrac{\partial V_i}{\partial X_i} & i = j \end{cases}$$

It should be mentioned that for the ideal liquid (μ = 0) the rheological equation has the following form:

$$P = -p\varepsilon$$

(all the parameters of the stress may be expressed through one scalar quantity, i.e. pressure) and in the simplest case of the straight layered motion of the viscous Newton liquid:

$$\tau = \mu\frac{\partial u}{\partial n}$$

where τ are the tangential stresses.

Solving these equations, Stokes calculated the drag coefficient:

$$F_s = 3\pi\mu vd$$

If it is assumed that $F_s = \frac{1}{2}\lambda\rho w^2\frac{\pi d^2}{4}$, then

$$\lambda = \frac{24}{\text{Re}}$$

It should be mentioned that this approach is not always successful in examining the flow around a long cylinder. It is therefore necessary to select linear-piecewise approximations of the Rayleigh curve of $\lambda = A/\text{Re}$ and $\lambda = A + B/\text{Re}$ form (for example, Ingebyu selected A and B for the entire curve), which are suitable in certain problems.

However, we shall not discuss them here. We shall only mention the interesting approximations for the entire section of the standard curve.

Here, the solution was obtained by Ozeen. Solving the equation together with the boundary conditions and the continuity equation, Ozeen obtained:

$$\lambda = \frac{24}{\mathrm{Re}}\left(1 + \frac{3}{16}\mathrm{Re}\right)$$

Thus, in contrast to the Stokes approximation, used at small distances from the sphere, the Ozeen approximation is used at large distances.

As regards experimental confirmation, the existing information results in a good agreement of the Stokes equation with the experiments for Re<0.1, and for the Ozeen équation 0.1<Re<10. Individual experimental data make it possible to use the Stokes equation for Re<1÷2.

In the case of relatively high velocities, the movement of the medium in flow around becomes unstable. The dependent variables (velocity, pressure) are not unambiguous functions of the spatial coordinates and time and should be described by stochastic laws. Therefore, regardless of the fact that the Navier–Stokes equations of motion may be regarded as applicable, unfortunately, in the present case, it is not possible to find the solution of these equations. A large number of experimental investigations, carried out to determine the dependence of the drag coefficient of the sphere on the Reynolds number resulted in the so-called standard Rayleigh curve obtained by statistical processing of the experimental data. In special literature, there is a large number of interpolation equations (approximations) of the Rayleigh curve in different sections, and also for the entire range of the values of the Reynolds number. The most interesting equations have been derived by:

Rozenbaum and Todes:

$$\mathrm{Re} = \frac{Ar}{18 + 0.61\sqrt{\mathrm{Re}}} \qquad \text{(XII-35)}$$

or

$$\lambda = 0.248 + 24\,\mathrm{Re}^{-1} + 0.248\sqrt{1 + 194\,\mathrm{Re}^{-1}}$$

Adamov

$$\mathrm{Re} = 21.2\left(\sqrt{1 + 0.038 Ar^{2/3}} - 1\right)^{3/2}$$

or $$\lambda = 24\,\mathrm{Re}^{-1}(1+0.065\,\mathrm{Re}^{2/3})^{3/2} \qquad \text{(XII-36)}$$

Brauer and Myus

$$\lambda = 0.4 + 24\,\mathrm{Re}^{-1} + 4\,\mathrm{Re}^{0.5} \qquad \text{(XII-37)}$$

The second method is based on the theory of the boundary layer. The concept of this method was developed for the first time by Prandtl. It is based on departure from the drag coefficient using a different method. In particular, in this approach it is possible to use the following approximation for solving the external flow-around problem. In the first stage, away from the sphere, the liquid may be regarded as ideal and, consequently, we can find the distribution of pressure and examine a viscous flow-around in the boundary layer. After the first step, it is possible to improve accuracy of the value of pressure and so on, i.e. use the iteration method. There are also other possible methods of solving the problem on the basis of this procedure. However, this method also has difficulties and shortcomings. It has been attempted to overcome them mainly on the basis of the application of the Rayleigh curve, because this is based on the similarity of the process. The friction coefficient $\lambda = \dfrac{F_s}{\frac{1}{2}\rho w^2 S}$ plays the role of the Euler criterion, since $\dfrac{F_s}{S}$ has the dimension of the pressure gradient and is the function $\lambda(\mathrm{Re})$ of the similarity criterion Re. As regards experimental data, it should be mentioned that all the available data have been obtained in flow-around (settling) of the spheres made from different materials and of different diameters, mainly in different liquids. The density of the liquids was varied in a limited range (the difference did not exceed a factor of 3–4). The experiments carried out with the settling of the particles in air in the normal conditions show that the values of λ also fit the Rayleigh curve. As regards the stationary flow around the spheres by a rarefied gas (the rarefaction range of 380÷10 mm Hg), the experimental data for the this range are not available.

The problem of non-stationary flow-around of the spherical particles is even more complicated. In the case of low values of the Reynolds number, the simplified Navier–Stokes equations in stationary flow-around were determined for the first time by Bussinesq and published in 1903.

Later, the equation was derived by Lourie and analysed by Ozeen and Chen. In addition to this, the evaluation of the contribution of different terms of the general equation has been presented in special literature.

Thus, the currently available methods of calculating the settling process are semi-empirical, approximate and are based on the experimental material. No experimental data are available for the settling of particles in rarefied gases. Consequently, it was interesting to carry out experiments that investigate the process of the flow-around of particles in the pressure range 380÷10 mm Hg. The Knudsen equation for the given degree of rarefaction is:

$$Kn = \frac{L}{d} << 1$$

where L is the free path length of the gas molecules; d is the particle diameter.

2. Experimental examination of the processes of settling of single particles in a rarefied gas

The aim of the experimental investigations was to examine the effect of gas density on the process of settling of spherical particles in a stationary medium using the method of stroboscopic recording. Figure XII-10 shows the diagram of experimental equipment which includes vacuum chamber *1*, photographic camera *2*, stroboscope *3*, vacuum pump *4*, mercury pressure gauge *5*, particle feeder *6*, photodiode *7*, frequency meter *8*, illumination lamps *9*. The vacuum chamber

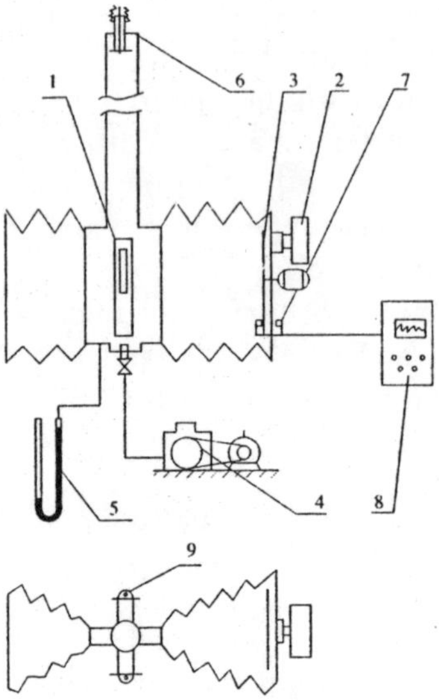

Fig.XII-10 Experimental equipment.

contains glass windows for illumination and photographic recording. Illumination is carried out using two 1 kW halogen lamps. The stroboscope is in the form of a non-transparent disc with slits, placed on a direct current drive. The photodiode and the frequency meter are used for measuring the time period of recording of the particles. The device for the feeding of the particles consists of a receiving pipe and a bar with a moving tray. In the experiments, we used polystyrene spheres with the density ρ_t = 1020 kg/m^3. The spheres of a regular shape were selected under a microscope from pre-selected narrow classes. The diameter of the sphere was measured using an indicating device with the accuracy of up to 1 μm. The axis of the chamber contained a measuring ruler and photographs were taken to determine the linear scale and this was followed by the determination of the required number of revolutions of the disc of the stroboscope, rarefaction in the chamber, and photography of the flight of the particles. The results were processed by the method of projecting of frames on a screen with appropriate measurements. The method of stroboscopic photographic recording makes it possible to ensure that the error of the measurements of the parameters of the particles is in the range not exceeding 2.5%. The height of throwing was selected in accordance with the size of the particles in order to ensure stationary settling. The gas density was calculated from the equation

$$\rho = \rho_n \frac{P(273 + t_n^\circ)}{P_n(273 + t^\circ)}$$

where ρ_n is the density of the gas in the normal conditions; P_n, t_n° are the barometric pressure and temperature, corresponding to the normal conditions; P is the absolute pressure in the chamber, mm Hg.

The Reynolds number was determined on the basis of the measured settling velocity and the drag coefficient λ was determined using the criterial dependence:

$$Ar = \frac{3}{4} \lambda \mathrm{Re}^2$$

where Ar is the Archimedes criterion.

The results of experiments and comparison with the calculated data on the basis of the standard Rayleigh curve and the Rosenbaum–Todes approximation at an air density of ρ = 0.12 kg/m^3 and ρ = 1.2 kg/m^3 are shown in Fig. XII-11. The data indicate that for the numbers Re>1.5 the experimental results are in good agreement with the standard Rayleigh curve and the Rosenbaum–Todes dependence

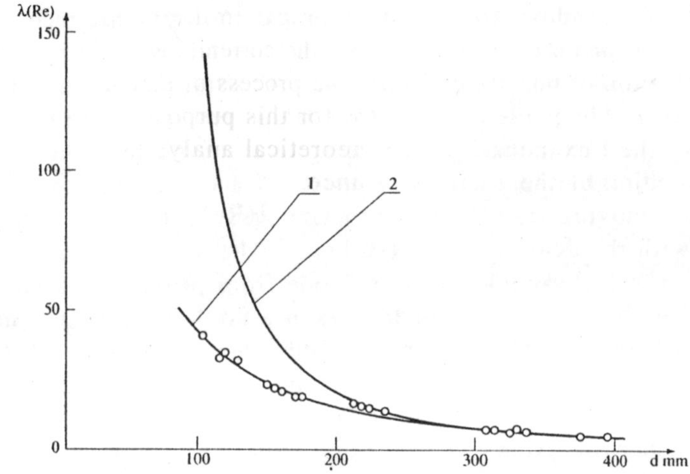

Fig.XII-11 Comparison of experimental results and the standard Rayleigh curve for Re>15: — ρ = 1.2 kg/m³; o – ρ = 0.12 kg/m³.

both for the case of the normal density of air ρ_n = 1.2 kg/m³ and for the rarefied gas ρ = 0.12 kg/m³. However, in the range of the small numbers Re < 1.5, the experiment gives low values of the drag coefficient in comparison with the available relationships. The graph presented in Fig. XII-12 shows that in the rarefied gas at ρ = 0.12 kg/m³ for polystyrene particles smaller than 200 μm, the experimental values of the drag coefficient will differ from the calculated values by more than 100%, and their settling velocities by almost

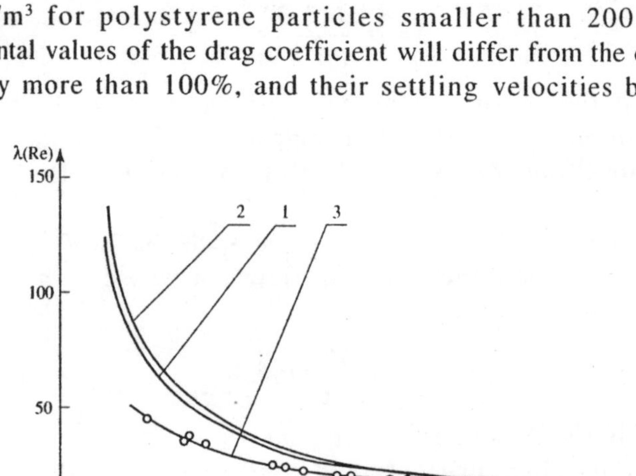

Fig.XII-12 Experimental results obtained in settling of spherical particle for Re<2: 1 – standard Rayleigh curve; 2 – Rosenbauer–Todes approximation; 3– experimental data obtained at ρ = 0.012 kg/m³.

50%. The results indicate the disruption of similarity $\lambda(\mathrm{Re})$ for the small Reynolds numbers and the fact that the currently available methods of determination of the parameters of the process of stationary settling of spherical particles are not suitable for this purpose. These results require detailed examination and theoretical analysis because they are of considerable applied importance.

Since any well-known relationships $\lambda(\mathrm{Re})$ at $\mathrm{Re}\to 0$ are the asymptotic approximation to the Stokes law, and all the experimental data, obtained previously (starting with $\mathrm{Re}\approx 1$) efficiently confirmed this, we shall turn directly to the solution of the Stokes equations. The value of the drag force $F_s = 3\pi\mu dV$, obtained by Stokes for the case of stationary flow-around of the spherical particle by a viscous incompressible Newtonian liquid, is the exact solution of the Navier-Stokes equations at $\mathrm{Re}<1$. The finite settling velocity according to this law is determined by the equation:

$$v=\frac{1}{18}\cdot\frac{\rho_T-\rho_0}{\mu}gd^2$$

For media in which ρ_0 is comparable with ρ_T (i.e. for the case of a liquid), the finite settling velocity will depend on the density of the liquid as a result of the effect of the Archimedes buoyancy force. For gases, the value of the buoyancy force is insignificantly small. Therefore, the small particles should settle in accordance with the Stokes law with the same velocity and should be subjected to the same drag force in the flow-around in gases characterised by different density but the same viscosity.

It should be mentioned that the viscosity of the gases does not depend and on the pressure up to the rarefaction of the order of $P = 10^{-7}$ mm Hg, i.e. up to the density of air of $\rho_0 = 0.012$ kg/m^3. Thus, varying the density of gas by two orders of magnitude, we should not obtain any changes in the finite settling speed or the drag force in flow-around. It is well-known that at $\rho_0 = 0$, the drag force is also equal to zero. The Stokes law does not ensure this limiting transition because in the case of strong rarefaction of the gas the latter should not be regarded as a continuous medium and, consequently, the initial hydromechanics equations are not applicable. The criterion of applicability of the model of the continuous medium is represented by the Knudsen number $K_n = L/d$, where L is the free path length of the molecules, and d is the characteristic size (the thickness of the boundary layer, particle diameter). For the case of the Knudsen numbers, approaching unity on the surface of the sphere, the gas velocity is not equal to zero and the correction introduced for the first time by Milliken, is added to the Stokes law:

$$\lambda = \frac{24}{\mathrm{Re}\left(1+1.2\,{}^{2L}\!/_{d}\right)}$$

We shall calculate the correction for the present case. It is well-known that the free path length of the molecules of the gas is inversely proportional to pressure. In the case of air at $t^0 = 15^0$, and at P = 760 mm Hg, $L = 6.2 \times 10^{-8}$ m, and at P = 76 mm Hg, $L = 6.2 \times 10^{-7}$ m. Consequently, for a particle with the diameter of 10 μm:

$$A = \left(1+1.2\cdot {}^{2L}\!/_{d}\right) = 1+1.2\frac{2\cdot 6.2\cdot 10^{-7}}{100\cdot 10^{-6}} = 1.015$$

i.e. negligibly small. Consequently, the results cannot be explained using this approach. The correction remains small up to the gas density of ρ_0 = 0.012 kg/m³. The rheology of the medium also changes only slightly with the decrease of the gas density. Therefore, it remains to be assumed that disruption of similarity may be associated with the variation of the physical pattern of the process in the boundary layer or with the disruption of the boundary conditions of the Navier–Stokes equations. For example, for the case of sliding on the surface of the sphere, the boundary conditions change. This problem was solved for the first time by Basse, who proposed that the tangential velocity of the liquid in relation to the solid body at the point on its surface is proportional to the tangential stresses, acting at this point. If the proportionality constant is denoted by β, the force acting on the solid is:

$$F_s = 3\pi\mu dV \frac{\beta\frac{d}{2}+2\mu}{\beta\frac{d}{2}+3\mu}$$

In the absence of sliding ($\beta = \infty$) we obtain the Stokes law, and in the case of complete sliding ($\beta = 0$), $F_s = 2\pi\mu dV$. Thus, the drag may decrease only by a factor of 2–3. It is insufficient, and in addition to this, it is not clear how β depends on ρ. In this respect, the Epshtein equation appears to be more suitable and was obtained by means of kinetic theory:

$$F_s = 3\pi\mu dV\left(1-\frac{\mu}{\beta\,{}^{d}\!/_{2}}\right) \approx \frac{3\pi\mu dV}{1+2\,{}^{\mu}\!/_{\beta}}$$

We examine whether it is possible to explain the disruption of similarity $\lambda(\mathrm{Re})$ from the position of the physical pattern in the boundary layer of the particle.

If the criterion is represented by the ratio l/d, then at Re = *idem* we obtain a similarity in respect of the parameter l/d, since $l \approx \frac{d}{\sqrt{\mathrm{Re}}}$ and, consequently $\frac{l}{d} = \frac{1}{\sqrt{\mathrm{Re}}}$. In general, at low values of Re the thickness of the boundary layer is comparable with the size of the particle and at small values of K_n the flow-around pattern hardly changes.

Thus, the resultant experimental data are difficult to explain on the basis of the currently available theory of the process and, consequently, they require detailed examination (both experimental and theoretical). This problem requires a solution. In conclusion, it should be mentioned that the application of the well-known theoretical relationships for the determination of the parameters of the settling process in the conditions of the rarefied gas may be regarded only as the first, very rough approximation.

3. Separation of powders in the rarefied gas

It is well-known that the process of gravitational separation of very fine powders takes place at low airflow rates ($w < 1$ m/s) and is accompanied by a reduction in the quality of fractionation and in the size of the separation boundary. The consequence of the low velocity of the airflow is the increase of the size of equipment and a decrease of productivity. Special investigations were carried out in order to determine positive effects resulting from a decrease of the gas density in the separation of fine powders.

The experiments were carried out on two types of classification equipment. Figure XII-13 shows the diagram of the first experimental equipment which includes the classifier *1*, the bunker with the coarse product *2*, the cyclone *3*, the bunker-feeder *4*, including the disc *5* for regulation of feed, valves *6* and *11* for maintaining the required rarefaction in the shaft of the classifier and the velocity of the airflow, the diaphragm *7*, and the pressure gauge *8* for the measurement of the airflow rate, the mercury pressure gauge *9* for inspecting the rarefaction in the shaft, the T-piece *10* for sampling the pressure in the chamber and equalisation of the pressure gradient in the feeder

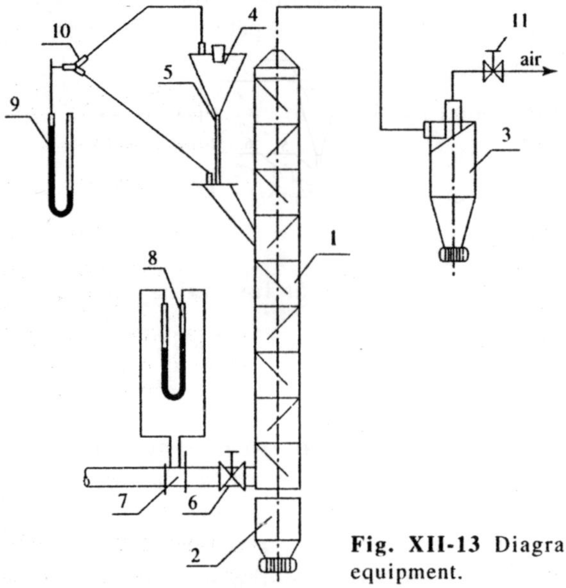

Fig. XII-13 Diagram of open experimental equipment.

in order to stabilise the initial material. The equipment was connected to a vacuum pump. In addition to this, in order to determine the rarefaction, a barometer was installed in the chamber of equipment. After every experiment, the classification products were weighed and subjected to sieve analysis and the results were used to determine $F_{f_i}(x_i)$ and other parameters of the process. Figure XII-14 shows the diagram of the second experimental equipment. The principal difference between the second and first equipment is that all the main sections of the second equipment (classifier, feeder, cyclone) are assembled in a hermetic metallic chamber. Therefore, in this case, there are less stringent requirements on the leak tightness of the classifier and its connection with the feeder and the cyclone, in comparison with the first case. In addition to this, the second equipment uses a cell feeder which makes it possible to regulate more efficiently productivity and maintain a constant consumption concentration of the material.

In the first experimental series, the density of air was normal (ρ = 1.2 kg/m^3), in the second and third series it was reduced. In each series it was possible to vary the speed of the airflow and, consequently, the separation boundary. The result of processing the experimental data were used to determine the dependences $F_f(x)$ and determine the efficiency of distribution according to the Eder-Mayer parameter χ.

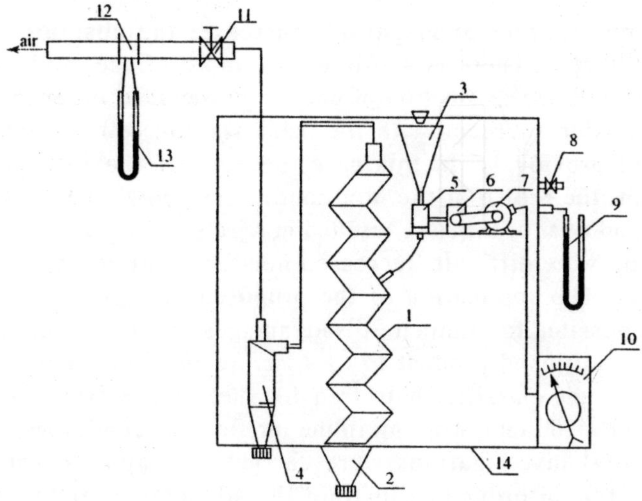

Fig.XII-14 Diagram of closed experimental equipment.

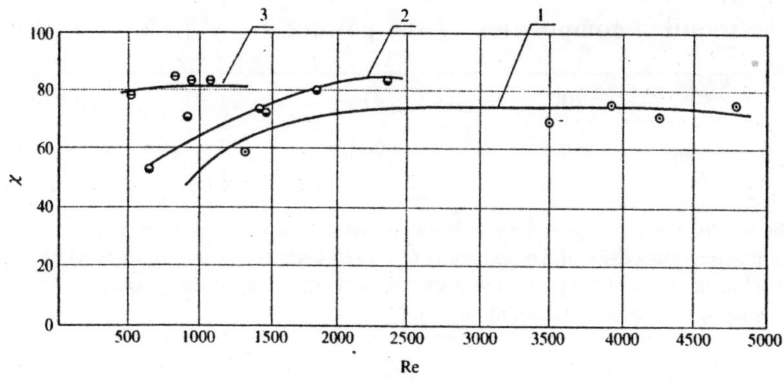

Fig. XII-15 Dependence of efficiency on the Reynold's number: 1) $\rho = 1.2$ kg/m^3; 2) $\rho = 0.6$ kg/m^3; 3) $\rho = 0.2$ kg/m^3.

The main experimental results are presented in Fig. XII-15. The resultant relationships show that the reduction of the density of air has a positive effect and increases the efficiency of separation.

The experimental results can be explained on the basis of several approaches.

The reduction in the density of the medium results, in addition to the improvement of the quality of separation, also in the intensification of the process as a result of the increase of the velocity of the airflow and the increase of productivity at the same concentration of the material in the flow as a result of increasing the main driving forces of the process, because the reduction of the density of the medium by an order of magnitude increases the absolute velocity by a factor of 3. There is a suitable example of the application of this method for the production of copper powders. In order to produce copper powders, screens with the mesh size of 100 μm and 71 μm were used previously. In this case, because of the low quality of separation, the return of the non-conditioned powder was 17% and breaking and wear of the sieve resulted in rejects. The working conditions in the shop were difficult (increased noise, vibration, contamination with dust). The application of the proposed pneumatic technology makes it possible to eliminate disadvantages and bring the return of the non-conditioned product to (4–6)%. Separation was carried out using a cascade classifier with transfer shelves $z = 10$, $i = 4$, cross-section 200 × 200 mm, working in the rarefied gas conditions. Detailed experimental investigations were carried out (approximately 100 experiments). In order to illustrate the advantages of the method, we present the data obtained in industrial tests of the first stage of separation.

The required composition of the powder

d, μm	100	71	45
Passes through sieve, mm%	99.5	90.0	65÷80

The results of separation in the conditions of the normal gas density $\rho_0 = 1.2$ kg/m³; the productivity of the apparatus $Q_n = 315$ kg/h; the flow rate of air $Q_a = 98$ m³/h; the yield of completed product $\gamma_m = 80\%$; the yield of the coarse product $\gamma_k = 20\%$, $\mu = 3.2$ kg/m³, $w = 0.68$ m/s.

The following results were obtained in these experiments:

	d, μm	100	71	45	Bottom	Yield of product
Partial	r_u	10.4	4.88	21.2	63.52	
residue,	r_m	0.1	3.5	18.8	77.6	80
%	r_k	52.0	10.0	30.8	7.2	20

In this case, the efficiency of separation according to the Eder-Mayer parameter was of the order of 69%.

The result of separation in the conditions of the rarefied gas: $\rho_0 = 0.6$ kg/m³, $Q_n = 756$ kg/h, $Q_a = 66.9$ m³/h (in normal air), $\mu = 11.3$ kg/m³ (in respect of normal air); $W = 0.93$ m/s.

The following results were obtained in these conditions:

r, %	d, μm	100	71	45	Bottom	Yield of product
r_{ii}		9.8	7.1	18.7	65.0	
r_{iii}		0.4	3.2	18.8	77.6	82
r_k		49.2	24.8	18.0	8.0	18

$\chi \approx 68\%$

The experimental results indicated that halving the density of gas makes it possible, whilst retaining the efficiency of separation, to increase the productivity of equipment 2.4 times and reduce the consumption of air 1.5 times. However, the application of the method has also a number of difficulties, with the main difficulty being to ensure the stability of feed of the initial material. This restricts the range of selection of the type of feeder because, for example, disc and belt feeders operate in an unstable manner in the rarefaction conditions. Evidently, cell (gate) feeders are most suitable for these conditions.

9. HOMOTHETIC TRANSFORMATION OF THE POWDERS BY FRACTIONATION METHODS

Recently, the problem of production of powders of a specific composition has arisen in different branches of technology. In most cases, it has been attempted to solve this problem by the currently available technological measures.

In this case, the initial powder is screened into narrow size classes in individual hoppers from which it is then supplied in accordance with the specific procedure, followed by mixing. All these processes are organised inefficiently and, consequently, the quality of powders is usually very low. There are considerable difficulties in improving the processes of this type. The main difficulty is that, up to now, the problem of the quantitative measure of the ratio of the produced and required composition of the powders has not been formulated.

The absence of this measure prevents optimisation of processes of this type.

In many cases, the problem can be solved by methods of classification, without preliminary separation of the initial product into narrow classes. As an example, there are principal diagrams of multiproduct separation (Fig. XII-2 to XII-4) in which the combination of the different output products into the same cyclone makes it possible to produce powders whose composition is close to the required composition.

It should be mentioned that the processes of this type cannot be regarded as classification processes because the problem is not limited to the separation of the powders, it is considerably wider. Therefore, we shall use a different term: *homothetic processing*, which describes the change of the composition of the powder towards required composition.

There are different method of controlling the composition of the final product in classification. In this case, there are wide possibilities in the previously mentioned methods of multiproduct separation. It is sufficient to have only two usual separately operating systems. This provides considerable possibilities for homothetic processing because it is possible to obtain different fine and coarse products and after mixing at a specific ratio to obtain the compositions close to the required composition.

Table XII-4 shows the results of homothetic processing of the powders in two independent classifiers as a result of mixing of the coarse product from one equipment with the fine product from the other one. The systems generated different velocities of the airflow, determined by the method of preliminary calculations.

As indicated by the table, the actual composition is relatively close to the required composition. We shall attempt to determine the degree of correspondence because optimisation of the processes of this type is possible only from this position.

The degree of correspondence between the required and actual composition of the powders can be determined most efficiently on the basis of the agreement of the corresponding parts of the grain size characteristics in partial residues because in this case we are concerned with similarity. For example, for the class +1.6 mm, the coincidence is 12.98%, and for the class +0.315 mm it is 10%. We shall construct the general characteristic of the quality of the powder from the comparison of the second and third lines of the table:

$$E = (12.98+23.26+17.22+19.04+10+10)=92.5$$

Table XII-4 Grain size distribution characteristics of powders in partial residues, %

Product	Mesh size of sieve, mm									
	2.5	1.6	1	0.63	0.4	0.315	0.2	0.160	Bottom	
Initial	3.93	34.06	26.85	12.32	11.38	3.31	3.9	1.24	1.66	1.41
Actual	0.1	12.98	23.26	17.22	19.04	11.96	10.98	4.5	0.940	0
Required	0	15	25	20	20	10	10	0	0	0

This result indicates that the agreement between the resultant and required compositions is quite satisfactory.

Thus, the efficiency of homothetic processing may be described as follows:

$$E_r = \sum_i r_c, \%$$

where E_r is the efficiency of homothetic processing, in%; i is the number of differentiated classes in the required composition; r_c is the coinciding part between the required and actual materials,%.

Using this rather simple parameter it is possible, from the number of powders, to choose quite unambiguously the composition which is closest to the required composition. Thus, the parameter can be used to solve the problem of selection of equipment for homothetic processing of the powders at the constant composition of the initial product.

If the composition of the initial powders is variable, this approach may be insufficient because in this case it is necessary to take into account the effect of this composition on the final products. In addition to this parameter, which remains the main parameter, it is necessary to accept another parameter, characterising the capacity of equipment for homothetic processing. The problem of the selection of such a parameter is relatively complicated. The solution requires carrying out extensive and detailed investigation.

To the first approximation, the problem can be solved using the entropy approach to the formation of quality criteria because this approach makes it possible to take into account the variation of the material of each class included in it. From this position, the homothetic processing capacity of the individual equipment or technological line as a whole may be expressed by the dependence:

$$E = \frac{|H_s - H_c|}{|H_s - H_r|} 100, \%$$

where H_s, H_c, H_r is the indeterminacy of respectively the initial, actual and required composition.

This dependence corresponds to the boundary conditions of the process:

If the actual material accurately corresponds to the required composition, then $E = 100\%$;

If the actual material does not differ from the initial material, then $H_c = H_s$, $E = 0$.

Any variation of the actual composition between these extreme cases gives, using this dependence, the parameter reflecting the degree of completion of the process. This will be shown on an example of Table XII-4. According to the table, H_s = 0.864; H_r = 0.754; H_c = 0.817. Consequently, one can write:

$$E = \frac{0.864 - 0.817}{0.864 - 0.754} \cdot 100 = 42.7\%$$

It should be mentioned that in the evaluation of the quality of homothetic processing there is a certain special feature by which the process differs from the separation processes. The point is that homothetic processing may occur in dual manner in each specific case: to the required composition is obtained both from the side of the excess and from the side of the shortage of the content of the narrow size class in the initial product. Therefore, deviations to either side from 100% parameter of homothetic processing indicates the degree of in completion of the process.

At $H_c > H_r$ we obtain $E < 1$ (approach from the side of the shortage); at $H_c < H_r$, we obtain $E > 1$ (approach from the side of the excess). It should be stressed that E = 42.7% and E_2 = 167.5% indicate the same degree of completion of the process of homothetic processing because they give the same deviation from 100%.

All these factors indicate that the given problem is very complicated. The processes of multiproduct separation and homothetic processing of the powders are obviously advanced in comparison with the currently available technology of preparation of powder material.

The detailed examination of this processes and the development of objective methods of optimising the processing makes it possible to raise the methods and technology of fractionation of powders to a higher, qualitatively new level.

References

1. Olevskii V.A., Design and calculation of mechanical classifiers and hydrocyclones, Gosgortekhizdat (1960).
2. Bond F., Theory of isotope separation, New York, No.4, 18-36 (1951).
3. Povarov F.I., Technological evaluation of the operation of classifiers, Obogashchenie Rud, No.3, 40–48 (1956).
4. Steinmetzer S., Windsichtung der Feinkohle, (1941), p.326.
5. Barsky M.D., Optimisation of the processes of separation of granular materials, Nedra, Moscow (1978), p.167.
6. Barsky M.D. and Barsky E., Criterion for separation of pourable material into

n components. 29th International Symposium on Computer Applications in Minerals Industries, Beijing, China (2001).

7. Barsky L.A. and Kozin V.Z., System analysis in enrichment of minerals, Nedra, Moscow (1978).
8. Plitt R., The analysis of solid-solid separations in classifiers. The Canadian Mining and Metallurgical Bulletin (April 1971).
9. Navrotski E., Graphical–analytical methods of evaluation of gravitational equipment, Nedra, Moscow (1980).
10. Rumpf H., Sommer K., Stiesse M., Berechnung von Trennkurven fur Gleichgewichtsichter, Verfahrenstechnik, Vol.8, No.9 (1974).
11. Loitsyanskii L.G., Mechanics of liquids and gas, Nauka, Moscow (1978).
12. Happel J. and Brenner H., Low Reynolds number hydrodynamics. Prentice-Hall (1965).
13. Sedov L.I., Methods of similarity and dimensions in mechanics, Nauka, Moscow (1981).
14. Fuks N.A., Mechanics of aerosols, Akademiya, Moscow (1955).
15. Boothroyd R., Flow of gas with suspended particles, Russian translation, Mir, Moscow (1975).
16. Gorbis Z.R., Heat exchanging hydrodynamics of dispersed continuous flows, Energiya, Moscow (1970).
17. Shishkin S.F., Intensification of the process of gravitation on pneumatic classification, Dissertation for the title of candidate of technical sciences, Sverdlovsk (1983).

Index